Intelligent Systems of Computing and Informatics

Since 2011, the Fourth Industrial Revolution (IR4.0) has played a significant role in education, and industry. Data technologies have also evolved rapidly to cater to the rapidly growing size of the data as well as to enhance the security of the shared data through multiple resources and platforms.

Intelligent Systems of Computing and Informatics aims to develop a new Intelligent Systems of Computing and Informatics (ISCI) to cater to the needs of industries in line with the United Nations' Sustainable Development Goals (SDGs) of affordable and clean energy and sustainable cities and communities.

Comprising 20 chapters by experts from all over the world, this book covers such topics as data technologies, machine learning, signal and image processing, software systems efficiency, computer networking, Internet of Things, and computational intelligence for real-life problems.

Key Features:

- Develops a new system for computing and informatics
- Provides the state of the art of current research and studies in intelligence systems
- Written by experts in the field of computing and informatics

This book is aimed at postgraduate students, researchers working in various research and development (R&D) agencies, and practitioners, as well as scientists that have an interest in ISCI.

Intelligent Systems of Computing and Informatics

Edited by Samsul Ariffin Abdul Karim,
Anand J. Kulkarni, Chin Kim On,
and Mohd Hanafi Ahmad Hijazi

CRC Press is an imprint of the
Taylor & Francis Group, an **informa** business

Cover image: Shutterstock

First edition published 2024
by CRC Press
2385 NW Executive Center Drive, Suite 320, Boca Raton FL 33431

and by CRC Press
4 Park Square, Milton Park, Abingdon, Oxon, OX14 4RN

CRC Press is an imprint of Taylor & Francis Group, LLC

ISBN: 978-1-032-50946-4 (hbk)
ISBN: 978-1-032-50949-5 (pbk)
ISBN: 978-1-003-40038-7 (ebk)

DOI: 10.1201/9781003400387

Typeset in Minion
by KnowledgeWorks Global Ltd.

Contents

Preface

Since 2011, the fourth industrial revolution (IR4.0) has played a significant role in education, research, and industrial growth. There are nine pillars under the framework of IR4.0 namely Big Data and Analytics, Robotics, Simulation, Horizontal and Vertical Integration, Internet of Things (IoT), Cybersecurity, Cloud Computing, Additive Manufacturing and Augmented Reality. Besides, data technologies have been evolving rapidly to cater the rapid growing size of the data as well as to enhance the security of the shared data through multiple resources and platforms. Machine learning, IoT, simulation and cloud computing are the mainstream tools to develop an up-to-date Intelligent Systems of Computing and Informatics (ISCI). This new system can be used to produce the desired outcome for the industries through efficient computational intelligence techniques as well as software development. The developments of ISCI are significant as they will enable the related agencies and policymakers to conduct the quick and reliable self-decision-making involving complex processes in the fields of engineering and computer sciences. In this book, a total of 20 chapters are contributed by the respective experts from all over the world. The main theme of the book is to develop a new Intelligent Systems of Computing and Informatics (ISCI) to cater the needs of industries that are in line with the United Nations Sustainable Development Goals (SDGs) No. 7: Affordable and Clean Energy; No. 9: Industry, Innovation, and Infrastructure; and No. 11: Sustainable Cities and Communities.

The main topics covered in this book are:

- Machine learning
- Big data and data analytics
- Computer networking and IoT

- Software systems efficiency
- Data processing algorithms and applications
- Signal and image processing
- Modeling and simulation via fast algorithm
- Computational intelligence and statistics

We would like to thank all contributors who provided excellent contributions for this book. Special thanks to the staff at Taylor & Francis/CRC Press for the publication of the book. Their superb support has made the publication process really a joyful and great experience for all of us!

The main editor is fully supported by the Ministry of Higher Education (MOHE), Malaysia, for the financial support received in the form of a research grant: **[FRGS/1/2023/ICT06/UMS/02/1]** (New Scattered Data Interpolation Scheme Using Quasi Cubic Triangular Patches for RGB Image Interpolation and fruits quality inspection) and Universiti Malaysia Sabah. Special thanks to the Faculty of Computing and Informatics, Universiti Malaysia Sabah, for the computing facilities support that made the completion of the book possible.

This book is highly suitable for postgraduate students, researchers working in various Research and Development (R&D) agencies, practitioners, policymakers as well as scientists that have an interest in Intelligent Systems of Computing and Informatics towards IR4.0.

Samsul Ariffin Abdul Karim
Kota Kinabalu, Malaysia

Anand J. Kulkarni
Pune, India

Chin Kim On
Kota Kinabalu, Malaysia

Mohd Hanafi Ahmad Hijazi
Kota Kinabalu, Malaysia

Editors

Samsul Ariffin Abdul Karim is an Associate Professor with Software Engineering Program, Faculty of Computing and Informatics, Universiti Malaysia Sabah (UMS), Malaysia. He obtained his PhD in Mathematics from Universiti Sains Malaysia (USM). He is a Professional Technologist registered with the Malaysia Board of Technologists (MBOT), No. Perakuan PT21030227. His research interests include numerical analysis, machine learning, approximation theory, optimization, science, and engineering education as well as wavelets. He has published more than 180 papers in journals and conference proceedings including five edited conference volumes and 80 book chapters. He was the recipient of Effective Education Delivery Award and Publication Award (Journal & Conference Paper), UTP Quality Day 2010, 2011, and 2012, respectively. He was certified WOLFRAM Technology Associate, Mathematica Student Level. He has also published 13 books with Springer Publishing including 6 books with Studies in Systems, Decision and Control (SSDC) series, 2 books with Taylor and Francis/CRC Press, 1 book with IntechOpen, and 1 book with UTP Press. He received the Book Publication Award in UTP Quality Day 2020 for the book Water Quality Index (WQI) Prediction Using Multiple Linear Fuzzy Regression: Case Study in Perak River, Malaysia, that was published by SpringerBriefs in Water Science and Technology in 2020.

Anand J Kulkarni holds a PhD in Distributed Optimization from Nanyang Technological University, Singapore; MS in Artificial Intelligence from University of Regina, Canada; Bachelor of Engineering from Shivaji University, India; and Diploma from the Board of Technical Education, Mumbai. He worked as research fellow at Odette School of Business, University of Windsor, Canada. Anand worked with Symbiosis International University, Pune, India, for more than six years. He is currently working as Professor and Associate Director of the Institute of Artificial Intelligence at the MITWPU, Pune, India. His research interests include optimization algorithms, multi-objective optimization, continuous, discrete and combinatorial optimization, swarm optimization, and self-organizing systems. Anand pioneered optimization methodologies such as Cohort Intelligence, Ideology Algorithm, Expectation Algorithm, and Socio Evolution & Learning Optimization Algorithm. Anand is the founder of Optimization and Agent Technology Research Lab and has published more than 70 research papers in peer-reviewed reputed journals, chapters, and conferences along with five authored and ten edited books. Anand is the lead series editor for Springer and Taylor & Francis as well as editor of several Elsevier journals. He also writes on AI in several newspapers and magazines. Anand has delivered expert research talks in countries such as the United States, Canada, Singapore, Malaysia, India, and France

Chin Kim On received his PhD in Artificial Intelligence with the Universiti of Malaysia Sabah, Sabah, Malaysia, in 2010, and he is currently working as an Associate Professor at the Universiti Malaysia Sabah in the Faculty of Computing and Informatics. His research interests are gaming AI, evolutionary computing, evolutionary robotics, artificial neural networks, image processing, agent technologies, sentiment analysis, evolutionary data mining, and biometric security system mainly focused on fingerprint and voice recognition. He has led several projects related to artificial neuro-cognition for solving real-world problems such as mobile-based number plate detection and recognition, off-line handwriting recognition, item drop mechanism, and auto map generation in gaming AI, to name a few. He has authored and co-authored more than 120 articles in the form of journals, book chapters, and conference proceedings. He is a senior member of IEEE and IAENG societies.

Mohd Hanafi Ahmad Hijazi is an Associate Professor of Computer Science at the Faculty of Computing and Informatics, Universiti Malaysia Sabah in Malaysia. He obtained his PhD from the University of Liverpool in 2012. His research work addresses the challenges in knowledge discovery and data mining to identify patterns for prediction on structured and/or unstructured data. His particular application domains are medical image analysis and understanding, sentiment analysis on social media data, and recently precision farming. He has authored/co-authored more than 50 journals/book chapters and conference papers, most of which are indexed by Scopus and ISI Web of Science. He also served on the program and organizing committees of numerous national and international conferences. He is currently the head of Data Technologies and Applications research group and the Dean at the faculty.

Contributors

Youssef Karim Ahmed Abdelkader
Clean Technology Impact Lab
Taylor's University
Subang Jaya, Malaysia

Adam Ahmad
Mathematical Sciences Department
Faculty of Science
Universiti Brunei Darussalam
Gadong, Brunei

Farzana Kabir Ahmad
School of Computing
Universiti Utara Malaysia
Changlun, Malaysia

Rainaf Akif
Clean Technology Impact Lab
Taylor's University
Subang Jaya, Malaysia

Suhana Mohezar Ali
Department of Decision Science
Universiti Malaya
Kuala Lumpur, Malaysia

Muhammad Zaki Almuzakki
Faculty of Science and Computer
University of Pertamina
Jakarta, Indonesia

Elayaraja Aruchunan
Department of Decision Science
Universiti Malaya
Kuala Lumpur, Malaysia

Mohammad Fadhli Asli
Optimization and Visual Analytics Research Lab
Faculty of Computing and Informatics
Universiti Malaysia Sabah
Kuala Lumpur, Malaysia

Julkifli Awang Besar
Department of Electrical Engineering
Politeknik Kota Kinabalu
Sabah, Malaysia

Kang Jia Cheng
Clean Technology Impact Lab
Taylor's University
Subang Jaya, Malaysia

Jackel Vui Lung Chew
Faculty of Computing and Informatics
University Malaysia Sabah Labuan
Labuan, Malaysia

Yee Wen Choon
Institute for Artificial Intelligence and Big Data
and
Department of Data Science
Universiti Malaysia Kelantan
Kelantan, Malaysia

Timothy Christyan
Computer Science, Faculty of Science and Computer
University of Pertamina
Jakarta Selatan, Indonesia

Darra Funna
Faculty of Science and Computer
University of Pertamina
Jakarta, Indonesia

Graham Edwin Gardner
Advanced Livestock Measurement Technologies project
Meat and Livestock Australia
Sydney, Australia

Adilah Abdul Ghapor
Department of Decision Science
Universiti Malaya
Kuala Lumpur, Malaysia

Habibollah Haron
School of Computing
Universiti Teknologi Malaysia
Johor, Malaysia

Mohamad Khatim Hasan
Faculty of Information Science and Technology
Universiti Kebangsaan Malaysia
Bangi, Malaysia

Mohammad Khatim Hasan
Centre for Artificial Intelligence Technology
Faculty of Information Science and Technology
Universiti Kebangsaan Malaysia
Bangi, Selangor, Malaysia

Rosilah Hassan
Faculty of Information Science and Technology
Universiti Kebangsaan Malaysia
Bangi, Malaysia

Faysal Hossain
Clean Technology Impact Lab
Taylor's University
Subang Jaya, Malaysia

Saiful Azmi Husain
Mathematical Sciences Department
Faculty of Science
Universiti Brunei Darussalam
Gadong, Brunei

Muhamad Irsan
School of Computing, Telkom University
Bandung, Indonesia
and
Faculty of Information Science and Technology
Universiti Kebangsaan Malaysia
Bangi, Malaysia

Sin Jie
Institute of Mathematical Sciences
Universiti Malaya
Kuala Lumpur, Malaysia

Chi Jing
School Information and Electrical Engineering
Hebei University of Engineering
Handan, Hebei, China

Wan Yu Jinq
Institute of Mathematical Sciences
Universiti Malaya
Kuala Lumpur, Malaysia

Nurul Syifa Kamarolzaman
School of Computing
Universiti Teknologi Malaysia
Skudai, Johor, Malaysia

Siti Sakira Kamaruddin
School of Computing, College of Arts and Sciences
Universiti Utara Malaysia
Sintok, Kedah, Malaysia

Norliza Katuk
School of Computing
Universiti Utara Malaysia
Sintok, Kedah, Malaysia

Muhammad Ashraf Khalid
Department of Decision Science
Universiti Malaya
Kuala Lumpur, Malaysia

Meng Chun Lam
Faculty of Information Science and Technology
Universiti Kebangsaan Malaysia
Bangi, Malaysia

Abdul Kadir Mahamood
Politeknik Sultan Abdul Halim Muadzam Shah
Jitra, Malaysia

Majid Khan Majahar Ali
School of Mathematical Sciences
Universiti Sains Malaysia
Penang, Malaysia

Muzdalini Malik
Politeknik Seberang Perai
Bukit Mertajam, Malaysia

Jayaseelan Marimuthu
School of Veterinary and Life Sciences
Mudoch University
Murdoch, Australia

Gan Jun Ming
Faculty of Computing and Informatics, Universiti Malaysia Sabah
Jalan UMS, 88400 Kota Kinabalu
Sabah, Malaysia

Mohd Saberi Mohamad
Health Data Science Lab, Department of Genetics and Genomics
College of Medical and Health Sciences
United Arab Emirates University
Abu Dhabi, UAE

Ervin Gubin Moung
Faculty of Computing and Informatics
Universiti Malaysia Sabah
Kota Kinabalu, Sabah, Malaysia

Muhammad Raidhy Mustafid
Faculty of Science and Computer
University of Pertamina
Jakarta Selatan, Indonesia

Mohana Sundaram Muthuvalu
Department of Fundamental and Applied Sciences
Universiti Teknologi PETRONAS
Seri Iskandar Perak, Malaysia

Kohilavani Naganthran
Institute of Mathematical Sciences
Universiti Malaya
Kuala Lumpur, Malaysia

Nurul Athirah Nasarudin
Health Data Science Lab
Department of Genetics and Genomics
College of Medicine and Health Sciences
United Arab Emirates University
Al Ain, UAE

Nurul Aida Muhamad Noor
Optimization and Visual Analytics Research Lab
Faculty of Computing and Informatics
Universiti Malaysia Sabah
Kota Kinbalu, Sabah, Malaysia

Mohd Shamshul Anuar Omar
HeiTech Padu Berhad
Selangor Darul Ehsan, Malaysia

Amutha Prabha
Department of Instrumentation
School of Electrical Engineering
Vellore Institute of Technology
Vellore, India

Nur Anisah Mohamed A. Rahman
Institute of Mathematical Sciences
Universiti Malaya
Kuala Lumpur, Malaysia

Muhammad Akmal Remli
Institute for Artificial Intelligence and Big Data, and Department of Data Science
Universiti Malaysia Kelantan
Kelantan, Malaysia

Adi Badiozaman Ruhani
Politeknik Seberang Perai
Bukit Mertajam, Malaysia

Azali Saudi
Faculty of Computing and Informatics
Universiti Malaysia Sabah
Kota Kinabalu, Sabah, Malaysia

Chai Soo See
Faculty of Computer Science and Information Technology
Universiti Malaysia Sarawak
Sarawak, Malaysia

Jumat Sulaiman
Faculty of Science and Natural Resources
Universiti Malaysia Sabah
Kota Kinabalu, Sabah, Malaysia

Andang Sunarto
Tadris Matematika
Universitas
Islam Negeri Fatmawati Sukarno
Bengkulu, Indonesia

Mohammed Ahmed Taiye
Department of Cultural Sciences
Linnaeus University
Vaxjo, Sweden

Tasmi
Faculty of Science and Computer
University of Pertamina
Jakarta Selatan, Indonesia

Murugan Thangiah
Clean Technology Impact Lab
Taylor's University
Subang Jaya, Malaysia

Chockalingam Aravind Vaithilingam
Clean Technology Impact Lab
Taylor's University
Subang Jaya, Malaysia

Atul Dsouza Victor Francis
Clean Technology Impact Lab
Taylor's University
Subang Jaya, Malaysia

Chin Pei Yee
Faculty of Computing and Informatics
Universiti Malaysia Sabah
Kota Kinabalu, Sabah, Malaysia

Shrley Chan Suet Yee
Institute of Mathematical Sciences
Universiti Malaya
Kuala Lumpur, Malaysia

Ling Hang Yek
Faculty of Computing and Informatics
Universiti Malaysia Sabah
Kota Kinabalu, Sabah, Malaysia

CHAPTER 1

Intelligent Systems of Computing and Informatics

An Overview

Samsul Ariffin Abdul Karim

1.1 INTRODUCTION

The fourth industrial revolution (IR4.0) has played a very significant role not just in Teaching and Learning (T&L) but also in Research and Innovation (R&I), and it has been used as a tool for commercial development as well as automation of the processes in the manufacturing and consumption. Basically there are nine pillars of IR4.0 (see Figure 1.1) namely Big Data and Analytics (BDA), Robotics, Simulation, Horizontal and Vertical Integration, Internet of Things (IoT), Cybersecurity, Cloud Computing, Additive Manufacturing, and Augmented Reality. BDA is a crucial technique to cater the huge size of the dataset worldwide. It can be used to improve the security of the data shared through multiple resources and platforms via multi-factor security. Machine learning that includes artificial intelligence, statistical learning, and BDA; IoT; simulation; and cloud computing are the main tools to develop and construct a new system called Intelligent Systems of Computing and Informatics (ISCI). This new system can be used to produce the desired outcomes for the industries through efficient computational intelligence techniques as well as software development. The developments of ISCI are significant since it will be able to help the related agencies and policy makers to conduct the quick and

DOI: 10.1201/9781003400387-1

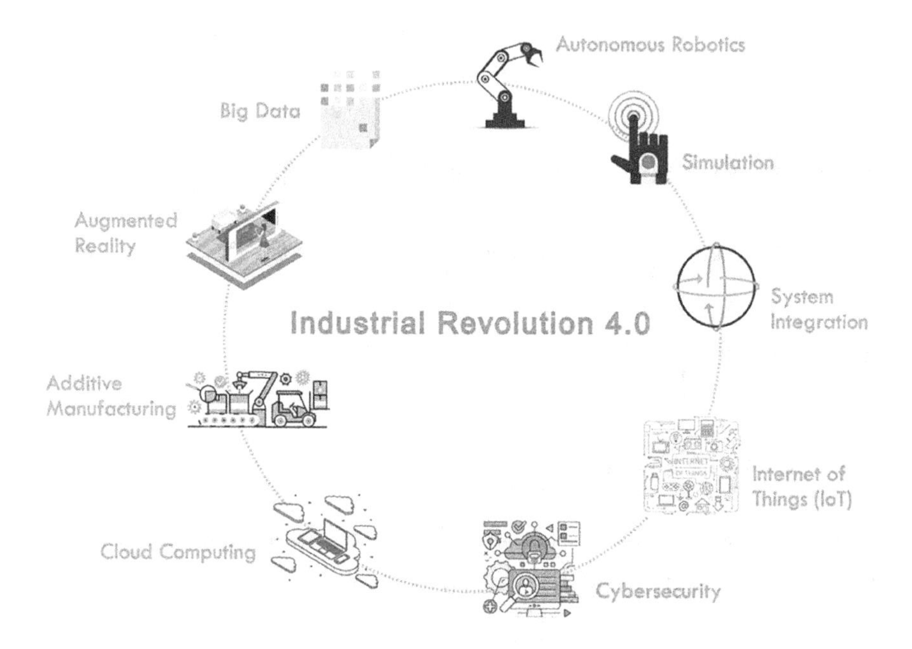

FIGURE 1.1 Nine pillars of IR4.0.

reliable decision-making involving complex processes in the fields of engineering, manufacturing, physical and life sciences, and computer sciences.

Malaysian National Fourth Industrial Revolution (4IR) policy was developed by the Malaysian government in 2021 [1]. It is human-centric (Figure 1.2) and focused on whole-of-nation approach. Besides, there are five foundational strategies for building capabilities based on 4IR. Figure 1.3 shows the foundational strategies in detail.

Furthermore, Malaysian Institute of Accountants (MIA) has developed their own MIA Digital Technology Blueprint on 13 July 2018. Their focus is on BDA, cloud computing, automation, and artificial intelligence to improve digital economy [2]. Notably, most of the government agencies in Malaysia have their own digitalization framework to meet the 4IR policy.

Besides 4IR, this book will also cater to the needs of industries that are in line with the United Nation Sustainable Development Goals (SDGs) No. 7: affordable and clean energy; No. 9: industry, innovation, and infrastructure; and No. 11: sustainable cities and communities. Figure 1.4 shows all 17 SDGs. We hope that by 2030, we can achieve SDGs No. 7, 9, and 11 [3]. Furthermore, the main agenda nowadays is also to achieve carbon net zero

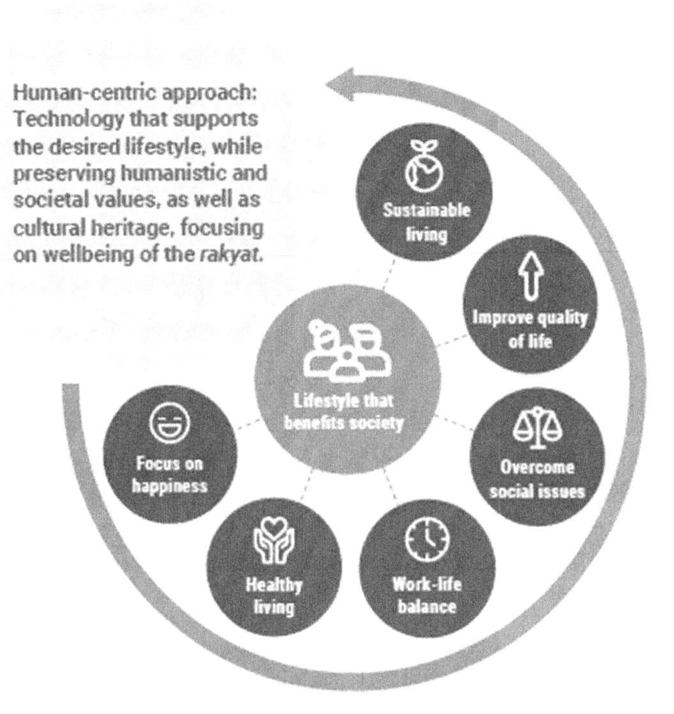

FIGURE 1.2 Human-centric approach under 4IR policy of Malaysia.

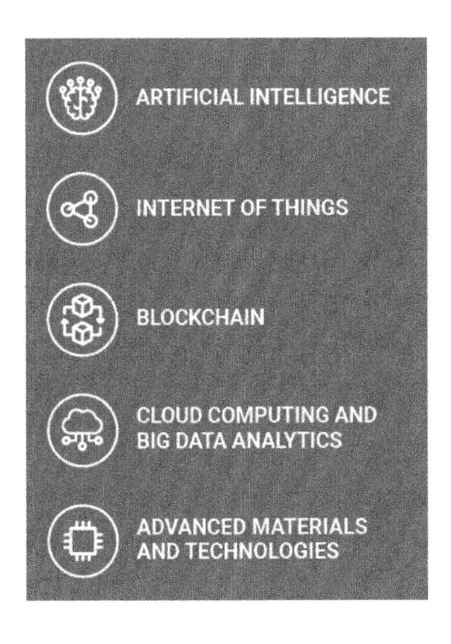

FIGURE 1.3 Five foundational technologies in Malaysian 4IR.

FIGURE 1.4 United Nations Sustainable Development Goals (SDGs).

emissions (CNZE) by 2050 [4,5]. The main objective of CNZE is to reduce carbon emissions up to zero, use low-carbon technologies as well as innovation towards circular economy.

1.2 SUMMARY

Following paragraphs provide the organization of the book and summarize each contributing chapter. Each chapter can be considered as a self-standing contribution.

Chapter 2 entitled "Intelligent Application of Partial Least Square Algorithm in Developing Model of Fat Depth Measurement" unveils a noninvasive, budget-friendly microwave system (MiS) for fat depth assessment. Through multiple linear regression and partial least square models, accurate predictions of organic animal c-site fat depth are achieved. Utilizing fivefold cross-validation and precision indicators like R-squared, both models demonstrate excellence. Tackling multicollinearity challenges from high-dimensional data, K-means clustering and principal component analysis streamline the explanatory variables. In a nutshell, the partial least square model shines, surpassing the accuracy parity due to its resilience against multicollinearity and unmet assumptions in the multiple linear regression model.

Chapter 3 entitled "Single Feature Imbalance Classification on Ensemble Learning Methods for Efficient Real-Time Malware Detection" presents ensemble machine learning techniques using imbalance classification approaches based on single feature for fast malware detection. Several classical machine learning algorithms and standard ensemble algorithms were tested, but they were not performing well due to imbalance class distribution. Thus, several ensemble bagging and boosting algorithms for imbalance classification were examined. It was found that the random forest with class weighting gave the highest accuracy, with true positive rate of 94.12% and false positive rate of 13.64%.

Chapter 4 entitled "Electroencephalogram-Based Emotion Recognition Using Binary Bat Algorithm and Least Square Support Vector Machine" presents binary bat algorithm (BBA) and least square support vector machine (LSSVM) in analyzing massive electroencephalogram (EEG) signals. The proposed model involved five stages: (a) data acquisition, (b) feature extraction, (c) feature selection, (d) classification, and (e) evaluation. The experimental results show that BBA has significantly reduced the features and provides outstanding results against other existing methods. The proposed algorithm achieved the maximum accuracy of 86.76% and 80.83%, and the mean accuracy obtained is 85.51% $\pm$ 0.07 and 80.97% $\pm$ 0.08 for valence and arousal emotions classification, respectively.

Chapter 5 entitled "Sequential Exception Technique for Text Anomalies" presents a deviation-based anomaly detection approach to identify anomalies in unstructured textual datasets. The technique which was originally developed for categorical dataset was adapted by replacing the variance-based dissimilarity function with the cosine similarity measure. The cosine similarity measure is proven to perform better for textual data similarity measurements. The adapted technique was tested on two textual datasets and the results show a relatively high F-score. The findings revealed the feasibility of the adapted technique and that it can be further explored to enhance its performance for textual anomaly detection problems.

Chapter 6 entitled "Intelligence Predictive Model for Lamb Carcass C-Site Fat Depth using Support Vector Machine" focuses on developing a robust predictive model for lamb carcass C-site fat depth through the application of support vector machine (SVM). This study systematically compares its efficacy against traditional multiple linear regression. Evaluation metrics encompass root mean square error (RMSE), mean absolute error (MAE), R-squared and adjusted R-squared, derived through fivefold cross-validation. Employing Python and the sci-kit-learn library, the analysis identifies the linear kernel in SVM, complemented by PCA transformed data, as the most proficient kernel. Ultimately, this investigation underscores SVM's superior predictive accuracy in contrast to multiple linear regression.

Chapter 7 entitled "Exploring the Power of Convolutional Neural Networks in Face Detection" presents a study that develops a neural network-based face detection system to count students in a learning environment accurately. Its primary goal is to provide an efficient attendance tracking method by capturing and detecting student faces with high precision. Object recognition and classification in digital images are achieved through a deep learning technique called neural networks. The training phase yielded an impressive accuracy rate of 99%, while testing yielded 98%. When implemented on a Raspberry Pi, facial recognition demonstrates robust performance within 1–5 m. However, it faces challenges when dealing with distances of 4 m. Nevertheless, accurate results were consistently obtained for distances ranging from 1 to 6 m when counting students. The main finding is related to UNSDG No. 9.

Chapter 8 entitled "Protecting Higher Learning Institutions from Phishing Attacks-A Staff Awareness Program" explores the growing menace of phishing attacks on higher education institutions, as cybercriminals relentlessly target staff and students to obtain sensitive data. Shockingly, educational organizations fell victim to 6.1 million malware attacks in just a month in 2021, with phishing being a prevalent method. The consequences are dire, including data breaches, financial losses, and irreparable damage to reputation. The chapter presents a comprehensive case study evaluating a multifaceted phishing awareness program that employs posters, infographics, videos, and interactive seminars. It also offers invaluable insights into the efficacy of the program and its potential for adoption by other institutions seeking to fortify their defenses against phishing threats.

Chapter 9 entitled "Intelligence Random Forest Application in Developing Regression Model from Lamb Carcass C-Site Fat Depth Data" presents a study that revolutionizes fat depth prediction by leveraging machine learning techniques. Addressing the limitations of conventional methods, the research employs random forest regression and multiple linear regression techniques to optimize accuracy. Notably, multiple linear regression with K-means clustering outperforms in key metrics (MSE, RMSE, R-squared, adjusted R-squared, and MAE). The findings establish random forest regression with K-means clustering as the superior model for accurate fat depth estimation. By showcasing the potential of machine learning in refining predictions, this chapter contributes to the advancement of fat depth estimation techniques in the realm of lamb carcass analysis.

Chapter 10 entitled "Intelligent Identification System for MOOC Security" presents a new idea to improve the safety and security of using the massive open online course (MOOC) platform. An intelligent standalone program with face recognition features was created to protect users' usage of the MOOC platform. In this study, we compare the performance of template matching, geometric-based methods, and convolutional neural networks (CNN) when using a learning management system. Tests were conducted under different conditions, including low resolution and excessive brightness. The results show that CNN slightly outperformed other methods. On the other hand, template matching and geometric-based methods may perform better if the hardware used can be improved or upgraded further.

Chapter 11 entitled "Low Illumination Surveillance for Object Detection and Recognition using Deep Learning Methods" introduces a novel approach to address the challenge of accurately detecting and recognizing objects in low illumination conditions through the application of deep learning algorithms. This research focuses on leveraging a specific low illumination dataset to train a deep neural network and compares its performance to that of images enhanced using image enhancement algorithms. The results of this investigation yield an object detection and recognition model capable of effectively detecting and recognizing objects in low illumination environments. Furthermore, the fine-tuned model developed in this study is implemented in a web-based surveillance system using a Raspberry Pi and its camera module, showcasing the practical application of the research findings.

Chapter 12 entitled "Intelligent System Design for the Solutions of Nonlinear Diffusion in the Two-Dimensional Porous Medium" designs an intelligent system for the solutions of nonlinear diffusion problems in the two-dimensional porous medium. First, we show the discretization of the problem using the finite difference method and the formulation of the explicit decoupled group successive over-relaxation iterative method. Then, we develop the computational algorithm for solving nonlinear diffusion problems in the two-dimensional porous medium setting. We evaluate the efficiency of the proposed numerical method based on the implementation of the method in a computational software. We compare its performance in terms of program and computational efficiency against some of existing methods. Next, this chapter presents the design of an intelligent system that uses the proposed method and algorithm to solve nonlinear diffusion problems subject to the prescribed conditions. The results of applying the proposed method to solve nonlinear problems demonstrate the efficiency level attained.

Chapter 13 entitled "Improved False Position Method Based on Slope (IFPMS)" develops an algorithm to search roots of nonlinear function. To find roots of function usually we use some methods such as bisection method, false position method, newton method, etc. Such methods only can find one root. Previously, researchers have built an algorithm to determine all roots in a given interval. This algorithm is called improved bisection method based on slope (IBMS). In this research, a new algorithm has been developed which is improved false position method based on slope (IFPMS). Simulation has been done to some case nonlinear function. The resulting of simulation has shown that IFPMS also can find some roots of nonlinear equation in given interval. Most of the relative errors of the IFPMS algorithm are smaller than the IBMS algorithm relative errors.

Chapter 14 entitled "Computing of Anxiety or Depression Symptoms Indicators Using Lagrange Exponential Modified Euler Method" presents a numerical scheme to compute anxiety or depression symptoms indicators data by using Lagrange and a new version of modified Euler methods. The scheme has proved to be more efficient than the fourth-order Runge–Kutta method for this dataset. Lagrange interpolating polynomial is employed to approximate the derivative function. Then three numerical methods are applied to predict the symptoms indicator. An efficient algorithm is constructed as the Lagrange exponential modified Euler algorithm. The resulting symptoms indicator have the smallest mean absolute percentage error compared to Runge–Kutta and modified Euler methods. The main finding is related to UNSDGs No. 9 and 11.

Chapter 15 entitled "Hybridization of Simulated Kalman Filter and Minimization of Metabolic Adjustment for Succinate and Lactate Production" presents a novel approach for classifying knockout genes. In addition, this chapter addresses the critical issue of augmenting succinate and lactate production in *Escherichia coli*. The proposed in silico method in this research is a hybrid technique combining simulated Kalman filter (SKF) and the minimization of metabolic adjustment (MOMA). SKF, being a population-based approach within metaheuristic optimization, draws inspiration from non-natural sources, while MOMA employs quadratic programming to pinpoint the fitness point in the flux space closest to the wild-type configuration after a gene knockout. The resulting hybrid method (SKFMOMA) provides a comprehensive output, including a list of gene knockouts, growth rates, and production rates for succinate and lactate. These findings serve as valuable resources for subsequent wet laboratory experiments, offering the potential to substantially boost the production of succinate and lactate in *E. coli*, with promising implications for various applications in biotechnology and industry.

Chapter 16 entitled "Intelligent Air Conditioning Systems: Enhancing Energy Efficiency and Indoor Air Quality through IoT and Air Conditioning Unit Using Machine Learning" proposes a comprehensive solution integrating IoT and air conditioning unit with machine learning technology that addresses the main issues with air conditioning (air conditioning unit) systems, such as high energy consumption and poor indoor air quality. The system uses a machine learning model for air conditioning to make more intelligent decisions and increase energy efficiency. A smartphone application gives customers up-to-date and historical information about their air conditioners. The main finding is related to UNSDGs No. 7B: Expand and upgrade energy services.

Chapter 17 entitled "A Comprehensive Analysis of Air Quality Data: A Case Study Approach with the OpenAir Package in R" considers scattered data interpolation scheme to visualize air quality data by using TheilSen, timeVariation, and scatterPlot functions for trend analysis, temporal variations, and linear correlation analysis, respectively. The OpenAir package model's data gathering, preprocessing, visualization, and statistical analysis tools are combined in the study to determine the level of air pollution and the resulting health hazards. The OpenAir model provides a comprehensive investigation of the intricate dynamics of air pollution, which aids in making wise decisions for efficient control measures and policy interventions. The main finding is related to UNSDG No. 7.

Chapter 18 entitled "A Systematic Review on Intelligent Mobile Beacon Systems: Applications and Features" presents a review on the recent advances of intelligent mobile beacon systems and applications. Major aspects like the essential components, interaction features, domain applications, and open issues are discussed in this chapter. Distinct characteristics of components and notification features inside a mobile beacon system and application are also deliberated in detail. Moreover, emerging trends in domain application of intelligent mobile beacon systems across many sectors are analyzed and discussed. Lastly, limitations and challenges by prior mobile beacon systems are highlighted for future undertaking. The main findings from this chapter offer insights on the fundamentals and future directions of intelligent mobile beacon systems.

Chapter 19 entitled "Enhancing the Use of Simulation Software with Artificial Intelligence Methods for Clean Energy System Performance" presents the use of ANSYS simulation software, which uses artificial intelligence methods to automatically find simulation parameters, in order to improve speed and accuracy simultaneously in the product design for clean energy system, which is a common topic of research which contributes to SDG No. 7, as to ensure access to affordable, reliable, sustainable, and modern energy for all. One of the product designs being simulated is for the hydrokinetic turbine system, typically used to drive power production from natural flow of a river, especially located in rural areas, and this system is usually not connected to national grid electricity services. Using artificial intelligence enhancement, the ANSYS simulation will help the engineers to quickly explore and predict whether or not the product design works in the real world by looking at its energy performance. Thus, this simulation software is not only important for research but also beneficial for teaching and learning in higher education, in which students, especially engineering students could explore their products design in more depths using AI enhancement simulation software packages.

Finally, in Chapter 20, we elaborate further on Intelligent Systems of Computing and Informatics: An Extension.

ACKNOWLEDGMENT

This research is supported by the Ministry of Higher Education (MOHE), Malaysia as financial support received in the form of a research grant: **[FRGS/1/2023/ICT06/UMS/02/1]** (New Scattered Data Interpolation Scheme Using Quasi Cubic Triangular Patches for RGB Image Interpolation) and Universiti Malaysia Sabah. Special thanks to Faculty of Computing and Informatics, Universiti Malaysia Sabah for the computing facilities support that have made the completion of the book possible.

REFERENCES

1. https://www.ekonomi.gov.my/sites/default/files/2021-07/National-4IR-Policy.pdf (Retrieved on 22 August 2023).
2. https://mia.org.my/knowledge-centre-resources/digital-economy/ (Retrieved on 22 August 2023).

3. https://sdgs.un.org/goals (Retrieved on 22 August 2023).
4. https://www.petronas.com/sustainability/getting-to-net-zero (Retrieved on 22 August 2023).
5. https://www.sustainabilitymatters.net.au/content/sustainability/article/net-zero-carbon-neutral-carbon-negative-what-do-they-mean-exactly-1597925690 (Retrieved on 22 August 2023).

CHAPTER 2

Intelligent Application of Partial Least Square Algorithm in Developing Model of Fat Depth Measurement

Shrley Chan Suet Yee, Elayaraja Aruchunan, Nur Anisah Mohamed A. Rahman, Kohilavani Naganthran, Adilah Abdul Ghapor, Jayaseelan Marimuthu, Graham Edwin Gardner, and Samsul Ariffin Abdul Karim

2.1 INTRODUCTION

In the pursuit of optimizing profits within the meat industry, it becomes imperative to minimize labor expenses associated with fat trimming. The Australian carcass trading landscape predominantly revolves around the weight of the carcass, wherein the amount of subcutaneous fat plays a pivotal role in influencing the yield of saleable meat from the carcass [1]. As the reduction in lean meat output occurs, the need to invest more labor into trimming excess fat becomes inevitable for meeting customer standards [2]. To mitigate the risk of substantial economic losses, a novel noninvasive and nondestructive technique has been developed for accurately assessing fat depth. Figure 2.1 shows the lamb carcass to assess the fat depth.

DOI: 10.1201/9781003400387-2

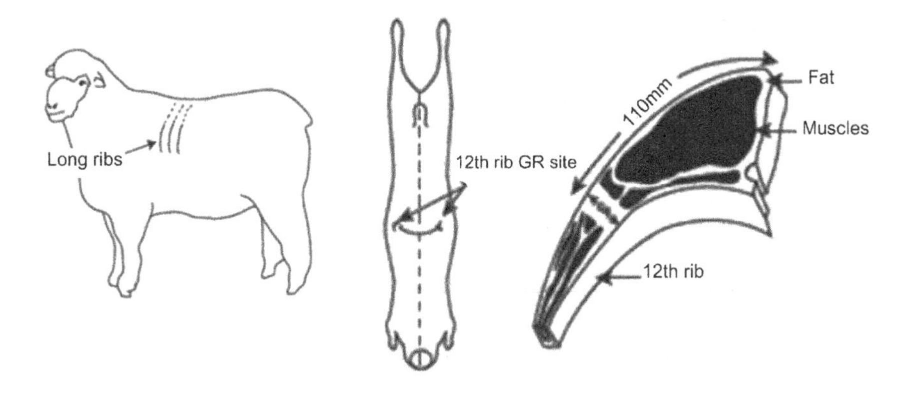

FIGURE 2.1 Lamb carcass to assess the fat depth.

This innovation involves the application of a microwave system (MiS) to gauge the fat content within the carcass. Unlike the invasive methods, this approach is favored for its minimal impact on animals and their tissues, as it employs low-power, nonionizing electromagnetic waves to quantify fat levels [3]. Currently, in the Australian context, the prevailing standards for assessing fat and tissue depth in lamb carcasses involve either subjective palpation estimates or invasive objective cut techniques (Anonymous, 2005). However, a more promising approach involves the use of MiS. This method capitalizes on the unique properties of microwave frequencies to distinguish between different layers, particularly biological tissues with varying dielectric characteristics. MiS, an active microwave measurement method based on inverse scattering, utilizes dielectric attributes to reflect microwave signals. These reflected and scattered signals are then collected at the same location [3]. Figure 2.1 shows the lamb carcass to assess the fat depth.

An innovative stride in this direction comes from Murdoch University, which has designed a portable MiS prototype at a reasonable cost. This prototype has been combined with several experimental broadband antennas to predict carcass fat depth [4]. For the acquisition of data and signal processing, the MiS incorporates integrated computer modules, an operating system, and automated Python-based programs. The experimentation apparatus involves a single broadband Vivaldi patch antenna (VPA) responsible for transmitting and receiving reflected signals [5].

The emergence of the noninvasive approach through the economical portable MiS offers a viable alternative to ultrasound measurements for forecasting C-site fat depth. The significance lies not only in the accuracy of this technology but also in its potential to facilitate well-informed decisions throughout the supply chain. Consequently, there is a clear need for further research in developing a predictive model for fat depth using this technology. The principal objective of this study, therefore, is to formulate a predictive model for fat depth by harnessing the capabilities of the noninvasive, cost-effective, portable MiS. Addressing the issue of multicollinearity is a significant concern in this high-dimensional dataset where the number of predictors amounts to 311, surpassing the available 120 observations. To tackle this, principal component analysis (PCA) stands out as a potent dimension reduction strategy. This technique effectively condenses the predictors into a smaller set of uncorrelated components [6]. Moreover, when grappling with the challenge of multicollinearity, the analysis of high-dimensional data through partial least squares (PLS) analysis demonstrates noteworthy utility and efficiency [7].

2.2 MATERIALS AND METHODS

This study employed various techniques and procedures to analyze the data and build regression models. The workflow chart shown in the Figure 2.2 provides an overview of the entire process.

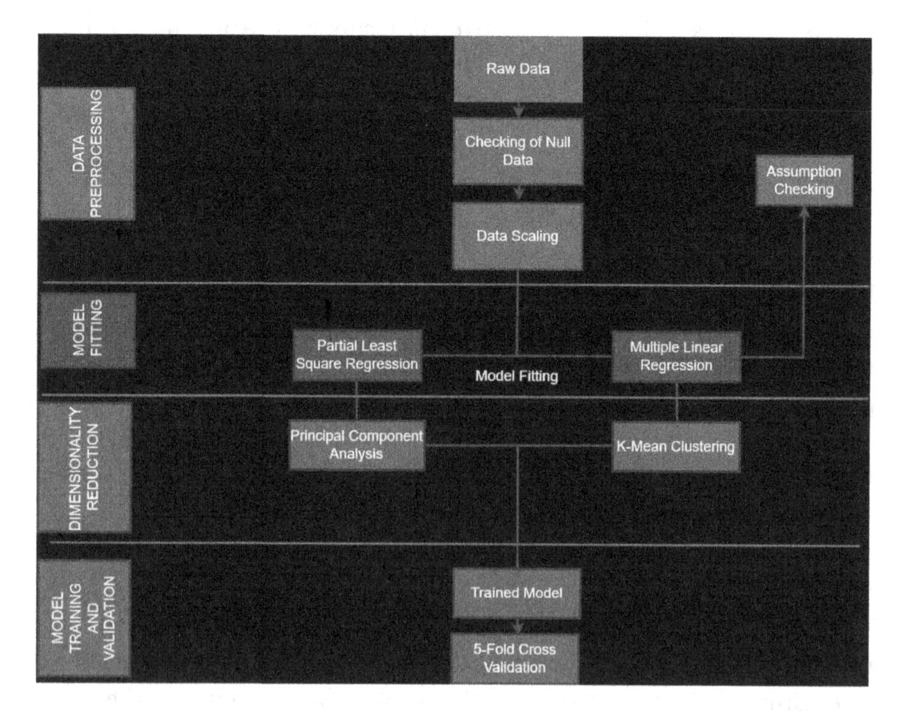

FIGURE 2.2 Flow chart.

In order to address the issue of multicollinearity in high-dimensional data, where the number of predictors (311) exceeds the number of observations (120), PCA was utilized as a dimension reduction strategy [6]. PCA helps reduce the number of predictors to a smaller set of uncorrelated components. In addition, PLS analysis was found to be useful and efficient, particularly when multicollinearity is a concern [7]. The performance of K-means clustering was compared with PCA in constructing the PLS model.

The variance inflation factor (VIF) in multiple linear regression was considered to determine the likelihood of rejecting a theoretically valid predictor as a non-significant variable. When the VIF is infinity, the chance of rejection increases. Therefore, PCA was performed prior to building the regression model.

The study followed a structured flow starting with data preprocessing. This step ensured that the data was suitable for model fitting. Subsequently, the data was scaled to prevent inaccuracies caused by large data ranges. Dimensionality reduction was then applied before fitting the model. Finally, the trained model was validated using fivefold cross-validation.

2.2.1 Principal Component Analysis

PCA was employed as a technique to reduce the dimensionality of the data. Scatterplots, two-dimensional graphs projecting multivariate data into a two-dimensional space, were used to identify correlations. The purpose of PCA was to capture the intrinsic variability in the data and prevent overfitting. PCA is commonly performed before classifying the data or addressing regression problems.

2.2.2 K-Means Clustering

K-means clustering, an unsupervised learning algorithm, was utilized to divide the dataset into K unique and non-overlapping clusters. The algorithm minimizes the dispersion of data to achieve this division in a straightforward and elegant manner. The number of clusters, K, needs to be defined before running the K-means algorithm. The algorithm assigns each observation to the closest centroid based on "closeness" distance, often measured using Euclidean distance. The process continues until the centroids no longer change.

2.2.3 Multiple Linear Regression

This technique was employed as a statistical supervised learning technique to build regression models. The goal was to design, analyze, and discover models based on a sample from a population. Multiple regression involves multiple independent variables (input) and one or more dependent variables (output). Several assumptions must be met when constructing a linear model, including the independence of observations, normally distributed and independent error terms, and constant variances. The formula for multiple linear regression is as follows:

$$y = B_0 + B_1 X_1 + \cdots + B_n X_n + \tag{2.1}$$

where

- y is the predicted value,
- B_0 is the y-intercept,
- $B_1 X_1$ is the regression coefficient of the first x, and
- ε is the model error.

2.2.4 Partial Least Square

PLS is a supervised way to say that the best explanation of predictors in that direction is the best directions and contributing to predict the response. It is one of the famous techniques to reduce the high dimensionality of data. The first step is to identify a new set of features $Z_1, \ldots, Z_M$ which are the linear combinations of the original features, and then a linear model will be fitted using a least square using M new features. The factors that are most closely related to the response are given the most weight in PLS. In other words, PLS regression is a technique whether the number of predictors will be reduced to a set of uncorrelated components and a least square regression is performed on those components. Often, PLS is helpful especially when the number of independent variables is high and even there is a present of high collinearity of the predictors. The aim of PLS is to build a predictive or explanatory model. PLS 1 and PLS 2 are distinct in some programs. When only one dependent variable is present, PLS 1 is used; when there are multiple dependent variables, PLS 2 is used.

2.2.5 R-Squared

It is the variation proportion that is explained by the predictor variables. In other word, it also explains how well the data fit the model. The better model will have higher R-squared.

$$R^2 = 1 - \frac{\Sigma(y_i - \hat{y}_i)^2}{\Sigma(y_i - \bar{y})^2} \tag{2.2}$$

where y_i is the observed value and $\hat{y}_i$ is the predicted value.

2.2.6 Mean Squared Error

It is a measure of mean error that explains the performance to predict the outcome of an observation.

$$Mean\ Squared\ Error = \frac{1}{n}\sum_{i=1}^{n}(y_i - \hat{y}_i)^2 \tag{2.3}$$

2.2.7 Root Mean Squared Error

It is the square root of the mean. squared error. In other word, it is an average squared difference between the predicted value and the observed value. The better model will have lower root mean squared error.

$$Root\ Mean\ Squared\ Error = \sqrt{Mean\ Squared\ Error} \tag{2.4}$$

2.2.8 Adjusted R-Squared

It adjusts the root squared when there are too many variables in a model.

$$Adjusted\ R - Squared = 1 - \left[\frac{(1 - R^2)(n-1)}{n-k-1}\right] \tag{2.5}$$

where n is the number of observations, R^2 is the R-squared, and k is the number of independent variables.

2.2.9 Mean Absolute Error

It measures the error of residual. In other words, it is the mean absolute difference between the predicted value and the observed value.

$$Mean\ Absolute\ Error = \frac{1}{n}\sum_{i=1}^{n}|\ y_i - \hat{y}_i\ | \tag{2.6}$$

where n is the number of observations and y_i is the observed value and $\hat{y}_i$ is the predicted value.

2.3 RESULTS AND DISCUSSION

The initial phase in constructing a multiple linear regression model involves verifying certain assumptions. Upon confirming the fulfillment of these assumptions, the subsequent course of action focuses on mitigating multicollinearity. The detection of infinite VIFs for all variables prompted the utilization of PCA to condense the feature set from 311 to 16. In the context of K-means clustering, following preprocessing, the application of the Elbow method determined the optimal number of clusters to be 3. This process effectively reduced the dimensionality from 311 features to 3. Subsequently, the PLS model was developed based on these three clusters. The ultimate model encompasses both PLS and PCA, resulting in a refined feature set shrinking from 311 to 15. Below, we delve into the assessment of these three models.

2.3.1 Assumption 1: Linearity Checking for Multiple Linear Regression

The residuals show a random pattern. Hence it can be concluded that relationship is linear between the independent variable and dependent variable. Figure 2.3 shows the plot of actual values and predicted values.

2.3.2 Assumption 2: Independence

As the distribution centered around zero, it can be concluded that distribution of residuals follows a normal distribution with a mean zero and a variance of σ^2. Figure 2.4 shows the distribution of residuals.

2.3.3 Assumption 3: Autocorrelation

As per Figure 2.5 which shows the Durbin–Watson test, the test result indicated 1.82 which is around 2. Hence, it can be concluded that there is no autocorrelation. The residuals are not dependent on each other.

2.3.4 Assumption 4: Homoscedasticity

From Figure 2.6 of the assumption of homoscedasticity, the results satisfied as there is an even spread of residuals. However, there is one outlier present from the above graph.

2.3.5 The Number of Clusters in K-means Clustering

In the above plot, we can see that there is a kink at $K = 3$. Hence $K = 3$ can be considered a good number of the cluster to cluster this data. Figure 2.7 shows the Elbow method.

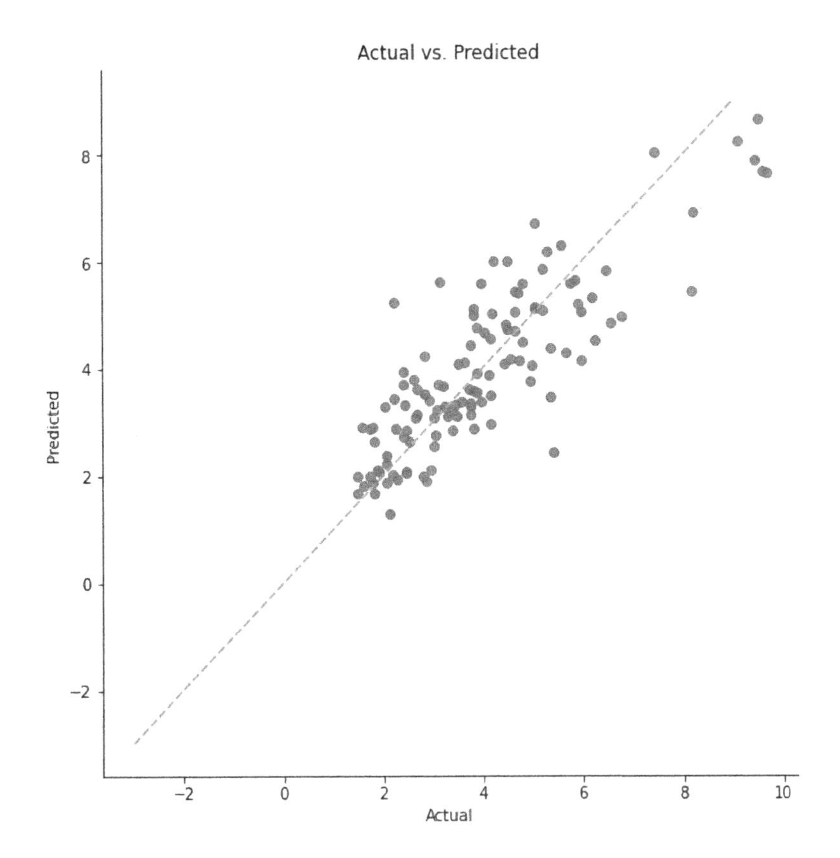

FIGURE 2.3 Assumption 1: Linearity.

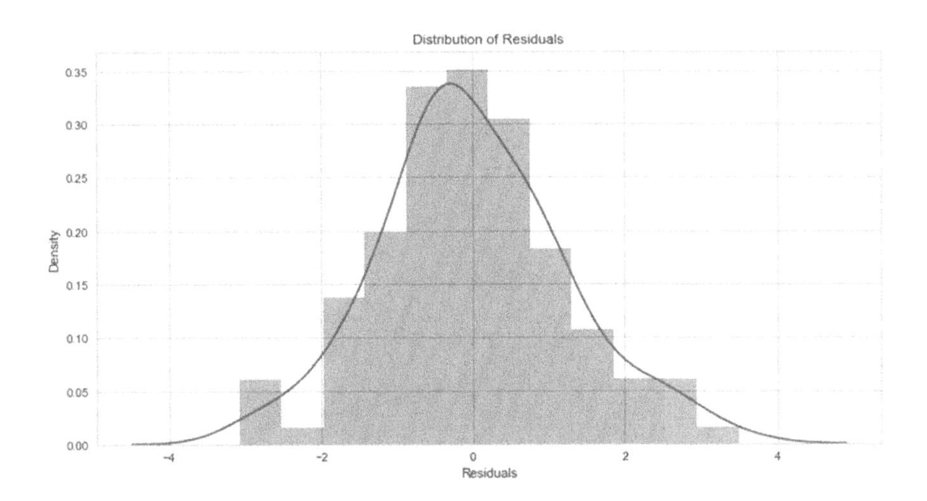

FIGURE 2.4 Assumption 2: The distribution of Residuals.

```
Assumption 3: No Autocorrelation


Performing Durbin-Watson Test
Values of 1.5 < d < 2.5 generally show that there is no autocorrelation in the data
0 to 2< is positive autocorrelation
>2 to 4 is negative autocorrelation
--------------------------------------
Durbin-Watson: 1.8181544450135014
Little to no autocorrelation

Assumption satisfied
```

FIGURE 2.5 Assumption 3: Autocorrelation.

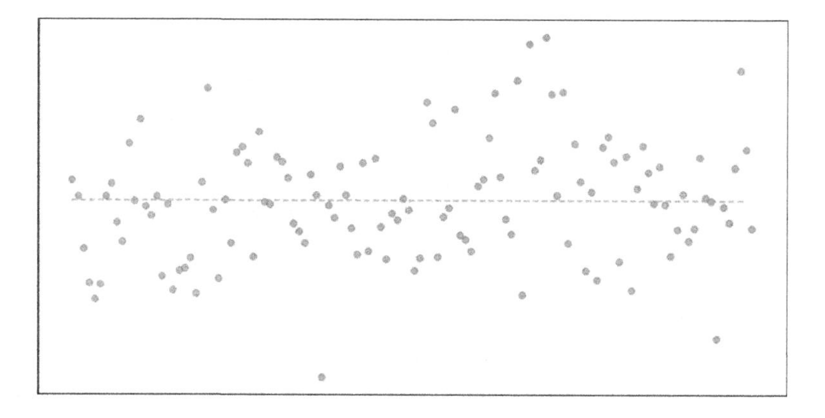

FIGURE 2.6 Assumption 4: Homoscedasticity.

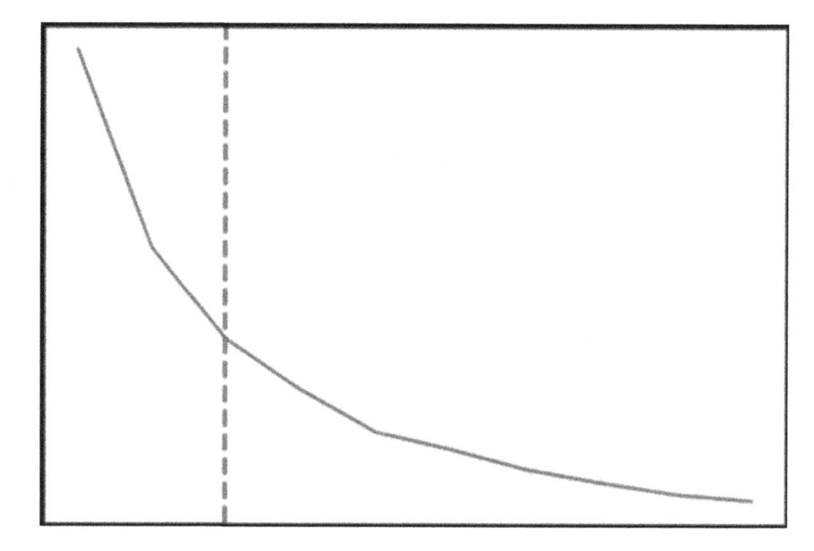

FIGURE 2.7 The Elbow method.

TABLE 2.1 Overview of Liquid and Gaseous Fuels Generated from Lignocellulose and Their Properties

Accuracy	MLR Model with PCA	MLR Model with *K*-Means Clustering	PLS Model with PCA	PLS Model with *K*-Means Clustering
R-squared	0.6018	0.5324	0.6018	0.5273
Mean square error	1.3357	1.5332	1.3357	1.5616
Root mean square error	1.1407	1.2277	1.1407	1.2408
Adjusted R-squared	0.5444	0.5203	0.5444	0.5151
Mean absolute error	0.8937	0.9884	0.8937	1.0000

2.4 MODEL COMPARISON

Table 2.1 summarize the results. The R-squared values obtained from both the PLS model and the multiple linear regression model are identical, standing at 0.6018. This value signifies a 60% explanation of the variation or 60% fitness to the model. The R-squared value for the multiple linear regression model, when utilizing K-means clustering, is 0.5324, explaining around 53% of the variation. Similarly, the R-squared value from the PLS model is 0.5273, accounting for approximately 53% of the variation. Notably, both the multiple linear regression model and the PLS model exhibit consistent outcomes across other accuracy metrics.

2.5 CONCLUSION

In summary, the primary objective of this study was accomplished, involving the construction of a predictive model for fat depth. The investigation centered around the PLS model. Despite the similarities in results between the multiple linear regression and PLS models, challenges such as multicollinearity and violations of the homoscedasticity assumption due to outliers were identified. Consequently, the accuracy of the multiple linear regression results could be compromised, necessitating a more robust statistical approach involving penalization techniques. Successful application of multiple linear regression hinges on the fulfillment of all underlying assumptions. In conclusion, the performance of the PLS model outshines that of the multiple linear regression model in terms of R-squared. This study establishes a foundational understanding of both the models. Future directions could involve advancing multiple linear regression model to incorporate Lasso regression and exploring avenues such as orthogonal projection to latent structures or nonlinear kernel PLS for enhancing the PLS approach. In addition, readers inclined toward further research could explore alternative shrinkage methods or variable selection techniques prior to embarking on the model construction process. This study has the potential for further advancement through the analysis of interactions and binary logic, adding an additional dimension to its findings.

ACKNOWLEDGMENT

This study was undertaken through the Advanced Livestock Measurement Technologies Project (ALMTech) and funded by the Department of Agriculture, Rural Research and Development for Profit program and Meat and Livestock Australia.

REFERENCES

1. Marimuthu, J., Hocking-Edwards, J., & Gardner, G. (2018). Non-invasive technique using low cost portable microwave system on carcase for fat depth measurement. ICoMST 2018. Melbourne, Australia.
2. Fowler, M., Qin, M., Fiscus, S. A., Currier, J., Flynn, P., Chipato, T., Mofenson, L. (2017). Benefits and Risks of antiretroviral therapy for perinatal HIV prevention. Obstetric Anesthesia Digest, 37, 154–155. doi:10.1097/01.aoa.0000521253.11653.d9.
3. Marimuthu, J., Konstanty S. Bialkowski, & Abbosh, A. (2016). Software-defined radar for medical imaging. IEEE Transactions on Microwave Theory and Techniques, 64, 663. doi:10.1109/TMTT.2015.2511013.
4. Marimuthu, J., & Gardner, G. E. (2019). P8 and rib fat depth measurement on beef carcase using a portable microwave system. ICoMST 2019. Potsdam, Germany.
5. Marimuthu, J., Loudon, K., & Gardner, G. (2021). Ultrawide band microwave system as a non-invasive technology to predict beef carcase fat depth. Meat Science, 179, 108455.
6. Campbell, A., & Ntobedzi, A. (2007). Emotional intelligence, coping and psychological distress: a partial least squares approach to developing a predictive model. E-journal of Applied Psychology, 3, 39–54.
7. Rosipal, R., & Krämer, N. (2005). Overview and recent advances in partial least squares. Paper presented as a part of the "Subspace, Latent Structure and Feature Selection". Lecture Notes in Computer Science series.

CHAPTER 3

Single Feature Imbalance Classification on Ensemble Learning Methods for Efficient Real-Time Malware Detection

Azali Saudi and Julkifli Awang Besar

3.1 INTRODUCTION

Malware include viruses, worms, Trojans, ransomware, spyware, adware, and rootkits, created for harmful purposes like stealing data or disrupting systems. Fighting malware is vital for cybersecurity due to its wide-ranging forms and impacts. These attributes help identify whether software is malware. In this chapter, the malware datasets are scrutinized through the use of virtual instances. To assess the impacts of these malware, they are executed securely within a virtual environment. The benign dataset is obtained by gathering executable programs designed for the running virtual Windows 10 environment. This dataset typically consists of harmless software or files that are used as a reference or comparison in the experiment. These programs are not considered malware and are used to establish a baseline for normal system behavior. By contrasting the behavior of

DOI: 10.1201/9781003400387-3

potentially malicious software with the benign dataset, malware program can be identified and detected more effectively.

Malware and benign program classification is widely carried out using classical machine learning algorithms like k-nearest neighbors (kNN), support vector machine (SVM), decision tree, Naive Bayes (NB), and stochastic gradient descent (SGD). In addition, ensemble algorithms based on Bagging techniques, including standard bagging with DT as the base estimator (bagging-DT), random forest, and extra trees, are employed. Furthermore, ensemble algorithms utilizing Boosting methods, such as AdaBoost, Gradient Boosting Machine (GBM), XGBoost, LightGBM, and CatBoost, are also applied. Ensemble algorithms based on Voting classifier and Stacking generalization are also employed. Given that the distribution of samples between benign and malware classes in malware classification often suffers from an imbalance, this study also explores modified ensemble algorithms designed to address such issues. These modified algorithms include bagging with random undersampling (bagging-RU), random forest with class weighting (RF-W), and random forest with random undersampling (RF-U).

3.2 LITERATURE REVIEW

Choi [1] introduced a vantage point tree with similarity hash and kNN for malware detection, achieving a 25% detection rate improvement, 67% faster detection, and 20% reduced search times. Yilmaz et al. [2] used SVM and NB to classify Android apps as malicious or benign, achieving high accuracy rates of 90.9% and 92.4% using 116 permission features. Maheswari et al. [3] introduced a malware classification system using SVM-SGD, achieving a remarkable accuracy of 99.13% on the CIC-IDS2017 dataset, reducing false alerts compared to signature-based detection. Smmarwar et al. [4] proposed a feature selection framework that significantly improves machine learning model performance. They tested it with RF, DT, and SVM classifiers on the CIC-InvesAndMal2019 dataset, achieving high accuracy rates, with RF reaching 91.32%.

Hussain et al. [5] developed a malware detection system using machine learning, with RF achieving a high accuracy of 99.44%. This suggests potential for a customizable Windows desktop malware scanning application. In ref [6], a study introduced a feature selection method for malware detection that reduced dimensionality, improving ensemble model training times while maintaining high accuracy. XGBoost performed the best among the models tested in terms of accuracy. Wu et al. [7] proposed a malware detection approach using a three-tier cascading XGBoost model with cost sensitivity to handle imbalanced data. They gather data by analyzing API calls in Portable Executable files. The experimental results highlight its high accuracy in detecting malicious code in unbalanced datasets. Al-kasassbeh et al. [8] developed an intelligent anti-malware system for Internet of Things (IoT) using LightGBM. It achieved nearly 100% accuracy in detecting and categorizing malware and effectively identified IoT botnet attacks, proactively stopping their spread in the network.

Agrawal and Trivedi [9] assessed supervised algorithms for Android malware detection. They created their dataset by collecting malware files from various repositories, resulting in 16,300 records with 215 features. CatBoost was the best performing classifier, achieving 93.15% accuracy.

Machine learning is widely used malware detection technique; however, relying on a single base classifier like kNN or SVM for improved detection is challenging. Therefore, ensemble classifiers such as RF, AdaBoost, and GBM are preferred. This study explores modified bagging and boosting algorithms in the context of a single-feature dataset with imbalanced class distribution, avoiding the need for multiple features and associated preprocessing, resulting in faster training times.

3.3 THE CLASSICAL MACHINE LEARNING ALGORITHMS

Classical machine learning methods typically involve the manual extraction of features from the data to address the problem at hand. This section describes the five classical machine learning algorithms examined in this study.

3.3.1 K-Nearest Neighbors

The kNN algorithm is a supervised machine learning method applicable for handling both classification and regression problems. It operates under the assumption that items sharing similarities are located near each other in the data space. The effectiveness of the KNN algorithm heavily depends on the validity of this assumption [10].

In kNN, the entire training dataset is stored, and during prediction, the algorithm retrieves the *k* most similar training patterns. It makes predictions based on the distances between data instances and often achieves good results. In classification, kNN selects the mode (most frequent class) from the *k* most similar training instances, as illustrated in Figure 3.1.

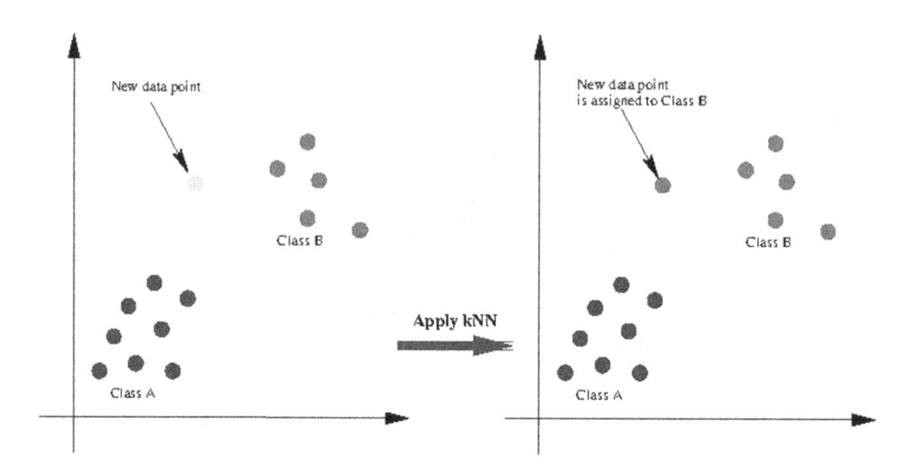

FIGURE 3.1 The kNN algorithm assumes that similar items are located near to each other.

3.3.2 Support Vector Machine

The SVM algorithm aims to identify the optimal boundary line that effectively segregates data points into their respective categories. This optimal boundary is referred to as a hyperplane. SVM, in this process, singles out the most critical data points that play a pivotal role in constructing this hyperplane. These crucial data points are known as support vectors, hence lending the algorithm its name, support vector machine [11].

SVM transforms data into a higher-dimensional feature space for better classification. It uses kernel functions to calculate dot products between transformed feature vectors, avoiding costly computations. SVM handles both linear and nonlinear data using various kernel functions. In training, it finds the optimal hyperplane in the higher-dimensional space, maximizing the margin between different class data points while minimizing errors, as shown in Figure 3.2.

3.3.3 Decision Tree

A decision tree serves as a valuable tool for providing clear and explicit explanations of decisions and choices made in decision-making processes. It adopts a model resembling a tree structure, as implied by its name. While its primary application was initially in data mining for devising strategies to achieve specific objectives, it has found extensive use in machine learning as well [12].

Decision trees represent a form of predictive modeling that assists in linking various decisions or courses of action to a particular outcome.

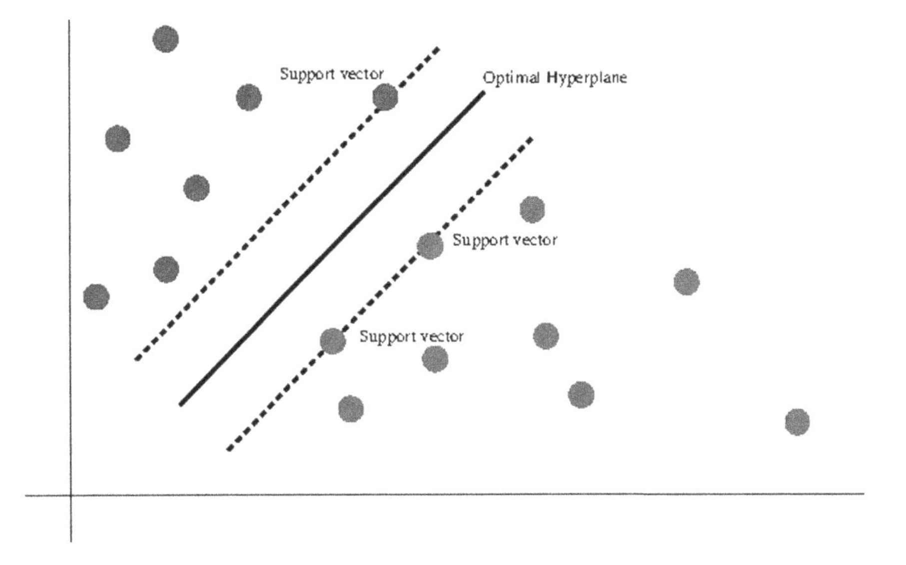

FIGURE 3.2 The support vector machine uses hyperplane to separate different class of items.

These trees consist of various nodes. The starting point of the decision tree is referred to as the root node, which in machine learning often represents the entire dataset. At the other end of a branch, representing the final result of a sequence of decisions, is what we call a leaf node. In the context of machine learning with decision trees, the data features are denoted as internal nodes within the tree, while the final result or outcome is represented by the leaf node as depicted in Figure 3.3.

3.3.4 Naive Bayes

The NB classifier, widely regarded as one of the simplest but efficient classification algorithms, holds good flexibility, making it a valuable tool in various domains of application. Its ability extends to scenarios involving large dataset, thus providing an excellent choice for handling large volumes of information.

NB excels in various practical applications like spam filtering, text classification, sentiment analysis, and recommender systems, demonstrating its adaptability and relevance. In malware detection [13], it achieved impressive accuracy of 93% with static characteristics and 85% with dynamic characteristics. This classifier relies on probability theory, particularly Bayes' theorem, for predictions, enabling informed decision-making and

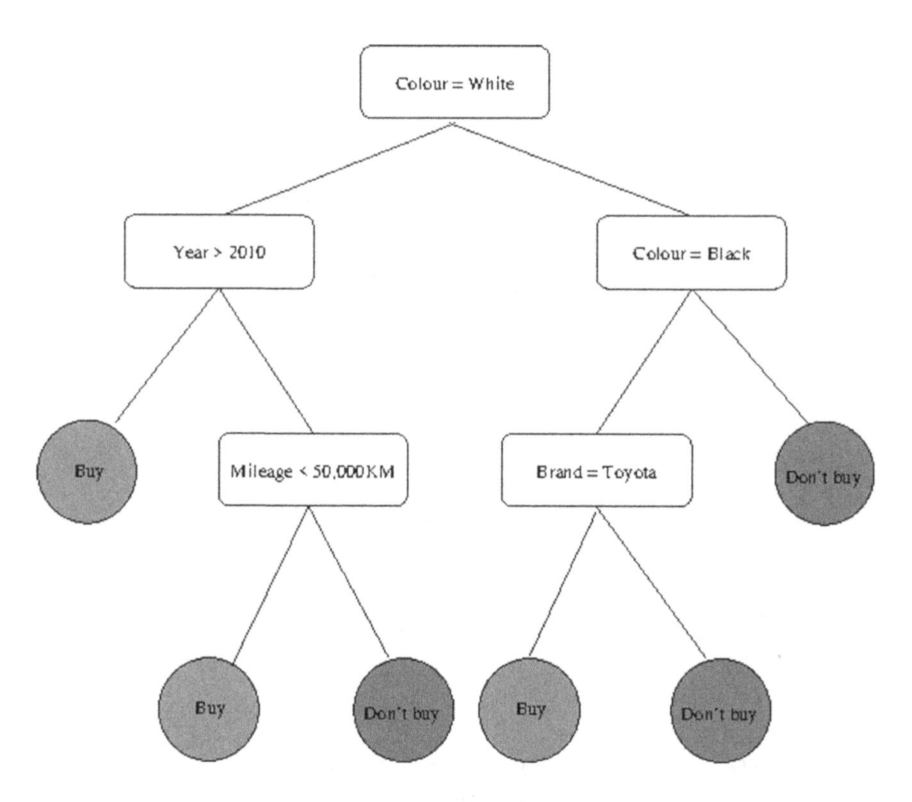

FIGURE 3.3 In a decision tree, the leaf node represents the final outcome.

classification. Its simplicity, computational efficiency, and reliability make it a crucial tool in machine learning and data analysis.

3.3.5 Stochastic Gradient Descent

SGD stands as a notable variant of the gradient descent algorithm, tailor-made for the optimization of machine learning models. This specialized approach addresses a pivotal concern: the computational inefficiency that arises when conventional gradient descent techniques are deployed to handle extensive datasets within the domain of machine learning. In ref [14], SGD was successfully applied in predicting harmful malware from two datasets.

SGD differs from standard gradient descent as it is more efficient by selecting a random training example or a small batch for gradient calculation and model updates, reducing computational overhead. This makes SGD highly efficient for large datasets compared to the resource-intensive standard gradient descent, making it a preferred choice for machine learning with big data.

3.4 THE ENSEMBLE LEARNING ALGORITHMS

Ensemble models in machine learning combine the decisions from multiple models to improve the overall accuracy of the classifier. In this study, we examine ensemble learning techniques based on bagging and boosting methods due to their light implementations that are suitable for low-powered computing devices.

Ensemble learning algorithms combine the outcomes of multiple models to enhance the overall accuracy of the classifier. In this work, we investigate ensemble learning methodologies that draw inspiration from bagging and boosting techniques. These approaches are chosen for their efficient implementations, making them particularly suitable for computing devices with limited processing power.

3.4.1 Bagging Algorithms

The bagging model combines the outcomes of multiple models to yield an improved result [15]. The bagging technique involves utilizing bags of observations from the original dataset, as illustrated in Figure 3.4. This approach serves as a means to diminish the variance associated with a base estimator, such as a decision tree, by introducing a degree of randomization during its construction and subsequently forming an ensemble from it. In many instances, the bagging method offers a straightforward

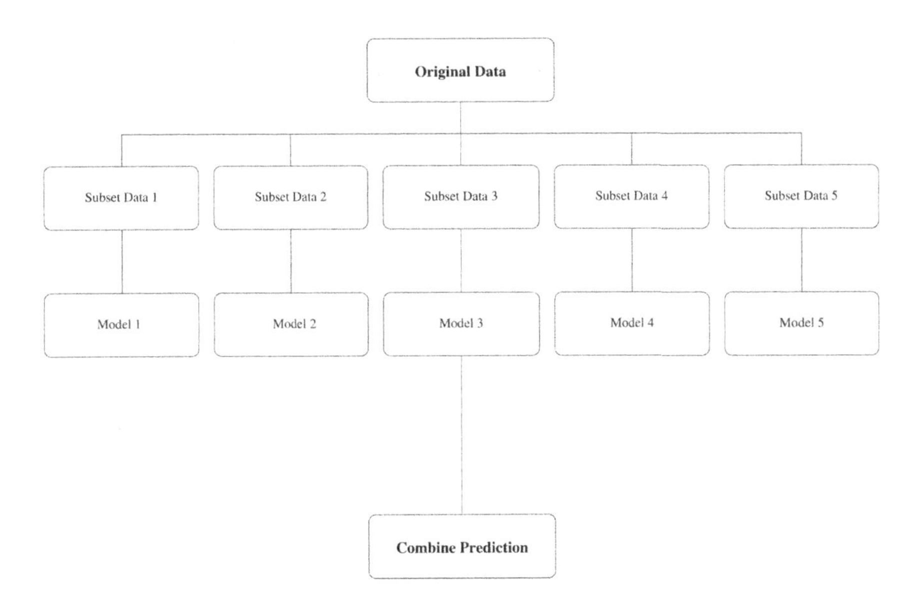

FIGURE 3.4 In bagging, multiple subsets are created from the original dataset.

approach to enhancing performance compared to a single model, all without necessitating adaptations to the underlying base algorithm.

Given its effectiveness in minimizing overfitting, bagging method performs optimally when applied to strong and complex models such as fully developed decision trees. This stands in contrast to the boosting method, which tends to excel when used with weak models. For each subset, a base weak model is generated, and these models operate independently and concurrently. The ultimate prediction is derived by consolidating the predictions made by all of these weak models.

3.4.1.1 Bagging Classifier

A bagging classifier serves as an ensemble meta-estimator. It operates by training individual base classifiers on random subsets extracted from the original dataset. These base classifiers' predictions are then combined, either through voting or averaging, to produce a final prediction as illustrated in Figure 3.5. This meta-estimator is often employed to minimize the variance of a black-box estimator, such as a decision tree. This is achieved by introducing a level of randomness into the construction process and subsequently creating an ensemble from it.

The bagging classifier initially involves the selection of random samples from a training dataset, allowing for replacement. Each of these data samples is then associated with a weak learner model, commonly employing decision trees. Ultimately, the predictions generated by all these weak learners are aggregated to produce a single prediction [16].

3.4.1.2 Random Forest

The random forest classifier, as shown in Figure 3.6, is a model formed by aggregating decision trees through the process of bagging [17]. A decision

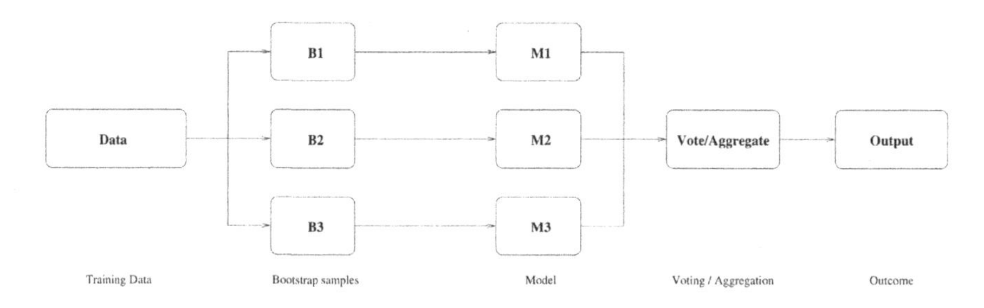

FIGURE 3.5 The standard bagging combines the predictions generated by weak learners to produce a single prediction.

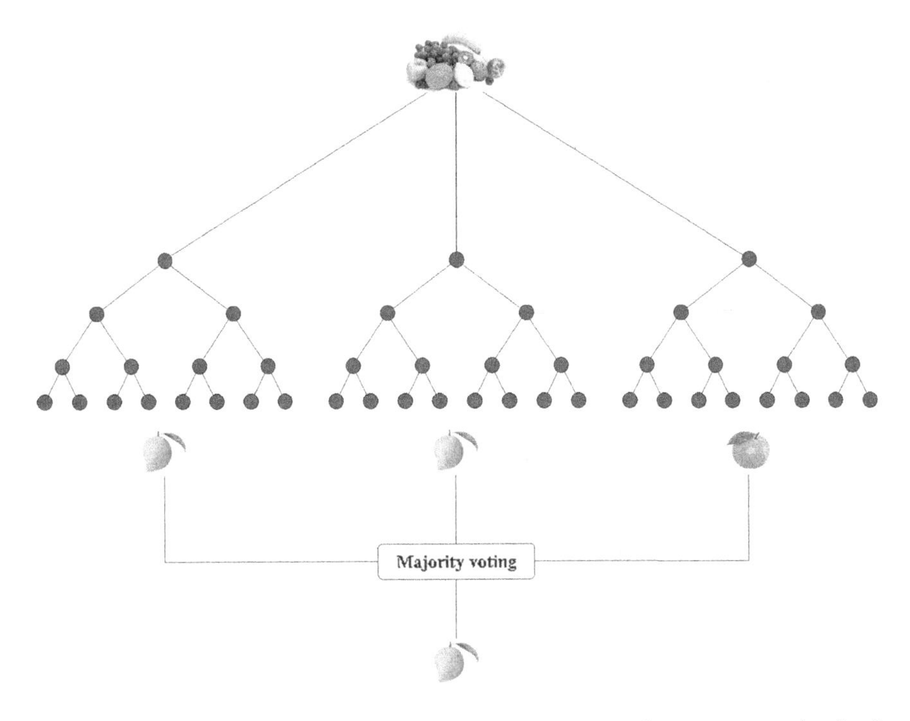

FIGURE 3.6 The random forest applies bagging approach to generate the final outcome.

tree is a graphical representation of actions and outcomes used for goal planning. It expresses information in a tree structure, enhancing clarity. It is a predictive model with nodes representing variables and branches splitting based on conditions. Edges connect nodes, defining paths by feature values, and leaf nodes represent predicted classes. It is a dynamic model for predictions based on variable values, not just a set of rules.

When creating a decision tree from a dataset, it is crucial to establish stopping criteria during the training phase to control its growth. An excessively branching tree can lead to high computational complexity with minimal gains in classification accuracy. Bagging, with data sampling without replacement, combines multiple models of the same type from the same dataset. Random forest consists of a collection of decision trees, each trained on a random subset of variables. The final classification is based on the most frequent outcome among the individual decision tree classifiers.

3.4.1.3 Extra Tree

The extra tree algorithm employs a high degree of randomization in both the selection of attributes and the determination of cut-points when dividing a tree node. In its most extreme manifestation, this algorithm constructs entirely randomized trees, whose structures remain entirely unrelated to the output values within the learning dataset. The degree of randomization can be tuned to the specific problem at hand by selecting an appropriate parameter [18].

Random forest introduces randomness by using random subsets of the dataset. Like in decision trees, the algorithm may initially choose a random threshold (e.g., 90) to split the dataset. If a student scores over 90, they pass; otherwise, the algorithm selects another random threshold (0–90) for further classification. This process results in a more extensive tree but significantly faster due to random threshold selection. Extra tree, which uses this random approach, is faster than random forest, making it advantageous for large datasets and speedy decision tree ensemble evaluation.

3.4.2 Boosting Algorithms

Boosting is a step-by-step process that aims to correct errors made by the previous model [19]. Each model builds on insights from the previous one, resulting in a more robust collective model, as depicted in Figure 3.7. While individual models may not perform well across the entire dataset,

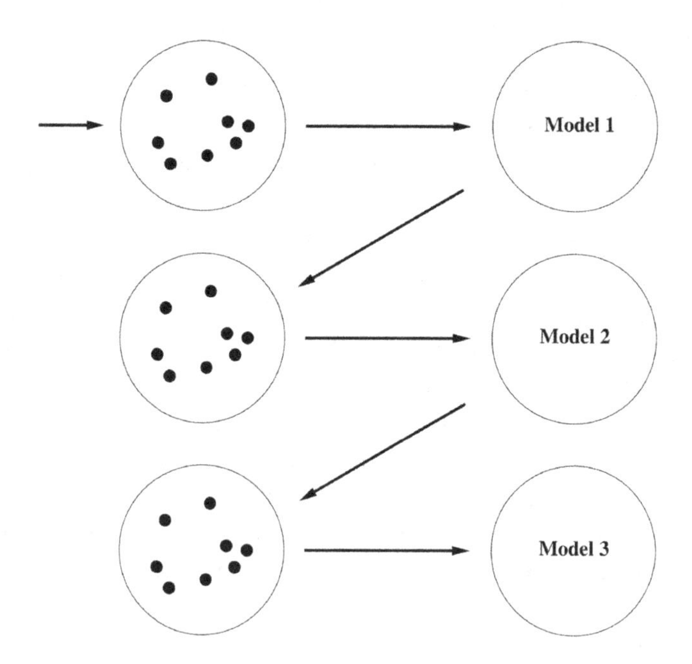

FIGURE 3.7 Boosting is like bagging but with the difference that it trains a sequence of weak learners that try to correct the mistakes of their predecessors.

they excel in specific portions, contributing to the overall performance of the boosting ensemble.

Boosting, like bagging, trains a series of weak learners using samples from the training dataset. However, unlike bagging, boosting selects samples without replacement, ensuring each instance is chosen only once. Weak learners are trained sequentially. Initially, a subset is drawn and a weak learner is trained. The process continues, with each subsequent subset including instances that the previous learner misclassified. Differences between learners are used to train the next one. The boosting model makes predictions through majority voting, similar to bagging. Boosting is better at reducing bias and variance compared to bagging in machine learning models.

3.4.2.1 AdaBoost

AdaBoost, short for adaptive boosting, is among the basic boosting algorithms [20]. Typically, modeling is performed using decision trees. Multiple sequential models are constructed, with each subsequent model aimed at rectifying the errors made by its predecessor as illustrated in Figure 3.8. AdaBoost introduces weights to the observations that are inaccurately

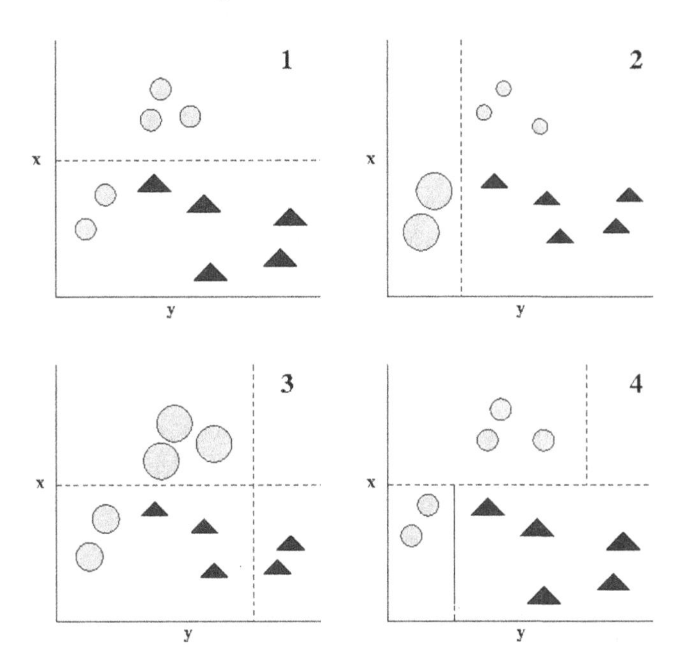

FIGURE 3.8 In AdaBoost model, each subsequent model attempts to correct the error made by the previous model.

predicted, prompting the following model to attempt for precise predictions on these instances.

Let us assume an adaptive boosting ensemble with three classifiers. Classifier 1 (clf1) initially classifies points (picture 1). Misclassified points gain higher weights, while correctly classified ones get lower weights. Classifier 2 (clf2) is trained with adjusted weights, shown in picture 2. This process continues with classifier 3 (clf3). Finally, an ensemble classifier (clf4) is formed by combining the outputs of these three classifiers, as depicted in picture 4. This ensemble, created through adaptive resampling, illustrates the AdaBoost classifier's functioning.

3.4.2.2 Gradient Boosting Machine

Gradient Boosting Machine (GBM) builds prediction models by combining smaller models, often represented as decision trees [21]. Like previous boosting methods, it uses an incremental approach but stands out for its ability to optimize differentiable loss functions. GBM is highly adaptable, making it suitable for diverse tasks, and it is compatible with various programming languages and operating systems, ensuring excellent portability.

Gradient boosting is a highly effective method for building predictive models. These classifiers are based on AdaBoosting and weighted minimization, with a focus on reducing the loss or variance between actual and predicted class values. In gradient boosting, the weights of prior learners remain unchanged when adding new ones, unlike AdaBoosting where values are adjusted. GBM's strength lies in its versatility, making it suitable for both binary and multi-class classification problems, as well as regression tasks.

3.4.2.3 XGBoost

XGBoost improves upon the GBM framework through system optimizations and algorithmic enhancements [22]. XGBoost employs cache-aware algorithms with internal buffers, prioritizes parallelization in tree construction through nested loops, and offers control over the tree split process using parameters like max depth. It has gained huge popularity in Kaggle competitions, where participants compete to create the most effective data prediction models.

XGBoost initially implemented in Python and R, and has expanded to include implementations in Java, Scala, Julia, Perl, and more. It integrates well with tools like scikit-learn for Python and caret for R, as well as distributed processing frameworks like Apache Spark and Dask. In 2019, XGBoost received the Technology of the Year award from InfoWorld.

3.4.2.4 LightGBM

The gradient boosting decision tree is a widely used artificial intelligence machine learning model that employs multiple elementary learners, iteratively training them to arrive at the optimal model. Within this domain, LightGBM emerges as a distributed and highly efficient framework [23]. It adopts a gradient boosted decision tree approach, refining a new decision tree by approximating the negative gradient of the loss function as the residual of the current decision tree.

In LightGBM, the gradient-based one-side sampling (GOSS) technique keeps data instances with significant gradients and randomly samples those with minor gradients. Gradients represent error slopes, crucial for finding optimal splits. Smaller gradients maintain accuracy in decision trees. This approach speeds up training by reducing data instances.

3.4.2.5 CatBoost

CatBoost is an open-source software package focused on providing top-tier performance in gradient boosting with decision trees [24]. During training, it constructs a sequence of trees, each with lower loss than the previous one, leading to unique tree structures. CatBoost automatically quantizes feature values, determining thresholds to create disjoint intervals for features and labels. It supports categorical data and offers a GPU-compatible version. It is known for its accuracy, robustness, practicality, and extensibility, and it is user-friendly. In addition, CatBoost can integrate seamlessly with deep learning frameworks like TensorFlow and Core ML.

CatBoost employs symmetric trees, ensuring consistent splitting conditions at the same depth, while LightGBM and XGBoost create asymmetric trees with varying conditions. Symmetric trees aim for the lowest loss at each depth, offering computational speed, efficient evaluation, and overfitting control. LightGBM uses horizontal growth, making more compact models compared to XGBoost's vertical growth. CatBoost and LightGBM outperform XGBoost, especially with larger datasets, despite all three generally delivering similar performance.

3.4.3 Voting Classifier

The voting classifier is a machine learning technique that combines predictions from multiple models to make a final prediction. Instead of relying on a single model, it creates an ensemble of various models and determines the output class by either selecting the highest probability or the majority vote from these individual models [25]. This approach simplifies model selection and boosts accuracy by leveraging the collective knowledge of the ensemble. Rather than training and assessing separate models independently, the voting classifier creates a single model that generates predictions based on the majority vote of the ensemble's models for each output class.

3.4.4 Stacking Generalization

The utilization of stacking ensemble models plays a pivotal role in reducing both variance and bias, thereby enhancing the predictive capabilities of the model. Stacking ensemble model can be utilized to improve not only the accuracy of predictions but also to minimize variance in our computational framework.

To construct this ensemble model, multiple different base-learners can be used, each with its unique strengths and capabilities. The base-learners include kNN, SVM, DT, RF, etc. Each of these base-learners brings a different perspective and approach to the predictive task, contributing to a rich and robust ensemble. In ref [26], multi-layer perceptron, SVM, and RF algorithms are used as low-level learners in stacking generalization model.

3.4.5 Bagging and Boosting for Imbalance Classification

Both bagging and boosting algorithms have demonstrated their effectiveness across a wide range of predictive modeling tasks. However, their suitability diminishes when applied to classification problems characterized by a skewed class distribution. Despite their inherent strengths, these algorithms struggle in the face of severe class imbalances. Nevertheless, various modifications have been suggested to adjust their behavior and render them more adept at handling significant class imbalances. Consequently, this study delves into the examination of enhanced bagging and boosting algorithms tailored for imbalanced classification scenarios.

3.4.5.1 Bagging with Random Undersampling

One approach to adapting bagging for use in imbalanced classification scenarios involves performing data resampling on the dataset prior to fitting the weak learner model. This resampling involves either oversampling the minority class or undersampling the majority class. Specifically, OverBagging refers to the technique of oversampling the minority class within a dataset, while UnderBagging involves undersampling the dominant class. The implementation of UnderBagging can be achieved through the utilization of the imbalanced learn package, which introduces a bagging variation designed to balance the two classes. This is achieved by applying a random undersampling method to the majority class within a bootstrap sample.

3.4.5.2 Random Forest with Class Weighting

One simple method for adapting a decision tree to address unbalanced classification is by modifying the weight assigned to each class when calculating the impurity score at a specified split point [27]. Impurity serves as a metric for assessing the degree of mixing within groups of samples for a given split in the training dataset. Typically, this is computed using metrics like Gini or entropy. By adjusting the formula, it is possible to bias it in favor of a minority class, which may result in some false positives for the dominant class. To achieve this, the class weight parameter within the RandomForestClassifier class can be employed. This parameter takes a dictionary containing a mapping between each class value (e.g., 0 and 1) and its associated weight.

3.4.5.3 Random Forest with Bootstrap Class Weighting

Since every decision tree is built from a bootstrap sample, wherein data points are randomly selected with replacement, the class distribution within each data sample can vary across different trees. This variability presents an opportunity to adapt class weighting by considering the class distribution within each bootstrap sample rather than the aggregate class distribution of the entire training dataset. This approach can potentially enhance the model's robustness to class imbalances by tailoring the class weights to the specific characteristics of each subset, thereby potentially minimizing issues like overfitting to the majority class. An extensive empirical assessment of random forest algorithm constructed from imbalanced data was reported in [28].

3.4.5.4 Random Forest with Random Undersampling

Another highly effective strategy to improve the performance of random forests is the deliberate modification of the class distribution within the bootstrap samples through a process known as data resampling [29]. This approach is elegantly implemented in the imbalanced learn library, specifically within the BalancedRandomForestClassifier class. Balanced random forest takes a unique approach by addressing class imbalance directly. Instead of leaving the class distribution within bootstrap samples to chance, it systematically employs random undersampling of the majority class in each bootstrap sample.

The technique ensures that each subset used for building individual decision trees in the random forest ensemble has a balanced mix of minority and majority classes. Balanced random forest introduces class distribution control in each bootstrap sample, reducing class imbalance's negative impact on performance. It is especially beneficial for skewed datasets, preventing the model from favoring the majority class excessively and leading to fairer, more accurate predictions across all classes.

3.4.5.5 Easy Ensemble

Easy ensemble handles class imbalance by selecting all minority class instances and a subset from the majority class, using boosted decision trees, specifically AdaBoost, to process these subsets [30]. Instead of relying on pruned decision trees, easy ensemble harnesses the power of AdaBoost, a boosting algorithm renowned for its ability to enhance model performance.

AdaBoost works by fitting an initial decision tree on the dataset, serving as the starting point for the ensemble. It then assesses the errors made by this tree, assigning weights to each example in the dataset based on these errors. This crucial step prioritizes misclassified instances, directing more attention to them during subsequent training iterations while reducing the influence of correctly classified ones. Next, a decision tree is trained on the weighted dataset to correct AdaBoost-highlighted errors. This process iterates for a set number of decision trees, progressively emphasizing challenging instances. Easy ensemble repeats this process separately for subsets from the majority class, generating diverse learners through multiple randomizations. Finally, the outputs of all these learners are thoughtfully combined to produce a robust and well-balanced ensemble model.

3.5 EXPERIMENTAL SETUP

The dataset under consideration consists of 46 samples classified as malware, while there are 336 samples labeled as benign. These samples are specifically stored in the Portable Executable (PE) format, commonly utilized in Windows operating systems. The PE format is essentially a data structure encompassing various elements such as executable code, Dynamic Link Library (DLL) files, Application Programming Interface (API) export and import tables, as well as data associated with resource management. To facilitate the classification process, the PE header of all these samples is extracted. Moreover, a collection of DLL file names is assembled from these PE headers, forming a corpus of terms for subsequent use in classification tasks. This corpus encompasses both benign and malware DLLs.

The next step involves transforming this corpus into a matrix of Term Frequency-Inverse Document Frequency (TF-IDF) features. This transformation is achieved by leveraging the TfidfVectorizer implementation within the Python Sklearn library. The resulting feature set, which essentially represents the corpus of DLL file names, is subsequently divided into two distinct datasets: a training set comprising 67% of the data and a testing set containing the remaining 33%. Note that the distribution of classes within this corpus is imbalanced, with the number of malware samples accounting for a mere 12% of the total samples. This inherent class imbalance is a noteworthy characteristic of this dataset.

Five classical machine learning algorithms, two bagging algorithms, four boosting algorithms, a voting classifier with three base estimators, and a stacking method with three base estimators and a decision tree as its final estimator are employed for the purpose of malware detection using the generated corpus, which consists of a single feature derived from both benign and malware datasets. Given the inherent class imbalance in the dataset, four modified ensemble algorithms designed to address such imbalance in classification are also investigated.

The performance assessment of the proposed models is conducted utilizing widely recognized evaluation metrics commonly applied to assess a variety of existing classifiers. These metrics encompass accuracy, true positive (TP), true negative (TN), false positive (FP), and false negative (FN). In our study, true positive (TP) represents the correct identification of benign applications, whereas TP and TN collectively measure the accuracy of classifying benign and malicious software. Conversely, FP and FN metrics measure the rate at which benign software is erroneously classified as malicious, and vice versa. The true negative rate (TNR), which measures the proportion of actual negatives correctly identified, is described by the following formula: $TNR = TN/(TN + FP)$. The false negative rate (FNR) quantifies the proportion of all benign apps that are mistakenly identified as malicious apps and is calculated as: $FNR = FN/(FN + TP)$. Finally, the overall accuracy measures the rate at which the actual labels of all instances are correctly predicted: $accuracy = (TP + TN)/(TP + FP + TN + FN)$.

3.6 RESULTS

Table 3.1 summarizes algorithm performance, including classical machine learning (kNN, SVM, DT, NB, and SGD), bagging (bagging-DT, RF, and ET), boosting (AdaBoost, GBM, XGBoost, LightGBM, and CatBoost), voting classifier, and stacking. Classical machine learning algorithms (kNN, SVM, DT) achieved 50–65% accuracy, with DT performing above 70%. Ensemble algorithms (boosting, voting, stacking) generally outperformed bagging. Boosting-based methods and voting/stacking achieved 66–70% accuracy, while bagging showed 63–65%. CatBoost and voting had the highest accuracy, exceeding 70%.

Initially, RF performed poorly due to dataset imbalance. This issue highlighted a common machine learning challenge: Bagging and boosting algorithms struggle with imbalanced class distributions. To address this and potentially enhance classification, we analyzed the dataset further. We explored modified algorithms like bagging-RU, RF-W, and RF-U, designed for imbalanced datasets. The goal was to see if these modifications could overcome class distribution challenges and improve performance.

As illustrated in Table 3.2, all three classifiers demonstrated good performance, achieving accuracy levels exceeding 80%. The output predictions generated by these algorithms were binary, falling into either the false or true categories.

TABLE 3.1 Performance of the Examined Algorithms

Classifier	Accuracy (%)
kNN	51.47
SVM	52.94
DT	65.48
NB	70.56
SGD	71.88
Bagging-DT	65.01
RF	65.46
Extra tree	63.87
AdaBoost	68.65
GBM	66.76
XGBoost	67.19
LightGBM	66.74
CatBoost	70.59
Voting (kNN, DT, SGD)	71.11
Stacking [(kNN, DT, SGD), DT]	68.03

TABLE 3.2 The Performance of the Imbalance Classifiers

Imbalance Classifier	Configuration	Accuracy (%)
Implementation with imblearn.ensemble library		
Balanced bagging-RUS	imblearn.ensemble; BalancedBaggingClassifier Classifier with default parameter	85.09
Balanced RF-RUS	imblearn.ensemble; BalancedRandomForest Classifier with default parameter	88.64
Easy ensemble (bag of balanced boosted learners)	imblearn.ensemble; EasyEnsembleClassifier with default parameter	83.56
RUS-AdaBoost (RUS integrated into AdaBoost)	imblearn.ensemble; RUSBoostClassifier with default parameter	82.18
Implementation with class weighting using class_weight or scale_pos_weight parameters		
RF-W	RandomForestClassifier; class_weight="balanced"	83.21
RF-BW	RandomForestClassifier; class_weight="balanced_subsample"	82.76
ET-W	ExtraTreesClassifier; class_weight="balanced"	84.57
ET-BW	ExtraTreesClassifier; class_weight="balanced_subsample"	88.28
XGBoost-W	XGBClassifier; scale_pos_weight=88	85.72
LightGBM-W	LGBMClassifier; class_weight="balanced"	83.24
LightGBM-W	LGBMClassifier; scale_pos_weight=0.88	82.27
CatBoost-W	CatBoostClassifier; class_weights=[0.12, 0.88]	86.18
Implementation of random oversampling using RandomOverSampler library with SMOTE		
kNN-ROS		85.00
SVM-ROS		84.55
DT-ROS		90.45
Bagging-ROS		90.00
RF-ROS		87.73
ET-ROS		88.64
AdaBoost-ROS		90.45
GBM-ROS		90.91
XGBoost-ROS		90.91
LightGBM-ROS		82.27
CatBoost-ROS		90.45
Implementation of random undersampling using RandomUnderSampler library		
kNN-RUS		85.45
SVM-RUS		84.55
DT-RUS		90.45
Bagging-RUS		91.36

(Continued)

TABLE 3.2 (Continued) The Performance of the Imbalance Classifiers

Imbalance Classifier	Configuration	Accuracy (%)
RF-RUS		89.09
ET-RUS		87.73
AdaBoost-RUS		90.45
GBM-RUS		90.91
XGBoost-RUS		90.91
LightGBM-RUS		82.27
CatBoost-RUS		90.45

However, in the real-world context of malware detection, the primary emphasis lies in minimizing the false positive rate (FPR) while simultaneously maximizing the true positive rate (TPR) of detection. These metrics are essential in evaluating the classifier's effectiveness. The FPR and TPR are computed using the following formulae:

$$\text{False Positive Rate (FPR): FPR} = \text{FP}/(\text{FP} + \text{TN})$$

$$\text{True Positive Rate (TPR): TPR} = \text{TP}/(\text{TP} + \text{FN})$$

Where FP represents the number of false positives, TN denotes the number of true negatives, TP signifies the count of true positives and FN corresponds to the number of false negatives. To obtain the optimal balance between the FPR and TPR, we employed a prediction probability ranging from 0 to 1. In this context, values approaching 0 denote true negatives, whereas values approaching 1 represent true positives. Leveraging this approach in conjunction with the aforementioned formulae, we observed encouraging results for the modified algorithms.

Specifically, the bagging-RU algorithm required an FPR of 7.27% to achieve a TPR of 52.94%. Notably, it managed to maximize the TPR at an impressive 82.35% while maintaining an FPR of 11.82%. The RF-W algorithm exhibited noteworthy performance, requiring an FPR of 5.45% to obtain a TPR of 70.59%. Remarkably, it reached its peak TPR, shoot up to 94.12%, at an FPR of 13.64%.

Lastly, the RF-U algorithm demanded an FPR of 10.9% to secure a TPR of 58.8%. Its optimal TPR, peaking at 82.35%, was achieved with an FPR of 14.55%. These findings underscore the adaptability of these modified algorithms in achieving a finely tuned balance between false positives and true positives, a crucial facet in real-world malware detection scenarios.

3.7 CONCLUSION

In this study, the task of classifying malware was accomplished by utilizing a single feature, specifically the list of DLL files present in the executable code of both the benign and malware datasets. This approach yielded an efficient and lightweight implementation, making it particularly well-suited for resource-constrained devices. The findings of our study revealed that, among the classical machine learning algorithms, the NB and SGD models exhibited the highest level of accuracy. Notably, algorithms founded on the boosting methodology consistently outperformed those based on the bagging technique. Standard bagging and RF, both based on bagging approach, performed poorly due to class imbalance between the benign and malware categories. On the other hand, boosting algorithms, including AdaBoost, GBM, XGBoost, and CatBoost, demonstrated strong performance. AdaBoost and CatBoost, in particular, achieved accuracy levels surpassing 70%. To address the class imbalance issue, we explored modified bagging algorithms, namely, bagging-RU, RF-W, and RF-U. Among these, RF-W stood out by delivering the highest accuracy, achieving a TPR of 94.12% at an FPR of 13.64%.

Collectively, the three imbalance classifiers consistently outperformed the other algorithms, showcasing their effectiveness in adapting to imbalance distribution of class in the dataset. In future works, it may be beneficial to explore other features to enhance detection accuracy while concurrently reducing the FPR and maximizing the TPR. In addition, the investigation of deep learning techniques such as convolutional neural networks, recurrent neural networks, and long short-term memory (LSTM) models holds promise for further advancing the field of malware detection.

REFERENCES

1. Choi, S. (2020). Combined kNN Classification and Hierarchical Similarity Hash for Fast Malware Detection. Applied Sciences, 10(15), 5173. https://doi.org/10.3390/app10155173.
2. Yilmaz, A., Taspinar, Y., & Koklu, M. (2022). Classification of Malicious Android Applications Using Naive Bayes and Support Vector Machine Algorithms. International Journal of Intelligent Systems and Applications in Engineering, 10(2), 269–274.
3. Maheswari, K. U., Shobana, G., Bushra, S. N., & Subramanian, N. (2021). Supervised malware learning in cloud through system calls analysis. In 2021 International Conference on Innovative Computing, Intelligent Communication and Smart Electrical Systems (ICSES) (pp. 1–8). Chennai, India. https://doi.org/10.1109/ICSES52305.2021.9633788.

4. Smmarwar, S. K., Gupta, G. P., & Kumar, S. (2022). A Hybrid Feature Selection Approach-Based Android Malware Detection Framework Using Machine Learning Techniques. In D. P. Agrawal, N. Nedjah, B. B. Gupta, G. Martinez Perez (Eds.), Cyber Security, Privacy and Networking. Lecture Notes in Networks and Systems. Vol. 370 Springer, Singapore. https://doi.org/10.1007/978-981-16-8664-1_30.
5. Hussain, A., Asif, M., Ahmad, M., Mahmood, T., & Raza, M. (2022). Malware Detection Using Machine Learning Algorithms for Windows Platform. In Proceedings of International Conference on Information Technology and Applications (pp. 619–632). Springer Nature, Singapore. https://link.springer.com/chapter/10.1007/978-981-16-7618-5_53.
6. Euh, S., Lee, H., Kim, D., & Hwang, D. (2020). Comparative Analysis of Low-Dimensional Features and Tree-Based Ensembles for Malware Detection Systems. IEEE Access, 8, 76796–76808. https://doi.org/10.1109/ACCESS.2020.2986014.
7. Wu, D., Guo, P., & Wang, P. (2020). Malware Detection based on Cascading XGBoost and Cost Sensitive. In 2020 International Conference on Computer Communication and Network Security (CCNS) (pp. 201–205). Xi'an, China. https://doi.org/10.1109/CCNS50731.2020.00051.
8. Al-kasassbeh, M., Abbadi, M. A., & Al-Bustanji, A. M. (2020). LightGBM Algorithm for Malware Detection. In K. Arai, S. Kapoor, & R. Bhatia (Eds.), Intelligent Computing (Vol. 1230, Advances in Intelligent Systems and Computing). Springer Nature Switzerland AG. https://doi.org/10.1007/978-3-030-52243-8_28.
9. Agrawal, P., & Trivedi, B. (2020). Evaluating Machine Learning Classifiers to Detect Android Malware. In 2020 IEEE International Conference for Innovation in Technology (INOCON) (pp. 1–6). Bangluru, India. https://doi.org/10.1109/INOCON50539.2020.9298290.
10. Altman, N. S. (1992). An Introduction to Kernel and Nearest-Neighbor Nonparametric Regression. The American Statistician, 46(3), 175–185. https://doi.org/10.1080/00031305.1992.10475879.
11. Cortes, C., & Vapnik, V. (1995). Support-Vector Networks. Machine Learning, 20, 273–297. https://doi.org/10.1007/BF00994018.
12. Quilan, J. R. (1988). Decision Trees and Multi-Valued Attributes. Machine Intelligence 11, 305–318. Oxford University Press, Inc., USA. https://dl.acm.org/doi/10.5555/60769.60782.
13. Ramadhan, B., Purwanto, Y., & Ruriawan, M. F. (2020). Forensic Malware Identification Using Naive Bayes Method. In 2020 International Conference on Information Technology Systems and Innovation (ICITSI) (pp. 1–7). Bandung, Indonesia. https://doi.org/10.1109/ICITSI50517.2020.9264959.
14. Smith, D., Khorsandroo, S., & Roy, K. (2023). Supervised and Unsupervised Learning Techniques Utilizing Malware Datasets. In 2023 IEEE 2nd International Conference on AI in Cybersecurity (ICAIC) (pp. 1–7). Houston, USA: IEEE. https://doi.org/10.1109/ICAIC57335.2023.10044169.
15. Breiman, L. (1996). Bagging Predictors. Machine Learning, 24, 123–140. https://doi.org/10.1007/BF00058655.

16. Maclin, R., & Opitz, D. (1997). An empirical evaluation of bagging and boosting. In Proceedings of the fourteenth national conference on artificial intelligence and ninth conference on Innovative applications of artificial intelligence (AAAI'97/IAAI'97) (pp. 546–551). AAAI Press. https://dl.acm.org/doi/10.5555/1867406.1867491.
17. Breiman, L. (2001). Random Forests. Machine Learning, 45, 5–32. https://doi.org/10.1023/A:1010933404324.
18. Geurts, P., Ernst, D., & Wehenkel, L. (2006). Extremely Randomized Trees. Machine Learning, 63, 3–42.
19. Drucker, H., & Cortes, C. (1995). Boosting Decision Trees. In Advances in Neural Information Processing Systems. MIT Press, Cambridge, MA, United States.
20. Freund, Y., & Schapire, R. E. (1997). A Decision-Theoretic Generalization of On-Line Learning and an Application to Boosting. Journal of Computer and System Sciences, 55(1), 119–139.
21. Friedman, J. H. (2001). Greedy Function Approximation: A Gradient Boosting Machine. The Annals of Statistics, 29(5), 1189–1232.
22. Chen, T., & Guestrin, C. (2016). XGBoost: A Scalable Tree Boosting System. In Proceedings of the 22nd ACM SIGKDD International Conference on Knowledge Discovery and Data Mining (pp. 785–794). Association for Computing Machinery.
23. Ke, G., Meng, Q., Finley, T., Wang, T., Chen, W., Ma, W., Ye, Q., & Liu, T. Y. (2017). LightGBM: A Highly Efficient Gradient Boosting Decision Tree. In Advances in Neural Information Processing Systems. Curran Associates, Inc.
24. Prokhorenkova, L., Gusev, G., Vorobev, A., Dorogush, A., & Gulin, A. (2018). CatBoost: Unbiased Boosting With Categorical Features. In Advances in Neural Information Processing Systems. Curran Associates, Inc.
25. Hussain, M. J., Shaoor, A., Baig, S., Hussain, A., & Muqurrab, S. A. (2022). A hierarchical based ensemble classifier for behavioral malware detection using machine learning. In 19th International Bhurban Conference on Applied Sciences and Technology (IBCAST), 2022 (pp. 702–706). Islamabad, Pakistan. https://doi.org/10.1109/IBCAST54850.2022.9990203.
26. Mahendran, N., Vincent, P. M. D. R., Srinivasan, K., Sharma, V., & Jayakody, D. K. (2020). Realizing a Stacking Generalization Model to Improve the Prediction Accuracy of Major Depressive Disorder in Adults. IEEE Access, 8, 49509–49522. https://doi.org/10.1109/ACCESS.2020.2977887.
27. More, A., & Rana, D. (2017). Review of random forest classification techniques to resolve data imbalance. In 2017 1st International Conference on Intelligent Systems and Information Management (ICISIM) (pp. 72–78).
28. Khoshgoftaar, T. M., Golawala, M., & Hulse, J. V. (2007). An Empirical Study of Learning from Imbalanced Data Using Random Forest. In 19th IEEE International Conference on Tools with Artificial Intelligence (ICTAI 2007) (pp. 310–317). Patras, Greece: IEEE. https://doi.org/10.1109/ICTAI.2007.46.

29. Yağcı, A., Aytekin, T., & Gürgen, F. (2016). Balanced random forest for imbalanced data streams. In 2016 24th Signal Processing and Communication Application Conference (SIU) (pp. 1065–1068).
30. Liu, X.-Y., Wu, J., & Zhou, Z.-H. (2009). Exploratory Undersampling for Class-Imbalance Learning. IEEE Transactions on Systems, Man, and Cybernetics, Part B (Cybernetics), 39(2), 539–550.

CHAPTER 4

Electroencephalogram-Based Emotion Recognition Using Binary Bat Algorithm and Least Square Support Vector Machine

Farzana Kabir Ahmad and Siti Sakira Kamaruddin

4.1 INTRODUCTION

Human emotion is mainly associated with thought, feeling, and behavioral responses, and it plays a vital role in decision-making, communication, and learning process. Emotion recognition is a fascinating research field that aims to get insight into human emotional states and gives the ability to recognize them properly. The need of computers to be able to detect and interact with user's emotion states is a growing research interest (Gannouni et al., 2021). In the past decades, many researchers have attempted to recognize human emotion through various approaches, such as facial expression, speech recognition, or gestures of the body. However, these approaches were considered unreliable since users can control their facial expressions, prosody, or body movements which could affect the ultimate emotion recognition results.

DOI: 10.1201/9781003400387-4

Human emotion recognition by using brain signals has become an emerging research area. EEG is one of the approaches used for recognizing human emotions through brain signals. Such signals are recorded and monitored based on the brain electricity that is generated from electrodes placed on human scalp. Signals obtained from EEG have been known to provide effective information on both mental and emotional activities as it measures brain waves that are produced when neurons communicate. In the past, EEG device had been widely used in medical domain typically to detect epilepsy and seizures (Shoeibi et al., 2022). Availability of portable, easy to operate, and inexpensive wireless EEG has attracted many recent researchers to use it in recognizing human emotion. In addition, several public repositories have been established to permit researchers to work on analyzing and enhancing techniques that could handle large scale of EEG signals data. Although EEG signals provide detailed information about human emotional states, it is not an easy task to infer the user's mental state from those same signals.

The nature of EEG signals is often complicated, unsettled, nonlinear, and random mainly due to the complex interconnections among neurons. Hence, the analysis of nonlinear and chaotic characteristics of EEG signals is a substantial problem. As a result, a thorough analysis is required to process and analyze such complex EEG signals. Moreover, the EEG signals are subject to high dimensionality features, which require further attention. The process of EEG classification starts with the stimulus that provokes brain signals which are later recorded. The recorded EEG signals undergo a preprocessing phase in which the noisy data is eliminated and resulting signals are ready to be used as an input for feature extraction phase. The features that are extracted are then used to classify the signals. However, as the number of features that are extracted is massive, obtaining results can be cumbersome. As a result, feature selection and channel selection methods are used. These play an essential role in enhancing the classification performance, reducing model complexity, and increasing the performance of subsequent process (Ahmad & Tuaimah, 2021; AL-Dyani et al., 2020).

Feature selection (FS) is one of the crucial steps in building machine learning model for real-world problems. The main objective of FS is to select informative features and remove irrelevant features from dataset to facilitate any data mining task. Hence, FS is essential to reduce the high dimensionality of space and address the overfitting problem. Moreover, adding redundant features reduces the generalization capability of the model and may also reduce the overall accuracy of a classifier. Besides, large number of feature space increases the overall complexity of the model and makes it difficult to understand the underlying information.

In a nutshell, FS can be categorized into two types: filter-based FS and wrapper-based FS. Filter-based FS generally ranks each feature according to univariate score such as signal-to-noise ratio (SNR), Euclidean distance (ED) and only the highest ranks are selected to be trained by classifier. Although filter-based FS is the simplest form of variable selection, this method is highly dependent on human intervention to determine the threshold. Moreover, the relationship between features is not taken into consideration in this underlying case. Wrapper-based FS, on the other hand, is a method that embed FS with classification process. In the wrapper-based method, a search is conducted in the space of features, evaluating the goodness of each found subset by the estimation of the accuracy percentage of the specific classifier to be used, then training the classifier only with the relevant subset of features. Nonetheless, these two FS approaches still suffer from some drawbacks such as the subset feature selected can be trapped in local optima and may trigger a high computational cost. Moreover, they tend to perform global searches to find the optimal features, yet it is impossible in most cases.

Swarm intelligence is a new and rising paradigm within bio-inspired computing which is able to find optimal features with global searchability (Slowik & Kwasnicka, 2018). These algorithms are based on the social behavior of organisms, not on their genetic adaptation. Swarm intelligence involves the execution of collective intelligence in groups of simple agents that depend on the behaviors of real-world animal or insect swarms as a tool for problem-solving. There are some popular swarm-based algorithms that have been used in EEG field studies. Genetic Algorithm (GA), Artificial Bee Colony (ABC), Particle Swarm Optimization (PSO), Ant Colony Optimization (ACO), and Firefly Algorithm (FA) have been successfully utilized to discover the optimal feature subset. However, despite the excellent findings, most of these algorithms have a poor convergence rate and are entrapped in local optima. Moreover, based on "No Free Lunch" (NFL) theorem, none of swarm intelligence methods can handle all optimization problems. Therefore, this research aims to explore the binary bat algorithm (BBA) in selecting the relevant EEG features and train these features using least square support vector machine (LSSVM). The structure of this chapter is as follows. Section 4.2 explores the swarm intelligence-based feature selection in EEG studies. The proposed methodology which includes BBA and LSSVM is described in detail in

Section 4.3. Section 4.4 on the other hand presents the experimental results and discussion. Finally, the conclusion is given in Section 4.5.

4.2 SWARM INTELLIGENCE-BASED FEATURE SELECTION IN ELECTROENCEPHALOGRAM STUDIES

Over past several years, swarm intelligence techniques have been proposed by many researchers to solve the optimization problem. As massive data has been generated due to the new technologies that were invented, the need for efficient algorithms has become inevitable. In the study of EEG data signals, several algorithms have been inspired to overcome the complexity of immense features, specifically related to the optimization problems which are multi-objective problems (Satchidananda Dehuri et al., 2015). Swarm intelligence is a new and rising paradigm within bio-inspired computing intended for employing adaptive systems. Swarm intelligence algorithms are based on the social behavior of organisms, not on their genetic adaptation. Swarm intelligence involves the execution of collective intelligence in groups of simple agents that depend on the behavior of real-world animal or insect swarms as a tool of problem-solving.

Several algorithms have been proposed in the study of EEG data and this chapter reviews some of the most popular swarm-based benchmark algorithms that have been used in EEG field studies and have shown significant performance, including PSO, ABC, ACO algorithms (Tzanetos & Dounias, 2020). Please take note that although GA is a heuristic algorithm, its description is given as it is among popular benchmark techniques used in EEG data analysis.

4.2.1 Genetic Algorithm

GA uses principles of natural evolution (Shon et al., 2018). This algorithm evolved to find the (close to) optimal solution for chromosomes to maintain survival based on stochastic optimization. It has been implemented in several research areas to find the most useful solutions (e.g., job scheduling, pattern recognition, networks, etc.). The algorithm carries a fixed number of chromosomes, which might be represented in binary code, with operators for crossover and mutation. Each of these binary chromosomes, along with a fitness function operator, is represented as a solution. These solutions are considered to produce a new solution within the searching procedure.

4.2.2 Practical Swarm Optimization

PSO is a strategy used for enhancing hard numerical functions in view of the analogy of social practices of herds of winged animals and schools of fish (Zhiping et al., 2010). It is a developmental calculation system in light of swarm insight. A swarm consists of people, called particles, which change their situations after some time. Every molecule represents a potential answer for the issue. In a PSO framework, particles fly around in a multidimensional seeking space. Amid its flight, every molecule alters its situation as indicated by its own experience and the experience of its neighboring particles to find the best solutions. The impact is that particles move toward the better arrangement regions while keeping up the capacity to look through a wide region around the better arrangement territories. The execution of every molecule is estimated by a pre-characterized wellness work, which is identified with the issue being unraveled. The PSO has been observed to be strong and quick at fathoming nonlinear, non-differentiable, and multi-modular issues. Mathematically, a molecule in PSO is a vector in a N-dimensional parameter space, and its position x and speed v change according to the following equations:

$$v_{i,j}^{t} = \omega v_{i,j}^{t-1} + c_1 \eta_1 \left(p_{i,j} - x_{i,j}^{t-1} \right) + c_2 \eta_2 \left(g_j - x_{i,j}^{t-1} \right), \tag{4.1}$$

$$x_{i,j}^{t} = x_{i,j}^{t-1} + v_{i,j}^{t}, \tag{4.2}$$

where $x_{i,j}^{t}$ and $v_{i,j}^{t}$ are the jth components of the ith particle's position and velocity, respectively for each iteration t. Therefore, $p_{i,j}$ is the jth component of p, the best position that the ith particle has found so far, and g_j is the jth component of g, the best position found by the swarm. The impacts of p and g on the particle's motion are controlled by two steady parameters, c_1 and c_2, and two autonomous irregular factors, η_1 and η_2, consistently distributed in [0, 1]. The particle's motion is additionally impacted by the speed at the previous cycle, and this impact is controlled by inactivity parameter ω. c_1 and c_2 are constants set by the experimenter that decide the harmony between the misuse of a potential arrangement (development toward g) and investigation for new arrangement (development toward p). In every cycle, p and g are refreshed if a situation with better fitness is found. This is the primary element of the PSO algorithm: Each particle utilizes the data of its own history and the swarm's history, together with irregular irritations, to look for the global optima. The PSO algorithm iteratively refreshes the speed and position of every particle, moving every one of them around the parameter space until the point when the best global arrangement g achieves the coveted fitness.

4.2.3 Artificial Bee Colony

ABC is a prevalent algorithm mimicking the real canny scrounging activities of a bumble bee swarm. In the ABC algorithm, the state of counterfeit honeybees consists of three arrangements of honeybees: utilized honeybees, passerby honeybees, and scout honeybees. The spectator honeybees play out a move to demonstrate the location of a nourishment source, and the utilized honeybees go to this sustenance source. The investigator honeybees convey a hunt to discover new sustenance sources haphazardly. The locations of new sustenance sources symbolize new likely solutions to the improvement issue, and the nectar amount of a nourishment source alludes to the quality (fitness) of the related arrangement.

A swarm of honeybees is made and afterward continues haphazardly inside a two-dimensional hunt space. At the point when the honeybees discover an objective nourishment source/nectar, the honeybees begin cooperating, and the arrangement of the issue can be acquired from these communications' force. An irregular introduction of populace solutions ($x_i = 1, 2, \ldots, D$) is prepared on the D-dimensional space of the issue. A utilized honeybee makes a modification for the location (arrangement) inside her memory space regarding the neighborhood data (visual data) and checks the real nectar amount (fitness esteem) of the new source (new arrangement).

Given that the nectar amount of the new source is more prominent than that of the earlier one, the honeybee remembers the most current location and overlooks the previous one. At last, after every single utilized honeybee finishes the hunt strategy, they share this nectar source data and their locations with the passerby honeybees inside the move region. Over the accompanying advance, proliferation, in light of the probability estimation of the nourishment source, the specific fake spectator honeybee chooses a wellspring of sustenance by following a detailed capacity:

$$P_i = \frac{fit}{\sum_{n=1}^{FN} fit_n} \tag{4.3}$$

where, P_i is the sustenance source; FN is the nourishment source number, which is equivalent to the quantity of utilized honey bees; and fit beta I is the arrangement fitness esteem I, which is corresponding to the nectar source amount inside the location I. Inside the last advance, substitution of honeybee and choice, if a location can't be improved further by means of a settled number of rounds, at that point that sustenance source is presumed to be relinquished.

After every chosen source location is generated and afterward evaluated by the fake honeybees, its viability is contrasted with that of the old one. In this way, if the updated one has a comparable or stunningly better nectar source than the old one, the old one is overlooked and supplanted with the updated one, or else the old sustenance source is put away in the memory. The nearby hunt usefulness of the ABC algorithm is needy upon neighborhood seek and in addition covetous choice mechanisms executed by utilized and passerby honeybees. The genuine worldwide pursuit usefulness of the algorithm relies on an irregular inquiry methodology executed by scouts and upon a neighbor arrangement generation mechanism executed by utilized and spectator honeybees. In the ABC algorithm, a neighboring sustenance source's position is dictated by changing one haphazardly chosen parameter while the rest of the parameters are unaltered, as in the accompanying expression:

$$x_{ij}^{new} = x_{ij}^{old} + u\left(x_{ij}^{old} - x_{kj}\right) \tag{4.4}$$

where $k \neq i$ and (k and i) $\in \{1,2, \ldots, Eb\}$. The multiplier u represents a random number between −1 and 1, and $j \in \{1,2, \ldots, D\}$. Hence, x_{ij} represents the jth parameter of a solution x_i that is selected to be customized. As the food source's position is abandoned, the employed bees related to it become scouts.

In the ABC algorithm, scouts produce a completely new food source position as follows:

$$x_i^{j(new)} = min_i^j + u\left(max_i^j - min_i^j\right) \tag{4.5}$$

where equation (4.4) applies to all j parameters.

4.2.4 Ant Colony Optimization

ACO is based on the foraging behavior of an ant seeking a path between its colony and source food. Initially, it was used to solve the well-known traveling salesman problem. Now, it is used for solving different hard optimization problems. Ants are social insects. They live in colonies. The behavior of the ants is controlled by the goal of searching for food. While searching, ants roam around their colonies. An ant repeatedly hops from one place to another to find food. While moving, it deposits an organic compound called pheromone on the ground. Ants communicate with each other via pheromone trails. When an ant finds some amount of food

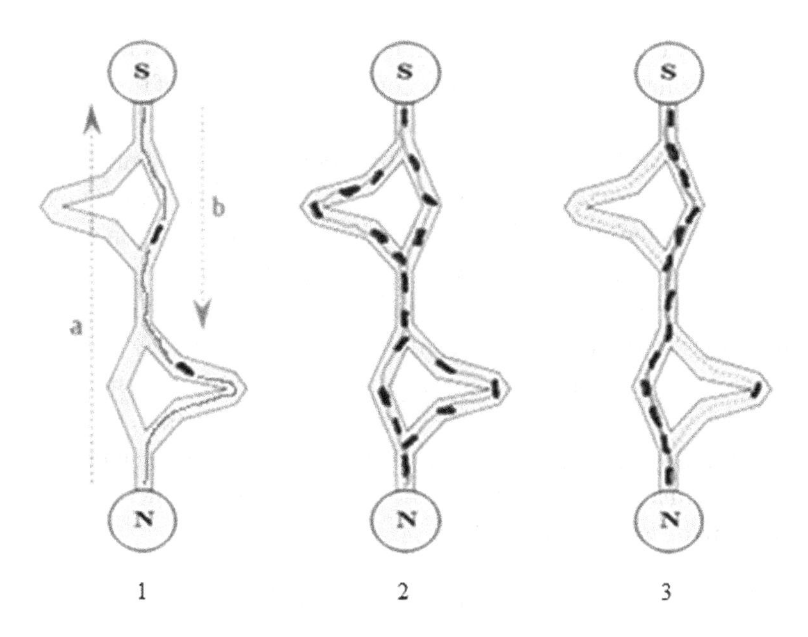

FIGURE 4.1 Mechanism of Ant Colony Optimization.

it carries as much as it can carry. When returning it deposits pheromone on the paths based on the quantity and quality of the food. Ants can smell pheromone. So, other ants can smell the pheromone trail and follow that path as depicted in Figure 4.1. The higher pheromone level has a higher probability of that path being chosen and as more ants follow the path, the amount of pheromone will also increase on that path. Figure 4.2 shows the flowchart of ACO.

4.3 BINARY BAT ALGORITHM AND LEAST SQUARE SUPPORT VECTOR MACHINE

This study has proposed the BBA for selecting the relevant EEG features and trained these features using LSSVM. The proposed methodology is illustrated in Figure 4.3, which consist of five phases namely, (a) EEG data acquisition, (b) data preprocessing, (c) features extraction, (d) feature selection and classification, and (e) evaluation. Each phase is explained in detail in the following sections.

4.3.1 Phase I: EEG Data Acquisition

Data acquisition is the process of sampling signals that measure real-word physical conditions, and converting the resulting samples into digital numeric values that can be manipulated by a computer. In emotion

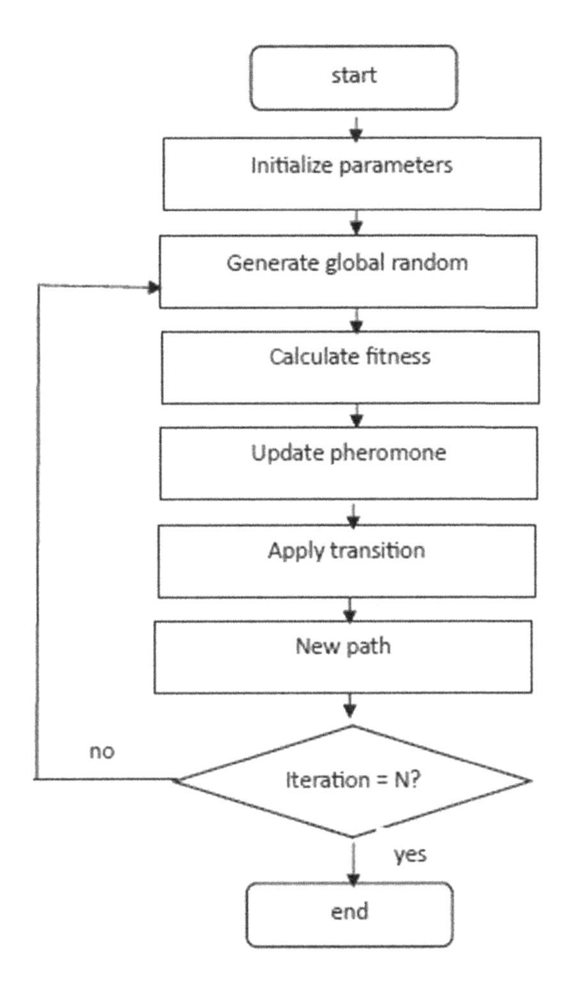

FIGURE 4.2 The flowchart of Ant Colony Optimization.

recognition research, several EEG public databases have been created to study the human emotion state in different situations/conditions. These datasets vary from one to another in terms of number of electrodes used, subjects, and series of trials. They also differ in terms of the kind of stimuli, as human emotions can be evoked by several kinds of stimuli, such as auditory, visual, and combined. The key idea of using public dataset is that researchers can use them as a benchmark to analyze and examine the effectiveness of their proposed methods.

In this study, DEAP dataset (Koelstra, 2011) has been examined. DEAP dataset is a multimodal dataset for humans full of feeling state examination. Both EEG and peripheral physiological signs of 32 subjects were

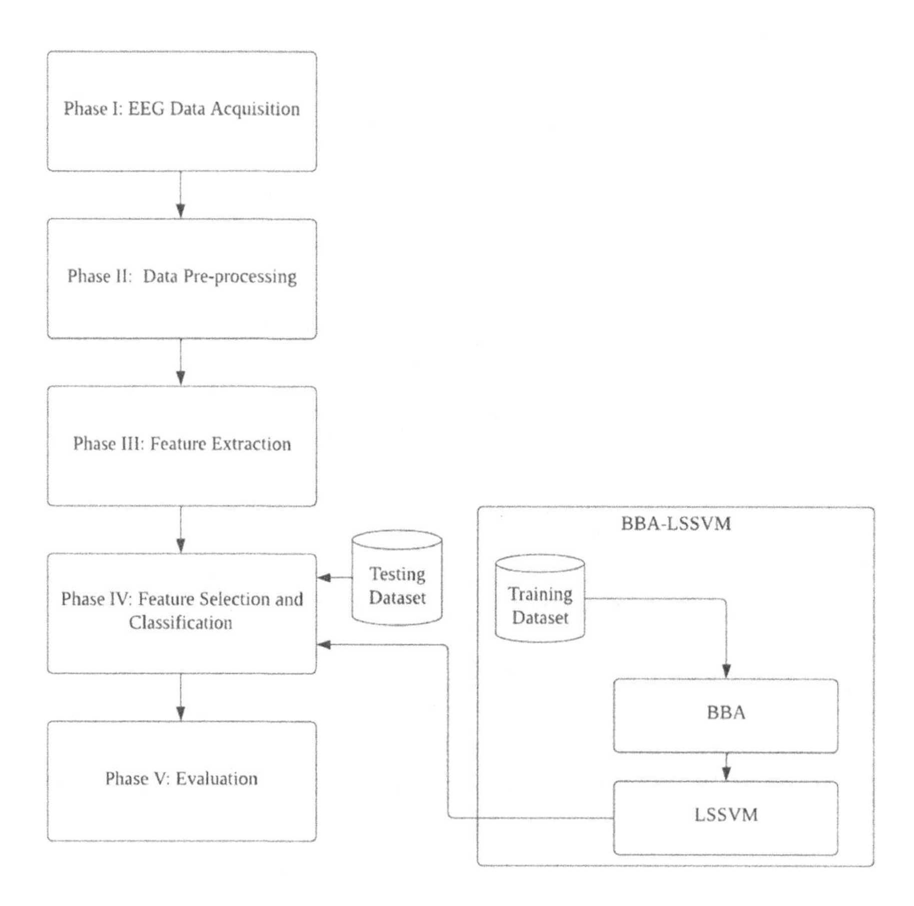

FIGURE 4.3 The Proposed BBA-LSSVM for EEG Signals Analysis.

recorded utilizing varying media boosts. Varying medias such as music or pictures were utilized to instigate subjects' emotions, thus prompts better data nature of EEG signals. Each subject was required to watch 40 music recordings for 1 minute each. After every moment of viewing the music-video, the subjects were solicited to round out the scale a self-evaluation of their emotion levels for example arousal, valence, predominance, and liking, based on the self-appraisal puppets (SAM) technique. Table 4.1 shows the summary description of DEAP dataset.

TABLE 4.1 Summary Description of DEAP Dataset

Dataset	Stimulus	Label	Channels	Participants	Frequency	Classes
DEAP	Music/Video	SAM	32	32	128	Stress/Calm

TABLE 4.2 Contents of Each Subject File

Array Name	Array Shape	Array Contents
Data	$40 \times 40 \times 8064$	Music-video/trial × channel × data
Labels	40×4	Music-video/trial × label (valence, arousal, dominance, liking)

4.3.2 Phase II: Data Preprocessing

The recorded EEG data were preprocessed and stored in 32.mat (MATLAB®) documents, in a shape of three-dimensional array (i.e., music-video/trial × channel × data), where each record has information chiefly for one subject; each document of the information incorporates two clusters (Data and Labels) as depicted in Table 4.2.

The 40 channels in each file include the 32 channels of the brain (EEG) as well as physiological signals from EOG, EMG, GSR, respiration belt, plethysmography, and temperature. In the current research, only the EEG signals are analyzed.

4.3.3 Phase III: Feature Extraction

In this study, discrete wavelet packet transform (DWPT) (Ong et al., 2017), which belong to time-frequency feature extraction techniques, is used. DWPT is a popular technique that decomposes signals in smaller packets, known as decomposing trees. The entropy values of these packets later form feature vectors that are used in subsequent analysis.

4.3.4 Phase IV: Feature Selection and Classification

The high dimensionality feature space generated over the large volume of extracted brain signals includes noise, sparse and uninformative features, which mislead the detection method and eventually reduce the overall accuracy of recognition models. As a result, the FS phase becomes a fundamental step to select the optimal feature subset and enhance the performance of the emotion recognition models. Wrapper-based FS methods based on binary swarm intelligence algorithms have shown promising performance. Subsequent sections discuss wrapper FS method based on the LSSVM and BBA algorithm.

4.3.4.1 Least Square Support Vector Machine

LSSVM was recently proposed to be the solution to the problem involving the equality limitation of the original SVM by solving a system of linear equations. In addition, LSSVM is easy to train and takes less computational effort compared with the original SVM. The essential instruction rules of whether to use SVM or LSSVM depends on determining the optimal hyperplane in which the predicted classification error regarding test samples can be minimized. The perfect hyperplane is usually the one that maximizes the margins. Maximizing margins can increase the capability of generalization. SVM employs a regularization parameter (C) that allows accommodation of outliers and enables error of test samples. Usually, the three common kernels related to SVM-based classifiers are used in order to avoid the optimization problem of hyperplane creation that usually happens in large-scale samples. These kernel functions are linear, multi-layer perceptron, and RBF. LSSVM, including a kernel function, has two essential parameters that play basic roles in identifying the performance of the classifier. These two parameters are the smoothing parameter σ and regularization parameter γ. Different values of these parameters continuously show different performance of the classifier. To tune these two values, numerous studies have depended on the default values of the classifier or the grid search method. In this study, these parameters will be tuned using BBA.

4.3.4.2 Binary Bat Algorithm

BBA is an algorithm developed on the basis of echolocation property of microbats. Echoes of bats are used as medium of communication to identify various types of insects, detecting the direction and distance of their prey as well as avoiding bumping with other close objects while moving in complete darkness. Generally, BBA consists of four main stages that are as follows:

i. Initializing bat population: The population of bats is initialized using randomly selected values to find an optimal solution for the given problem in the D-dimensional search area. The solution that is found by the population is evaluated using Equation (4.6)

$$x_{ij} = x_{min} + \varphi(\ x_{max} - x_{min}) \tag{4.6}$$

where x_{min} and x_{max} are lower and higher borders for the dimension space $j = 1, 2, \ldots, D$, while $i = 1, 2, 3, \ldots, N$, is the population number of BA. φ is a randomly generated value from [0, 1].

ii. Updating frequency (f), velocity (v), and new solution (x_i): the position x_i and velocity v_i of every bat in a D-dimensional search space are updated based on Equations (4.7), (4.8), and (4.9):

$$f_i = f_{\min} + (f_{\max} - f_{\min})\beta, \tag{4.7}$$

$$v_i^t = v_i^{t-1} + (x_i^{t-1} - x_*) f_i \tag{4.8}$$

$$x_i^t = x_i^{t-1} + v_i^t, \tag{4.9}$$

iii. Updating r and A: In reality, when a bat finds its prey, the A normally decreases, while r increases. Indeed, both A and r are updated using Equation (10) and Equation (11),

$$A_i^{t+1} = \alpha \ A_i^t, \tag{4.10}$$

$$r_i^{t+1} = r_i^0 \left[1 - exp(-\gamma t)\right] \tag{4.11}$$

where γ and α are constants; α is the cooling aspect of a cooling schedule in simulated annealing algorithm.

iv. Evaluation, saving, and ranking best solutions: After updating both A and r, an evaluation process is carried out to evaluate newly generated solutions for all bats. If the obtained solutions satisfy the given condition, then they will be archived conditionally as the best solutions. Finally, a ranking process will be performed on all bats to find the current best solution (x*). Figure 4.4 shows the flowchart of BBA.

4.3.5 Evaluation

This study has used several evaluation methods to assess the performance of their investigated methods: for example, accuracy, specificity, sensitivity, and F-score. Table 4.3 shows the definitions of some evaluation criteria: *TP* and *TN* are the true positive and true negative points, respectively, and *FP* and *FN* are the false positive and false negative points.

4.4 EXPERIMENTAL RESULTS AND DISCUSSION

This section presents the experimental results of EEG-based emotion recognition that are obtained when tested on BBA-LSSVM techniques. In this study, the proposed technique is benchmarked with several other swarm intelligence techniques such as GA, PSO, ABC, and ACO. These techniques were applied to the DEAP dataset as explained in previous section. Several metrics are tested to determine the performance of the proposed technique. Figure 4.5 shows the execution time that are recorded for BBA and other techniques. Based on this results, BBA has achieved the execution time of 9.92 second while ACO and ABC have less computational time

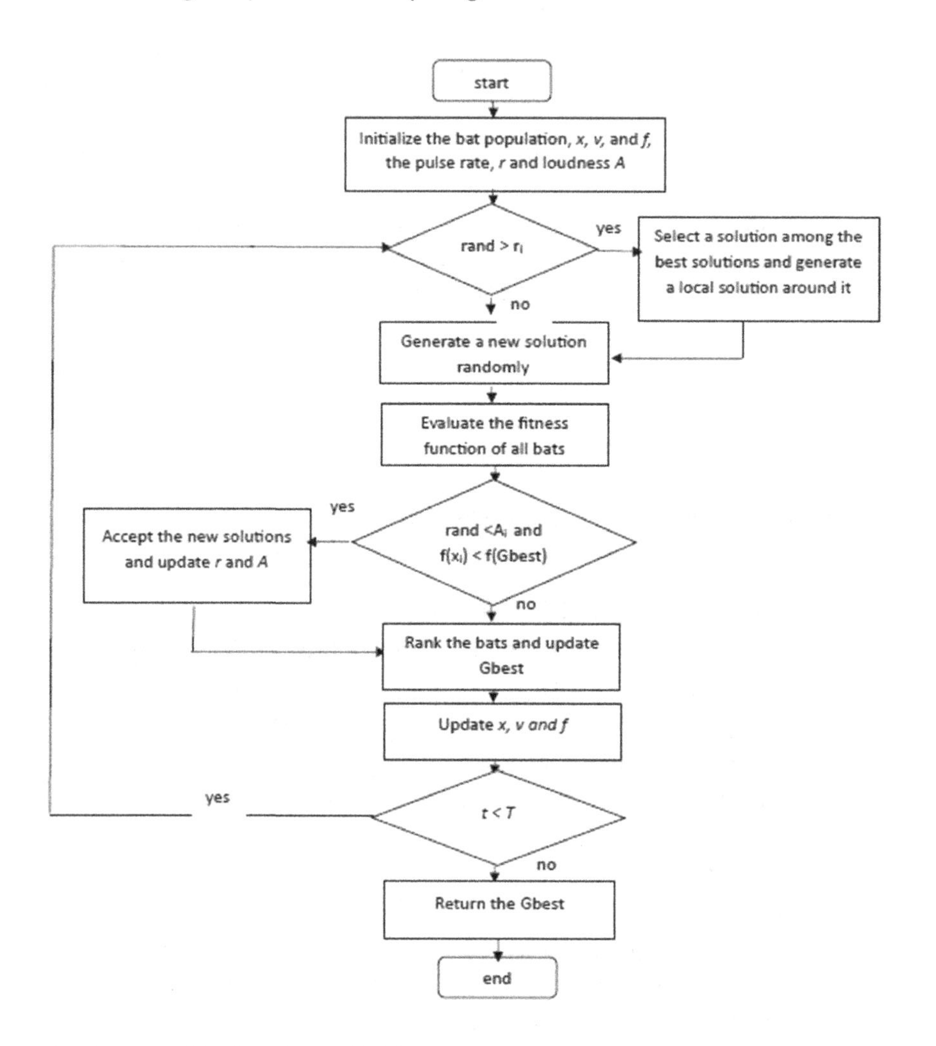

FIGURE 4.4 The flowchart of BBA.

with 8.67 and 8.84 seconds, respectively. These results reveal that BBA has the capacity to produce meaningful information even with few extra seconds. More explanation on this point can be described by Figure 4.6.

Figure 4.6 depicts the accuracy as well as the number of selected feature achieved by different swarm intelligence techniques. The highest accuracy is obtained by BBA, i.e., 89.8%, with 102 features selected from the 188 features. PSO on the other hand has achieved 88.1% of accuracy and 100 features being selected. These results show the ability of BBA to balance the exploration and exploitation in comparison to PSO. Other techniques such as ABC and ACO tend to fall in local optima and are not able to

TABLE 4.3 Evaluation Metrics

Formula	Description
$Accuracy = \frac{TP+TN}{TP+TN+FP+FN}$	Accuracy is calculated by dividing the number of correct decisions by the total number of cases
$Specificity = \frac{TN}{FP+TN}$	Specificity is calculated by dividing the number of true negative decisions by the number of actual negative case
$Sensitivity = \frac{TP}{TP+FN}$	Sensitivity is calculated by dividing the number of true positive decisions by the number of actual positive cases
$Precision = \frac{TP}{TP+FP}$	Precision (P) is the ratio of the number of relevant classes retrieved to the total number of irrelevant and relevant classes retrieved
$Recall = \frac{TP}{TP+FN}$	Recall (R) is the ratio of the number of relevant classes retrieved to the total number of relevant classes
$Fscore = 2.\frac{P.R}{P+R'}$	F-score is a measure of a test's accuracy. Both precision and recall are considered to measure F-score. The F-score can be interpreted as a weighted average of the precision and recall. F-score reaches its best value at 1 and its worst value at 0

perform global search. Even though BBA and PSO have more computational times relative to ABC and ACO, findings in Figure 4.6 offers promising results. The proposed algorithm achieved the maximum accuracy of 86.76% and 80.83%, and the mean accuracy obtained is 85.51% ± 0.07 and 80.97% ± 0.08 for valence and arousal emotions classification, respectively.

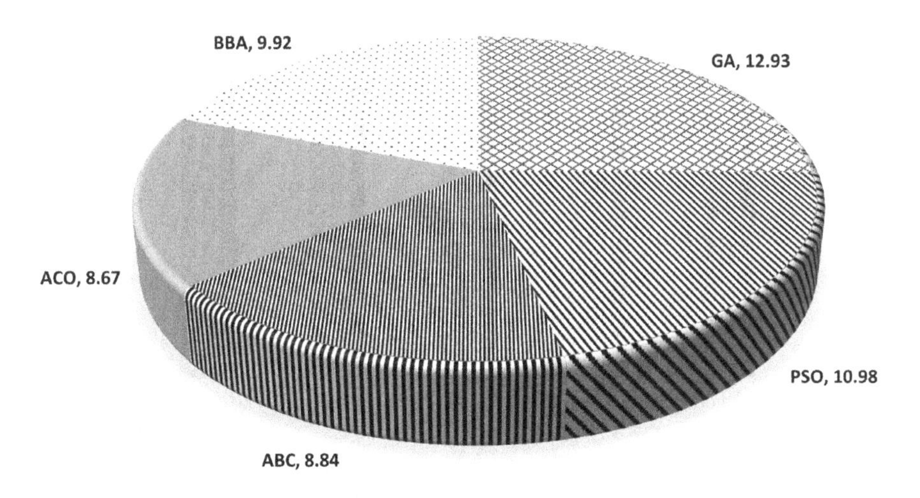

FIGURE 4.5 The execution time of BBA and other benchmarked algorithms.

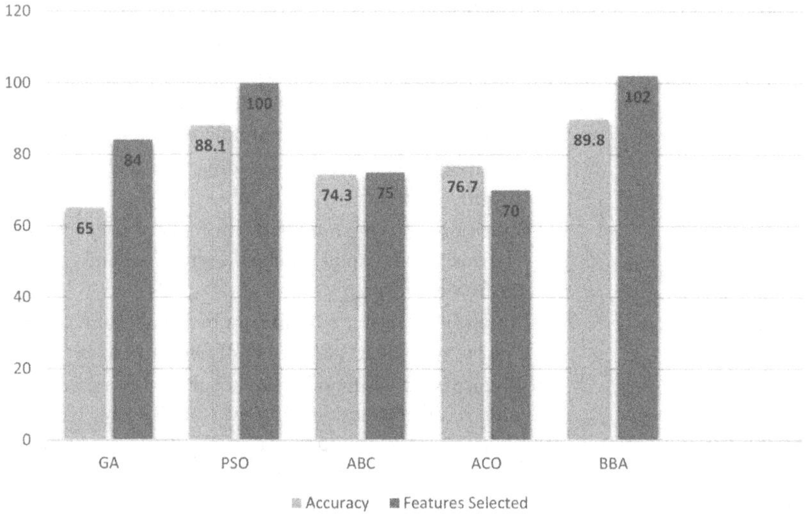

FIGURE 4.6 The accuracy and number of selected features obtained using different swarm intelligence techniques.

To verify the performance of proposed techniques, BBA and LSSVM, this study has measured the performance by testing it with several metrics as illustrated in Figure 4.7. Five metrics are measured such as specificity, sensitivity, precision, recall, and f-score. The BBA-LSSVM has achieved an average performance of 92%, 90%, 90%, 99%, and 96% of specificity,

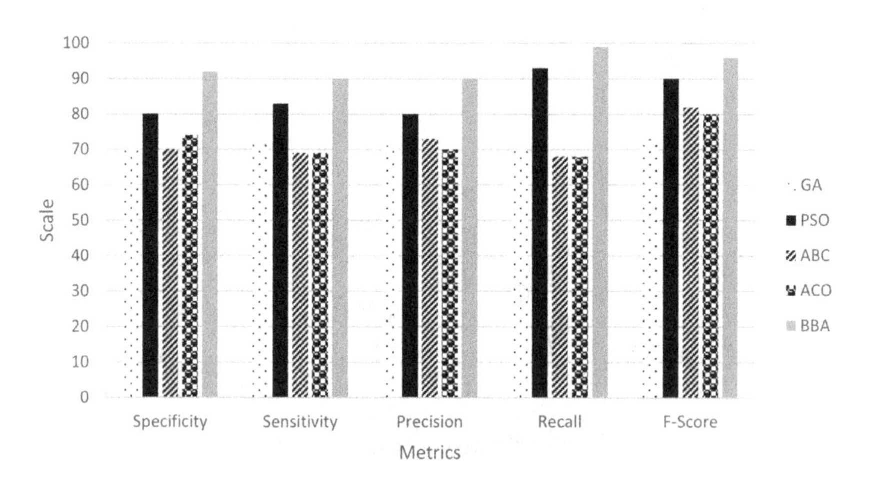

FIGURE 4.7 Performance metrics of BBA and other techniques trained using LSSVM.

sensitivity, precision, recall, and f-score. These results have outperformed other benchmarks techniques, such PSO, and GA being in second and third rank.

4.5 CONCLUSION

EEG is a remarkable tool to measure brain waves that help in determining human emotion. With various public data sources available on internet, the exploration of human emotion using recent swarm intelligence techniques is feasible. EEG signals are often complex, nonlinear, and chaotic in nature, which makes it difficult to extract features from such high-dimensional signals. Hence, this study aims to propose the BBA-LSSVM to detect informative features in emotion recognition using EEG signals. Rigorous analysis has been conducted on DEAP dataset and several metrics such as execution time, accuracy, number of selected features, specificity, sensitivity, precision, recall, and f-score have been measured to determine the performance of proposed technique. The experimental results have shown that BBA-LSSM has achieved highest accuracy of 89.8%, with 102 features selected from the 188 features. The maximum accuracy of 86.76% and 80.83%, and the mean accuracy of 85.51% $\pm$ 0.07 and 80.97% $\pm$ 0.08 for valence and arousal emotions classification are recorded respectively. These results have shown the ability of BBA to balance the exploration and exploitation in selecting informative features for better emotions classification. Regardless these promising results, in the future more swarm intelligence techniques need to be tested on different datasets to determine the quality of selected features.

REFERENCES

Ahmad, F. K., & Tuaimah, A. T. (2021). A comparative analysis of time-frequency feature extraction techniques for large scale electroencephalogram data. *International Journal of Advanced Trends in Computer Science and Engineering (IJATCSE)*, *10*(1), 14–24.

AL-Dyani, W. Z., Ahmad, F. K., & Kamaruddin, S. S. (2020). A survey on event detection models for text data streams. *Journal of Computer Science*, *16*(7), 916–935. https://doi.org/10.3844/jcssp.2020.916.935.

Gannouni, S., Aledaily, A., Belwafi, K., & Aboalsamh, H. (2021). Emotion detection using electroencephalography signals and a zero-time windowing-based epoch estimation and relevant electrode identification. *Scientific Reports*, *11*(1). https://doi.org/10.1038/s41598-021-86345-5.

Koelstra, S. (2011). DEAP: A Dataset for Emotion Analysis using Physiological and Audiovisual Signals. *Qmul.ac.uk*, 2011. https://www.eecs.qmul.ac.uk/mmv/datasets/deap/

Ong, P., Zainuddin, Z., & Lai, K. H. (2017). A novel selection of optimal statistical features in the DWPT domain for discrimination of ictal and seizure-free electroencephalography signals. *Pattern Analysis and Applications, 21*(2), 515–527. https://doi.org/10.1007/s10044-017-0642-7

Satchidananda Dehuri, Alok Kumar Jagadev, & Panda, M (2015). *Multi-objective Swarm Intelligence*. Springer: Berlin, Heidelberg.

Shoeibi, A., Ghassem i, N., Khodatars, M., Moridian, P., Alizadehsani, R., Zare, A., Khosravi, A., Subasi, A., Rajendra Acharya, U., & Gorriz, J. M. (2022). Detection of epileptic seizures on EEG signals using ANFIS classifier, autoencoders and fuzzy entropies. *Biomedical Signal Processing and Control*, 73, 103417. https://doi.org/10.1016/j.bspc.2021.103417.

Shon, D., Im, K., Park, J.-H., Lim, D.-S., Jang, B., & Kim, J.-M. (2018). Emotional stress state detection using genetic algorithm-based feature selection on EEG signals. *International Journal of Environmental Research and Public Health, 15*(11), 2461. https://doi.org/10.3390/ijerph15112461.

Slowik, A., & Kwasnicka, H. (2018). Nature inspired methods and their industry applications: Swarm intelligence algorithms. *IEEE Transactions on Industrial Informatics, 14*(3), 1004–1015. https://doi.org/10.1109/tii.2017.2786782.

Tzanetos, A., & Dounias, G. (2020). A comprehensive survey on the applications of swarm intelligence and bio-inspired evolutionary strategies. *Learning and Analytics in Intelligent Systems*, 337–378. https://doi.org/10.1007/978-3-030-49724-8_15.

Zhiping, H., Guangming, C., Cheng, C., He, X., & Jiacai, Z. (2010, November 1). A new EEG feature selection method for self-paced brain-computer interface. IEEE Xplore. https://doi.org/10.1109/ISDA.2010.5687156.

CHAPTER 5

Sequential Exception Technique for Text Anomalies

Siti Sakira Kamaruddin, Farzana Kabir Ahmad, and Mohammed Ahmed Taiye

5.1 INTRODUCTION

An increasing amount of natural language text, as opposed to structured databases, is overwhelming the repository of world knowledge. Humans are generating vast quantities of textual data due to advancements in Information Technology and the widespread adoption of social media platforms. The rapid increase in textual data has generated significant research interest in the field of text mining. However, the high dimensionality of text and the lack of formal structure in documents pose challenges in extracting and mining essential information from large volumes of text data. Mining text to detect anomalies is a field that is gaining relevance due to its ability to unearth intriguing, uncommon patterns concealed in the vast amount of textual data. Text anomalies refer to implicit knowledge that differs from the comprehensive information found in textual data. A diverse array of disciplines, including statistics, machine learning, data mining, information theory, and natural language processing (NLP) (Chandola et al., 2009), contribute to the assortment of approaches utilized in text anomaly detection, also known as outlier detection.

Statistics-based methods are deemed inappropriate due to the large dimensionality of text data and the requirement for prior knowledge of

DOI: 10.1201/9781003400387-5

data distribution. The distance-based approach, investigated by Agyemang et al. (2005) and Miller and Myers (2001), is adequate, but as the dimensionality of datasets increases, distance loses some of its significance. On the other hand, the classification-based method, as explored by Manevitz and Yousef (2001), is only appropriate if there is a clear distinction between the anomaly classes and the normal classes. In the clustering-based approach, anomalies are data objects that do not belong to any clusters (as described in Miller and Browning (2003), Montes-y-Gómez et al. (2002), and Cooley et al. (1997)). These methods appear to be slow because it is unclear how the data are grouped and because anomalies are typically clustering by-products. As a result, in comparison to other specialized techniques, clustering algorithms are not tuned to discover anomalies. Furthermore, a distance calculation between the data items is used by the majority of cluster-based algorithms.

However, a deviation-based detection method offers a significant benefit because it linearly processes high-dimensional data (Xie et al., 2006). The deviation-based method examines the key properties of the items in a group and discovers anomalies using a dissimilarity function. Anomalies are objects that differ from these properties. Hence, the deviation-based approach is more favorable when dealing with datasets in which the differentiation between normal and abnormal data is not as apparent as in the subjective text underpinning this study. As mentioned in Arning et al. (1996), Xie et al., (2006), and Zhang and Feng (2009), this technique offers linear complexity. However, the dissimilarity function, which is the method's central component, must be generally applicable to all types of data, which is a challenging requirement.

This study focuses on the discovery of knowledge through anomaly detection in text using the sequential exception technique (SET). SET was adapted by employing a cosine similarity function to replace the variance calculation. The adapted SET was tested on two datasets namely the ENRON email corpus and the 20 Newsgroup (20NG) dataset. Results are promising and indicate that the method can be further explored and enhanced to increase the performance in identifying anomalies in textual data.

This chapter is organized as follows: In Section 5.1, we gave some basic introductions to the topic. Section 5.2 discusses the existing text anomaly detection methods. Section 5.3 is devoted to describing SET. Section 5.4 details the methodology of this study followed by Section 5.5 which discusses the results. The conclusion is presented in Section 5.6.

5.2 TEXT ANOMALY DETECTION METHODS

As reviewed in Chandola et al. (2009) and Pang et al. (2021), the anomaly detection methods proposed for data which are structured or categorical might not be suitable for handling unstructured text. The high dimensionality and the sparseness of text cause the development of appropriate text anomaly detection methods more challenging. Machine learning methods are typically used to overcome these challenges as it involves the development of algorithms to enable computers to automatically learn the anomalies from existing example data. There is very little work that addresses anomaly detection in text. Among them, the classification and clustering-based methods are more prevalent.

Methods based on classification involve building a model from provided examples or training data. The classification-based approach seeks to teach the system from pre-classified training examples. In solving the text anomaly detection problem, these methods often assign text to one of the known anomaly or normal classes, i.e., in supervised learning; a learning algorithm is developed to learn the relationship between text documents with known training data. In this case, the training data may contain a pre-classified set of example anomalies. Neural network (NN) and support vector machine (SVM) are the most common classification-based methods used to solve the text anomaly detection problem. NN has been explored by Manevitz and Yousef (2001) for the reason that during the training of the network, only a limited number of parameters involve optimization. Another reason is no prior assumptions of the data characteristics are required. However, the main limitation that hinders the wide application of NN is the time it takes to process and converge especially when high-dimensional text data are considered. SVM has been used to detect anomalies in the works of Manevitz and Yousef (2001) and Srivastava et al. (2006). Within SVM, the training data is structured into two predefined categories, namely normal and anomaly. The SVM algorithm functions by constructing a model that maps these training instances onto a high-dimensional vector space, effectively differentiating between the normal and deviating data points. Classification-based method for anomaly detection is very efficient if a predefined anomaly category is available, nevertheless, in real-life cases, preparing a correctly annotated corpus is challenging. In most cases, the anomalies are unknown therefore methods based on text classification are considered not suitable for this situation. Furthermore, the computation cost can be high for a very large dataset.

The clustering-based method is an unsupervised learning method that involves learning patterns without any examples. In unsupervised learning, training data and predetermined classes are not provided. Applied to the text anomaly detection problem, the clustering-based method typically groups text automatically into different clusters based on their differences. Clustering is a popular method for anomaly detection among researchers especially when categorical data are considered. Items that do not belong to groups are considered anomalies. Clustering typically involves the calculation of a distance measure such as Euclidean distance to measure the similarity between data instances (Chandola et al., 2009). Implementation of the clustering method on textual documents involves developing a function to automatically divide a text collection into several different categories based on their text contents. Several researchers have employed clustering-based text anomaly detection. Among them are Srivastava and Zane-Ulman (2005) who analyzed flight readiness reports and discrepancy reports for the space shuttle, Srivastava et al. (2006) who attempted to use the statistical Von Mises Fisher (VMF) algorithm for clustering text to detect anomalies, and Zhang et al. (2004) who performed document clustering to detect anomalies using a hierarchical probabilistic model. One advantage of the clustering-based text anomaly detection method is that prior knowledge of data distribution is unnecessary if clustering techniques are employed. Another benefit of the clustering method is, that it can be employed incrementally. In other words, the method can detect anomalies as new data instances are fed into the system. However, clustering techniques involve distance computation which has its limitations and anomalies are a by-product of the clustering tasks.

Other approaches rely on external contextual data to identify anomalies using the topic modeling method (Mahapatra et al., 2012). The non-negative matrix factorization method was explored by Kannan et al. (2017) which is based on the block coordinate descent optimization technique. Besides that Fadhel and Nyarko (2019) explored the generative adversarial network (GAN) which learns the distribution of the normal data. GAN aims to generate realistic samples that are similar to real data by learning the data distribution (Fadhel & Nyarko, 2019). The challenge of using GAN-based method is that the data distribution needs to be known in advance as in the statistical-based methods and this method also needs a predefined anomaly category to learn from samples. Furthermore, GAN is challenging to train and frequently experiences convergence problems.

5.3 SEQUENTIAL EXCEPTION TECHNIQUE

The SET introduced by Arning et al. (1996) is a deviation-based anomaly detection method which does not employ any distance-based measures or statistical analysis to detect anomalies. Rather, it analyzes the key properties of the objects which are in a subset. An object that deviates from the identified properties is regarded as anomaly.

SET mimics how humans can pick out unusual objects from a group of apparently similar ones. Given n objects in as set, a sequence of subset will be constructed $\{D_1, D_2, D_3, \ldots, D_m\}$ of these objects with $2 \leq m \leq n$ such that shown in Equation (5.1).

$$D_{j-1} \subset D_j \; where \; D_j \subseteq D \tag{5.1}$$

The subsets are analyzed for dissimilarities. Instead of determining how different the present subset is from its complimentary set, the algorithm chooses a subset sequence from the group for analysis. It examines the sequence to establish the differences between each subset and the subset that precedes it. The objective is to identify an exception set that will be identified as the set of anomalies. The building block of this technique is the calculation of three functions, namely dissimilarity, cardinality, and smoothing factor.

A dissimilarity function is employed to identify the exception set. When given a set of items, it calculates the variance of the items in the set and yields a small number if the items are similar to one another and a higher value is returned when the items are more dissimilar to one another. Equation (5.2) is the dissimilarity function of SET.

$$\frac{1}{n}\sum_{i=1}^{n}\left(X_i - \bar{X}\right)^2 \tag{5.2}$$

Then a cardinality function is used to count the elements in the set as shown in Equation (5.3)

$$I_1 \subset I_2 \rightarrow C\left(I_1\right) < C\left(I_2\right) \; for \; all \; I_1, I_2 \subseteq I \tag{5.3}$$

The smoothing factor is used to analyze the effect of removing a subset D_j of elements from the set D. The purpose is to reduce the dissimilarity. It helps to remove noise. Equation (5.4) shows the smoothing factor of SET.

$$\mathrm{SF}(\mathrm{I_j}) := \mathrm{C}(\mathrm{I} - \mathrm{I_j}) * (\mathrm{D}(\mathrm{I}) - \mathrm{D}(\mathrm{I} - \mathrm{I_j})) \tag{5.4}$$

SET has been used for anomaly detection in categorical datasets (Xie et al., 2006; Settino et al., 2006; Zhang & Feng, 2009; Graaff & Engelbrecht, 2011). Although SET was used on categorical data, the method is applicable for sequence data represented as a set of strings which are the prominent form of textual data. SET is dedicated to finding potential anomalies in a set-based representation of categorical data by determining the subset of data to be pruned.

One major advantage of this method is it can run in linear time and it is able to process large amounts of data; however, the dissimilarity function which is based on variance calculation does not work for all types of data. The method became not popular in categorical data for the reason that it is non-trivial to derive a universal dissimilarity function; however, for textual datasets, this method is worth exploring since text can be represented as a set of vectors and cosine similarity can be used to find vector similarities.

5.4 METHOD

Figure 5.1 shows the method used in this study. Two datasets, namely the ENRON email corpus and 20 Newsgroup datasets were tested. The first step is to preprocess the text. Text preprocessing techniques such as stopword removal, stemming and lemmatization, were performed on the dataset. Then the text was converted into vectors using the count-vectorizer function before implementing the adapted SET. The adaptation is done by employing the cosine similarity function to replace the variance-based dissimilarity function as highlighted in Figure 5.1. The performance was evaluated in terms of precision, recall, and F-score. The next subsections provide a detailed description of the dataset, the adaptation of SET to process text data, and the evaluation measures used in the experiments.

5.4.1 Dataset

The datasets used in the experiment are the ENRON e-mail corpus and 20 Newsgroup datasets. ENRON email is a corpus that contains a large

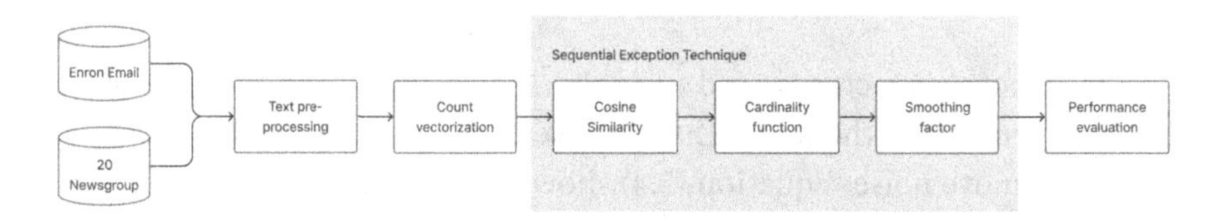

FIGURE 5.1 Text anomaly detection method with adapted SET.

collection of corporate email messages that were made available to the public in 2002 as a result of the ENRON Corporation's bankruptcy. ENRON Corporation, a company involved in the energy, commodities, and services sector, gained widespread attention for its bankruptcy in December 2001 due to engaging in fraudulent business practices. The dataset is available in various public versions. It consists of more than 1.2 million emails with attachments and metadata from 151 different users (Klimt & Yang, 2004; Gloor et al., 2006; Miz et al., 2019). Machine learning practitioners have the opportunity to utilize this dataset to develop models capable of identifying persons of interest (POIs) based on the data's attributes. The ENRON investigation targeted several high-level executives, who were eventually prosecuted for their involvement in fraudulent activities. In this study, we aim to identify the email messages from the identified POIs as anomalous text.

The second dataset used in this study is the 20 Newsgroup dataset (20NG) which is approximated to have 20,000 newsgroup documents. The news datasets used in this context are evenly divided into 20 different newsgroups, initially gathered by Lang (1995). Some articles within these groups exhibit close relationships, such as "comp.sys.ibm.pc.hardware" and "comp.sys.mac.hardware," while others are significantly unrelated, like "misc.forsale" and "soc.religion.christian." In this study, the aim is to identify topics that differ from the rest. In other words, the proposed method should accurately identify anomalous text. Both of these datasets have been extensively utilized by the machine learning community for a diverse range of applications, including text classification, topic modeling, text clustering, sentiment analysis, and text anomaly detection.

5.4.2 Adaptation on SET

SET was developed to find anomalies in categorical and sequential datasets from log files of large databases. In this study, SET was adapted to enable it to analyze textual data. The adaptation that was made to SET is in the calculation of the dissimilarity function. Instead of using variance as in Equation (5.2), we propose the cosine similarity function as shown in Equation (5.5).

$$\mathrm{SIM}(t_1, t_2) = \frac{\sum_1^n t_1 t_2}{\sqrt{\sum t_{1_i}{}^2} \times \sqrt{\sum t_{2_i}{}^2}} \tag{5.5}$$

The cosine similarity function is proven to perform well for text similarity measurement (Pradhan et al., 2015; Bakarov et al., 2018; Szoplák, & Andrejková, 2021). One of the reasons is that the relative frequency of terms is taken into account rather than their actual frequency. Equation (5.5) returns a value between 0 to 1 where 0 denotes dissimilarity and 1 denotes similarity. A threshold of 0.5 can be applied to determine whether an item is similar or otherwise. The cardinality function and the smoothing factor were used as proposed in Arning et al. (1996).

a. *Implementation of adapted SET on the ENRON email dataset.*

   The implementation of the adapted SET on the ENRON dataset is concerned with extracting information relating to the identified POI's fraudulent email messages. The aim is to identify the anomalous sender and receiver's email messages to and from one of the POI: Kenneth Lays who was the ENRON CEO. Before the adapted SET can be applied to the dataset, the dataset needs to go through a series of steps as shown in Figure 5.2.

   As depicted in the steps outlined in Figure 5.2, the parsing and counting methods are employed on the email messages exchanged between Kenneth Lay and the most frequent senders and receivers. The resulting texts are subsequently fed into the adapted SET. The obtained cosine similarity score is then recorded with the objective of identifying POIs who share similar textual information with Kenneth Lay based on the type of information exchanged. These findings are compared to the research conducted by Gloor et al. (2006), who identified and labeled certain employees as potential suspects in the ENRON company using temporal link analysis.

Step1. Initialize email messages as input files
Step2. Extract the email body of most to and from Kenneth Lay emails messages then append each mail.
Step 3. The extracted email body is then parsed as well and appended
Step 4. Appended files are then written into the root directory file of the systems
Step 5. The numbers of the list of emails sent to and received from messages with email addresses are then displayed on the written files saved in the root directory of the system.

FIGURE 5.2 Steps of parsing extracted mail messages.

comp.graphics comp.os.ms-windows.misc comp.sys.ibm.pc.hardware comp.sys.mac.hardware comp.windows.x	rec.autos rec.motorcycles rec.sport.baseball rec.sport.hockey	sci.crypt sci.electronics sci.med sci.space
misc.forsale	talk.politics.misc talk.politics.guns talk.politics.mideast	talk.religion.misc alt.atheism soc.religion.christian

FIGURE 5.3 20NG topic grouping.

b. *Implementation of adapted SET on the 20NG dataset.*
The utilization of the adapted SET on the 20NG dataset involves the detection of topics that demonstrate the highest textual similarity to other topics, as well as topics that exhibit the most dissimilar text. Figure 5.3 presents the categorized list of the 20 newsgroups, organized according to their respective subject similarities.

There are six groups: computers, recreational, science, politics, religion, and miscellaneous. In this regard, the adapted SET should identify the anomalous text from the group of computers as these group contains computer related discussion which are generally different from the discussion contains in other groups from the 20NG dataset.

5.4.3 Performance Evaluation

Once the anomalies have been identified by the adapted SET, an evaluation of the results was performed. The evaluation measurements that were employed are precision, recall, and F-score. For the ENRON email dataset, the precision and recall are calculated as shown in Equations (5.6) and (5.7), respectively.

$$\text{Precision} = \frac{\text{Retrieved} \cap \text{Relevant}}{\text{Retrieved}} \tag{5.6}$$

$$\text{Recall} = \frac{\text{Retrieved} \cap \text{Relevant}}{\text{Relevant}} \tag{5.7}$$

Precision is defined as the ratio of relevant material correctly retrieved to the total retrieved material, while recall is the ratio of relevant material correctly retrieved to the total relevant material.

For the 20NG dataset, the precision and recall were calculated using a confusion matrix that records the number of true positives (TP), true negatives (TN), false positives (FP), and false negatives (FN). TP represents the count of correctly predicted instances that are positive, FP denotes the number of positive instances that were incorrectly predicted, TN represents the number of negative instances that were correctly predicted, and FN signifies the number of negative instances that were incorrectly predicted. Equations (5.8) and (5.9) are used to calculate the precision and recall score using TP, TN, FP, and FN.

$$\text{Precision} = \frac{\text{TP}}{\text{TP} + \text{FP}} \tag{5.8}$$

$$\text{Recall} = \frac{\text{TP}}{\text{TP} + \text{FN}} \tag{5.9}$$

The F-score is computed using Equation (5.10) based on the precision and recall results.

$$\text{F} - \text{Score} = 2 \times \left[\frac{(\text{precision} \times \text{recall})}{(\text{precision} + \text{recall})} \right] \tag{5.10}$$

5.5 RESULTS AND DISCUSSION

The implementation of the adapted SET on the ENRON dataset has identified the anomalous sender and receiver of email messages to and from ENRON CEO Kenneth Lays. The cosine similarity score obtained is as shown in Figure 5.4. In our work, a threshold is set for the cosine similarity score. A similarity score above 0.5 shows that the text in the email messages is similar. Figure 5.4 shows the cosine similarity scores of the emails to and from Kenneth Lay plotted as a scatter plot. The recorded similarity score is higher (0.7 and above) for POIs such as K-Lay, J-Skillings, J-Steffes, J-Dasovich, R-Shapiro, S-Kean, K-Mann, and M-Taylor which indicates the closest information with Kenneth Lays mail messages.

The comparison between the relatedness of POIs in their email messages with Kenneth Lay, as illustrated in Figure 5.4, was conducted in comparison to the study by Gloor et al. (2006). Our method successfully identified 9 out of the 10 POIs identified by Gloor et al. (2006). To assess the effectiveness of the proposed technique, precision and recall were calculated using Equations (5.6) and (5.7), respectively, based on the relevance

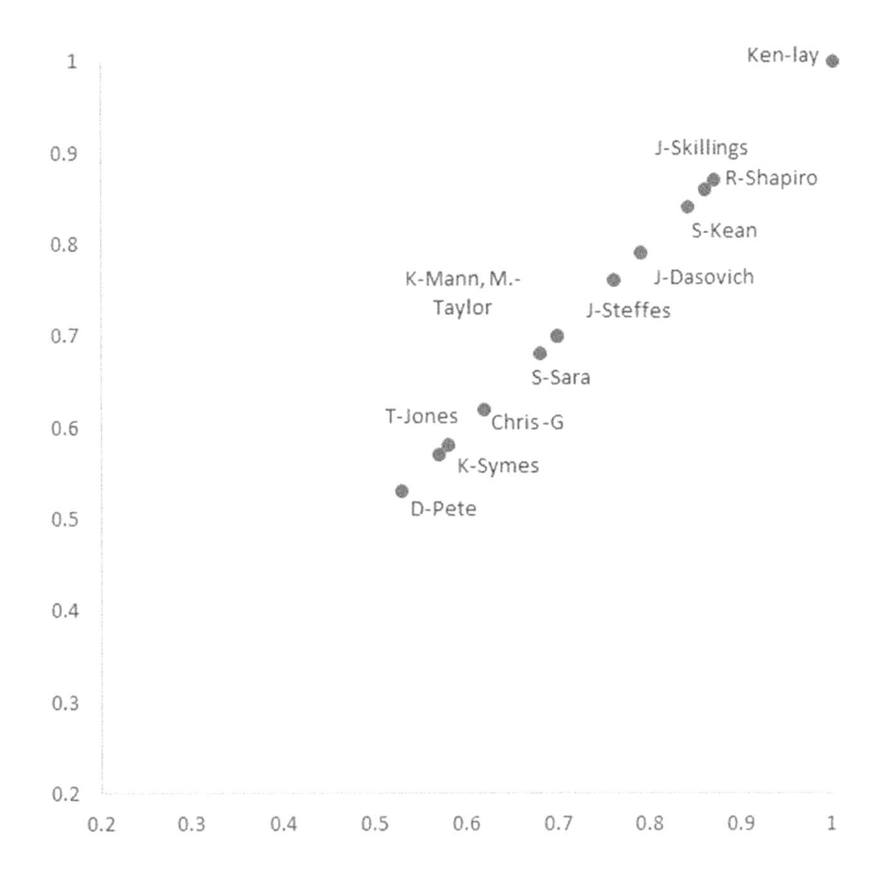

FIGURE 5.4 The similarity scores of the email messages of the POIs of the ENRON dataset.

and retrieved text results obtained from the proposed method. Table 5.1 presents the results.

A precision score of 69.2% was achieved, along with a recall score of 90% and an F-score of 78.1%. An acceptable precision score and a higher recall score indicate that the proposed method has successfully identified all the required anomalies. However, it should be noted that some additional

TABLE 5.1 Precision, Recall, and F-Score for the Retrieved POI's Email Messages from the ENRON Dataset

	# Items Retrieved by the Proposed Method (Ret)	# Items Relevant as Identified by Gloor et al., 2006 (Rel)	Ret ∩ Rel	Precision	Recall	F-Score
Number of POIs	13	10	9	69.2%	90%	78.1%

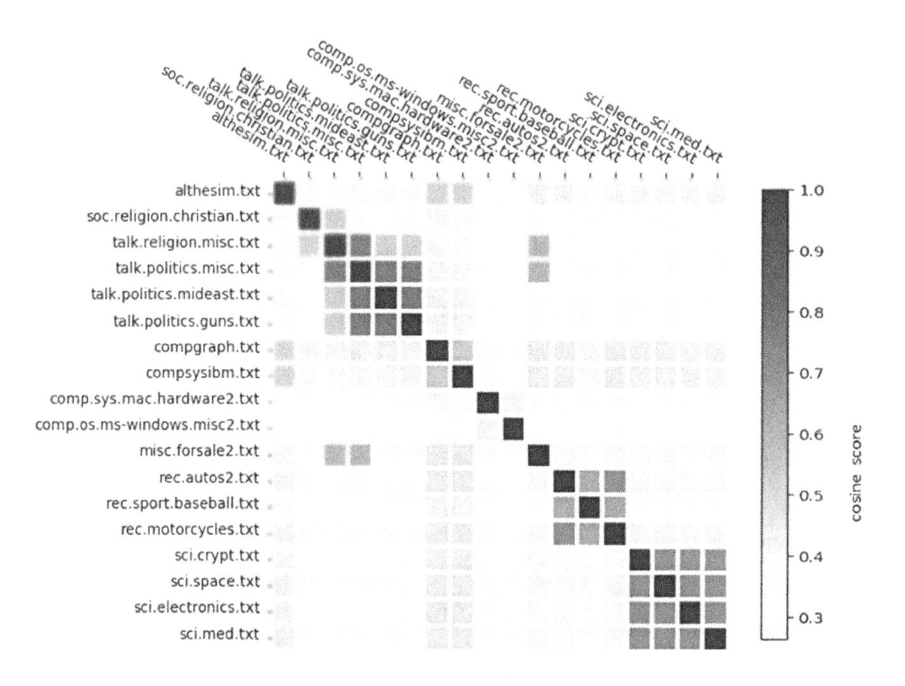

FIGURE 5.5 The similarity scores of the 20NG dataset.

items were also retrieved, which are not truly anomalous. Further exploratory analysis can be applied to this dataset to improve results and more effectively extract anomalous information.

For the 20NG datasets, the adapted SET has identified the anomalous topic of "comp.os.ms-windows.misc" and "comp.sys.mac.hardware." Figure 5.5 shows the similarity score calculated by comparing text documents from each of the topics in the 20NG dataset.

Figure 5.5 shows that the topic "comp.os.ms-windows.misc" and "comp.sys.mac.hardware" with lowest similarity score indicating they are entirely different from other topics. The performance evaluation was conducted by calculating the precision, recall, and F-score using Equations (5.8), (5.9), and (5.10) correspondingly. Table 5.2 shows the scores for the ENRON dataset, 20NG dataset, and the average scores for both datasets.

TABLE 5.2 Performance of Adapted SET on ENRON and 20NG Datasets

	Performance Measures		
Datasets	**Precision**	**Recall**	**F-Score**
ENRON	69.2%	90%	78.1%
20NG	92%	100%	95%
Average	80.6%	95%	86.55%

The results suggest that the adapted SET exhibited better performance on text datasets characterized by specific topics or focused areas of discourse (such as 20NG) compared to text datasets (like ENRON) that encompass a more generalized topic of discussion.

5.6 CONCLUSION

This study focuses on discovering knowledge through anomaly detection in text using SET. The SET was adapted by employing a cosine similarity function instead of variance calculation. The adapted SET was tested on two datasets: the ENRON email corpus and the 20 NG dataset. Promising results indicate the feasibility of using the cosine dissimilarity function as a significant adaptation in the SET to identify text anomalies. Although the average F-score of 86.55% obtained from the two datasets is relatively high, further exploration and enhancement of the method for identifying anomalies in textual data can be pursued. In future endeavors, it is imperative to prioritize the enhancement of document feature vectors, term-document matrices, and the semantic relationships of words through thorough human and corpus analysis. This approach will contribute to improving the overall effectiveness and accuracy of the analysis process. These improvements can help analyze and detect semantic-based text anomalies in documents, providing a foundation for further examination and refinement of the study findings.

ACKNOWLEDGMENT

This research was supported by the Universiti Utara Malaysia through the University Centre of Excellence (COE) Research Grant.

REFERENCES

Agyemang, M., Barker, K., & Alhajj, R. S. (2005). Mining web content outliers using structure oriented weighting techniques and n-grams. In *Proceedings of the 2005 ACM symposium on Applied computing* (pp. 482–487).

Arning, A., Agrawal, R., & Raghavan, P. (1996). A linear method for deviation detection in large databases. *KDD*, *1141*(50), 972–981.

Bakarov, A., Yadrintsev, V., & Sochenkov, I. (2018). Anomaly detection for short texts: Identifying whether your chatbot should switch from goal-oriented conversation to chit-chatting. In *Digital Transformation and Global Society: Third International Conference, DTGS 2018, St. Petersburg, Russia, May 30–June 2, 2018, Revised Selected Papers, Part II 3* (pp. 289–298). Springer International Publishing.

Chandola, V., Banerjee, A., & Kumar, V. (2009). Anomaly detection: A survey. *ACM Computing Surveys (CSUR)*, *41*(3), 1–58.

Cooley, R., Mobasher, B., & Srivastava, J. (1997). Web mining: Information and pattern discovery on the world wide web. In *Proceedings ninth IEEE International Conference on Tools with Artificial Intelligence* (pp. 558–567). IEEE.

Fadhel, M. B., & Nyarko, K. (2019). GAN augmented text anomaly detection with sequences of deep statistics. In *2019 53rd Annual Conference on Information Sciences and Systems (CISS)* (pp. 1–5). IEEE.

Gloor, P. A., Niepel, S., & Li, Y. (2006). Identifying potential suspects by temporal link analysis. In *MIT CCS working paper.*

Graaff, A. J., & Engelbrecht, A. P. (2011). Using sequential deviation to dynamically determine the number of clusters found by a local network neighbourhood artificial immune system. *Applied Soft Computing, 11*(2), 2698–2713.

Kannan, R., Woo, H., Aggarwal, C. C., & Park, H. (2017, June). Outlier detection for text data. In *Proceedings of the 2017 Siam International Conference on Data Mining* (pp. 489–497). Society for Industrial and Applied Mathematics.

Klimt, B., & Yang, Y. (2004). The enron corpus: A new dataset for email classification research. In *European Conference on Machine Learning* (pp. 217–226). Berlin, Heidelberg: Springer Berlin Heidelberg.

Lang, K. (1995). Newsweeder: Learning to filter netnews In *Machine Learning Proceedings 1995* (pp. 331–339). Morgan Kaufmann.

Mahapatra, A., Srivastava, N., & Srivastava, J. (2012). Contextual anomaly detection in text data. *Algorithms, 5*(4), 469–489.

Manevitz, L. M., & Yousef, M. (2001). One-class SVMs for document classification. *Journal of machine Learning research, 2*, 139–154.

Miller, D. J., & Browning, J. (2003). A mixture model and EM-based algorithm for class discovery, robust classification, and outlier rejection in mixed labeled/unlabeled data sets. *IEEE Transactions on Pattern Analysis and Machine Intelligence, 25*(11), 1468–1483.

Miller, R. C., & Myers, B. A. (2001). Outlier finding: Focusing user attention on possible errors. In *Proceedings of the 14th Annual ACM Symposium on User Interface Software and Technology* (pp. 81–90).

Miz, V., Ricaud, B., Benzi, K., & Vandergheynst, P. (2019). Anomaly detection in the dynamics of web and social networks using associative memory. In *The World Wide Web Conference* (pp. 1290–1299).

Montes-y-Gómez, M., Gelbukh, A., & López-López, A. (2002). Detecting deviations in text collections: An approach using conceptual graphs. In *MICAI 2002: Advances in Artificial Intelligence: Second Mexican International Conference on Artificial Intelligence Mérida, Yucatán, Mexico, April 22–26, 2002 Proceedings 2* (pp. 176–184). Springer Berlin Heidelberg.

Pang, G., Shen, C., Cao, L., & Hengel, A. V. D. (2021). Deep learning for anomaly detection: A review. *ACM Computing Surveys (CSUR), 54*(2), 1–38.

Pradhan, N., Gyanchandani, M., & Wadhvani, R. (2015). A review on text similarity technique used in IR and its application. *International Journal of Computer Applications, 120*(9), 29–34.

Settino, M., Ruffolo, A., & La Regina, F. (2006). "Mining" the sky from data mining. *Acta Astronautica, 59*(6), 499–502.

Srivastava, A., Akella, R., Diev, V., Kumaresan, S. P., McIntosh, D. M., Pontikakis, E. D., ... & Zhang, Y. (2006, January). Enabling the discovery of recurring anomalies in aerospace system problem reports using high-dimensional clustering techniques. In *IEEE Aerospace Conference.*

Srivastava, A. N., & Zane-Ulman, B. (2005, March). Discovering recurring anomalies in text reports regarding complex space systems. In *2005 IEEE Aerospace Conference* (pp. 3853–3862). IEEE.

Szoplák, Z., & Andrejková, G. (2021). Anomaly Detection in Text Documents using HTM Networks. In *ITAT* (pp. 20–28).

Xie, C., Chen, Z., & Yu, X. (2006, September). Sequence outlier detection based on chaos theory and its application on stock market. In *International Conference on Fuzzy Systems and Knowledge Discovery* (pp. 1221–1228). Berlin, Heidelberg: Springer.

Zhang, J., Ghahramani, Z., & Yang, Y. (2004). A probabilistic model for online document clustering with application to novelty detection. *Advances in Neural Information Processing Systems, 17,* 1617–1624. *17.*

Zhang, Z., & Feng, X. (2009, August). New methods for deviation-based outlier detection in large database. In *2009 Sixth International Conference on Fuzzy Systems and Knowledge Discovery* (Vol. 1, pp. 495–499). IEEE.

CHAPTER 6

Intelligence Predictive Model for Lamb Carcass C-Site Fat Depth Using Support Vector Machine

Wan Yu Jinq, Elayaraja Aruchunan, Nur Anisah Mohamed A. Rahman, Kohilavani Naganthran, Mohana Sundaram Muthuvalu, Jackel Vui Lung Chew, Jayaseelan Marimuthu, Graham Edwin Gardner, and Samsul Ariffin Abdul Karim

6.1 INTRODUCTION

Forecasting is extremely important in any field in the modern era. As the world is becoming more advanced, the forecasting method is slowly adapting from conventional statistical methods such as multiple linear regression, basic time series prediction methods, etc. to modern forecasting using machine learning and deep learning. The improvisation of these methods allows us to optimize the prediction model and obtain a more accurate output. The term "carcass" pertains to the deceased body of an animal, particularly a larger one destined for meat consumption or as sustenance for wildlife [1]. Accurate determination of fat depth in meat-producing animal carcasses is pivotal for optimizing productivity and profitability throughout the supply chain [2]. Traditionally, carcass fatness assessment has relied on

DOI: 10.1201/9781003400387-6

manual methods such as dissection or chemical analysis, which can lead to inaccuracies. The microwave system (MiS), utilizing low-power nonionizing electromagnetic waves, emerges as a suitable alternative to reduce human errors during data collection. MiS effectively gauges fat depth and carcass composition, boasting precision. Moreover, MiS requires cost-effective, portable equipment, adding to its appeal. The dataset originates from researchers affiliated with Murdoch University in Perth, Western Australia. These researchers conducted a study centered on a noninvasive microwave technique for measuring lamb's C-site fat depth. Their objective was to develop an effective predictive model utilizing microwave signals to forecast fat depth accurately. Figures 6.1 and 6.2 show the lamb carcass to assess the fat depth and measuring of carcase C-site fat depth using microwave device, respectively. The dataset consists of 120 observations, encompassing a solitary dependent variable and 311 independent variables. The dependent variable pertains to the C-site fat depth, whereas the independent variable corresponds to the magnitude of the frequency domain microwave signals.

Machine learning and statistical methods share similarities yet serve distinct purposes. Machine learning primarily centers on prediction, identifying patterns within complex data, whereas statistical methods

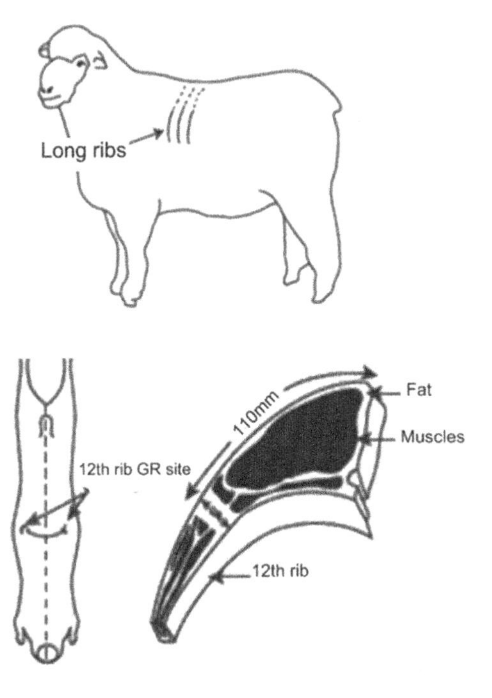

FIGURE 6.1 The Lamb carcass.

FIGURE 6.2 Measuring Carcase C-site fat depth using microwave device.

prioritize inference [3]. Despite their differences, machine learning is rooted in a statistical framework, making direct efficiency comparison challenging. The choice between the two hinges on our objective: Machine learning excels in prediction, while statistical models are optimal for elucidating variable relationships [4]. In the context of the ongoing COVID-19 pandemic, machine learning finds valuable application in the healthcare sector, alongside artificial intelligence, as potent tools combatting the crisis [5]. The classical support vector machine (SVM) model is frequently employed in weather forecasting, stock market analysis, and real estate pricing. Yet, inherent limitations of this model sometimes lead to inaccuracies in predictions. A quantum-assisted model was introduced in 2021 to enhance the regression performance of SVM, particularly demonstrated in the case of facial landmark detection. Comparison between the quantum-assisted and classical SVM models highlighted empirical advantages of quantum-assisted algorithms based on quantum annealing [6]. To optimize SVM prediction outcomes, hyperparameter tuning is crucial. The SVM, serving as a fundamental predictor for blasting mean fragment size, synergizes with five optimization algorithms: grey wolf optimization (GWO), grid search (GS), genetic algorithm (GA), salp swarm algorithm

(SSA), and particle swarm optimization (PSO). Among these, GWO-v-SVR emerged as the superior predictor for blasting mean fragment size [7].

The strengths of support vector regression (SVR) are noteworthy. It exhibits resilience against outliers, enables straightforward decision model updates, and boasts high generalization capability. In addition, SVR's practicality stands out as it requires no preliminary assumptions, rendering it easily applicable across various scenarios [8].

6.2 MATERIALS AND METHODS

6.2.1 *K*-Means Clustering

Clustering stands as an unsupervised learning technique, particularly valuable for managing unlabeled datasets. Its primary application involves identifying meaningful groupings within data. Clustering aims to segment a population into distinct groups, where each group shares similar characteristics among its members. In essence, clustering categorizes objects based on their shared and distinct attributes. *K*-means clustering is a technique rooted in vector quantization, originally from signal processing. Its purpose is to allocate n observations to k clusters, with each observation joining the cluster whose mean (centroid) is closest, effectively representing the cluster prototype. This method computes the distance between each data point and centroids to assign them to clusters. The objective is to distribute observations into clusters with similar characteristics, achieved through iterative steps where data points are gradually clustered based on common attributes. The ultimate goal is to minimize the sum of distances between data points and cluster centroids. *K*-means partitions data space into K clusters, assigning each a mean value. Data points align with clusters where they exhibit the shortest distance to the cluster's mean. This technique is widely utilized in dimensionality reduction within machine learning.

As previously mentioned, K-means clustering strives to minimize total intra-cluster variance, guided by the following objective function:

$$J = \sum_{j=1}^{k} \sum_{i=1}^{n} \left\| x_i^{(j)} - c_j \right\|^2 \tag{6.1}$$

where

k = number of clusters,

n = number of cases, and

c_j = centroid for cluster j

6.2.2 Principal Component Analysis

Principal component analysis (PCA) is one of the most efficient approaches used in the data field. It is often utilized for dimensionality reduction, where each data point is projected onto only the first few principal components to produce lower-dimensional data while preserving as much variability as possible. PCA tries to reduce the variables (principal component) by solving an eigenvalue/eigenvector problem and looking for new variables that are linear functions of those in the raw data that successively maximize variation and have no autocorrelation with each other. Further to the above, it helps us solve the multicollinearity issue in the dataset as the linear combinations that are formed are always uncorrelated with each other [9].

6.2.3 Support Vector Machine

SVM analysis, initially introduced by Vladimir Vapnik and colleagues in 1992, gained renown as a prominent machine learning method for classification [10]. Over time, this approach underwent refinement to accommodate regression tasks, leading to its incarnation as SVR, also used for prediction. Thanks to its reliance on kernel functions, SVR can be categorized as a non-parametric algorithm. With minor deviations, SVR and SVM adhere to a shared classification principle. In regression, SVR accommodates a tolerance margin approximation. However, the overarching concept remains consistent: optimizing error reduction while maximizing margin through personalized hyperplane positioning.

When applying SVR, several key hyperparameters come into play, including the hyperplane, kernel, and boundary line.

i. Type of kernel:

a. Linear kernel

$$y=\sum_{i=1}^{N}\left(\alpha_i-\alpha_i^*\right)\cdot\langle x_i, x\rangle+b \tag{6.2}$$

b. Nonlinear kernel

$$y=\sum_{i=1}^{N}\left(\alpha_i-\alpha_i^*\right)\cdot\langle\varphi(x_i),\varphi(x)\rangle+b \tag{6.3}$$

$$y = \sum_{i=1}^{N} \left(\alpha_i - \alpha_i^*\right) \cdot K(x_i, x) + b \tag{6.4}$$

c. Kernel functions

$$k\left(x_i, x_j\right) = \left(x_i, x_j\right)^d \text{ (Polynomial)} \tag{6.5}$$

$$k\left(x_i, x_j\right) = \exp\left(-\frac{\left\|x_i, -x_j\right\|^2}{2\sigma^2}\right) \text{(Gaussian Radial Basis Function)} \tag{6.6}$$

ii. Boundary lines: Two lines that are used to create a margin between the data points.

6.2.4 Multiple Linear Regression

A traditional statistical model, multiple linear regression (MLR), serves to establish the connection between a single dependent variable and multiple independent variables. In addition, MLR evaluates the strength of relationships within the variables across the entirety of the regression model. Furthermore, MLR finds applicability in predictive contexts. The fundamental structure of MLR can be depicted as:

$$y = B_0 + B_1 X_1 + \ldots + B_n X_n + \varepsilon \tag{6.7}$$

where

$Y =$ Dependent variable,
$X_i =$ Independent variable,
$\beta_0 =$ Intercept,
$\beta_n =$ Slope coefficients for each independent variable, and
$\varepsilon =$ is the model error.

6.2.5 Assumption Checking

There are a few preliminary assumptions that need to be met before fitting the data into the MLR model. It is significant as the result might not be valid if the data used does not fulfill every single assumption, and the outcome might not be able to capture the actual characteristic, causing the outcome to be totally out of form. Hence, assumption checking is needed to ensure the data used is appropriate to be fitted into the MLR model.

Otherwise, transformation of the data is needed so that the data can manage to achieve all of the assumptions.

6.2.6 Normality

The normal probability plot or normal quantile plot of the residuals can be seen as a good indicator of normality. The points in the plot should bounce along the diagonal line or show a normal distribution shape to ensure the residuals of the model are normally distributed.

6.2.7 Homoscedasticity Test

Equally spread points along with the horizontal line from the residual plot are a good indication of homoscedasticity. It allows us to ensure the residuals of the model have constant variance.

6.2.8 Autocorrelation Test

A Durbin–Watson test can be used to detect autocorrelation among variables and ensure that the residuals of the model are independent. A Durbin–Watson statistics, d which is in between 1.5 and 2.5 shows that there is no autocorrelation in the data. A d value of 0–2 and 2–4 shows that the data has positive autocorrelation and negative autocorrelation, respectively.

6.2.9 Linearity Test

Nonlinearity is usually most obvious in a plot of observed values versus predicted values or a plot of actual values versus predicted values. In such a plot, the points should be symmetrically distributed around the diagonal line with a roughly constant variance to determine if there exists a linear relationship between the dependent variables and independent variables.

6.2.10 Root Mean Square Error

$$RMSE = \sqrt{\frac{\sum_{i=1}^{N}\left(x_i - \hat{x}_i\right)^2}{N}} \tag{6.8}$$

where

N = Sample size,

x_i = actual observation, and

$\hat{x}_i$ = estimated observation.

According to Wikipedia, root mean square error (RMSE) is an often-used measure of the difference between predicted values and observed values. It can be said to be the square root of the average of squared errors. It is sensitive to outliers, and hence it is a commonly used measure in evaluating model accuracy, where the lower the RMSE, the closer the estimated value is compared to the actual value. A model that minimizes the RMSE will result in a higher accuracy of the predicted value.

6.2.11 Mean Absolute Error

$$MAE = \frac{\sum_{i=1}^{N} \left| \hat{y}_i - y_i \right|}{N} \tag{6.9}$$

where

N = Sample size,
y_i = actual observation, and
$\hat{y}_i$ = estimated observation.

The mean absolute error (MAE) is a measure of errors between paired observations. It can be seen as an arithmetic average of absolute errors. MAE is widely used as it is easy to understand and compute.

6.2.12 *R*-Squared (R^2)

It is the variation proportion that is explained by the predictor variables. In other words, it also explains how well the data fit the model. The better model will have higher *R*-squared.

$$R^2 = 1 - \frac{\Sigma\left(y_i - \hat{y}_i\right)^2}{\Sigma\left(y_i - \bar{y}\right)^2} \tag{6.10}$$

where

y_i = actual observation,
$\hat{y}_i$ = estimated observation, and
$\bar{y}$ = average actual observation.

Coefficient of determination, or *R*-squared, is the proportion of the variation in the dependent variable that is predictable from the independent variables. An *R*-squared of 1 tells us that all of the variations from dependent variable can be explained by the features.

6.2.13 Adjusted *R*-Squared

It adjusts the root squared when there are too many variables in a model.

$$Adjusted\ R-Squared = 1-\left[\frac{\left(1-R^2\right)(n-1)}{n-k-1}\right] \tag{6.11}$$

where n is the number of observations, R^2 is the R-squared, and k is the number of independent variables.

6.3 RESULT AND DISCUSSION

6.3.1 Dimensionality Reduction

6.3.1.1 K-Means Clustering

Utilizing the Elbow method, our analysis has ascertained that the optimal quantity of clusters, suitable for k-means clustering, is determined to be three. This outcome signifies that our assemblage of 311 independent variables will be partitioned into three distinct clusters. These clusters are characterized by sizes of 122, 91, and 98, respectively. Figure 6.3 shows the Elbow method results.

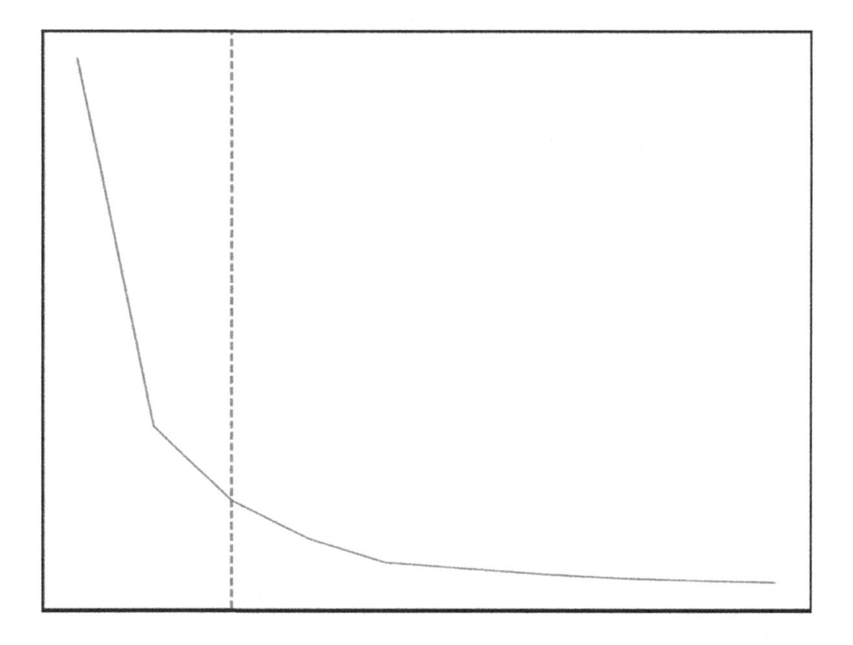

FIGURE 6.3 The Elbow method result.

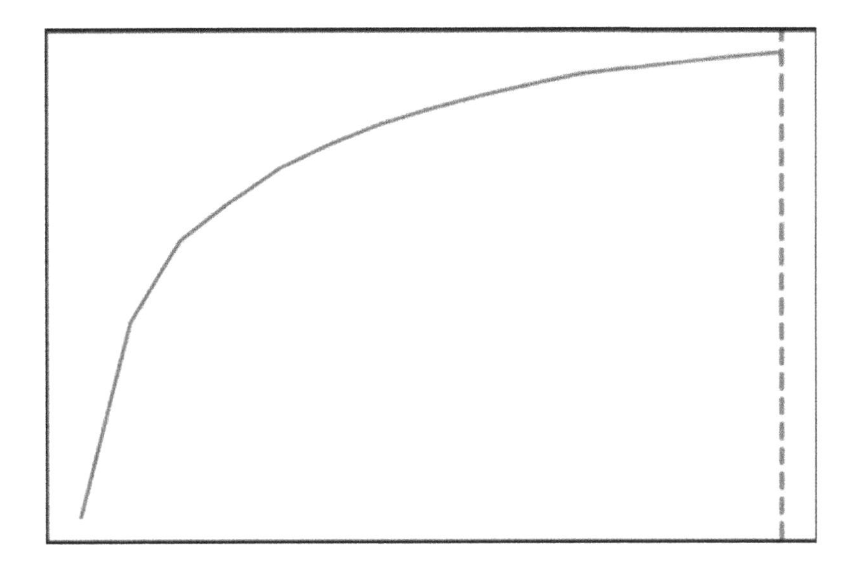

FIGURE 6.4 Principal Component Analysis results.

6.3.1.2 Principal Component Analysis

In our investigation, Q-mode PCA is employed to address the emphasis on variables. Following the execution of the PCA, the multitude of independent variables is streamlined to 15, effectively mitigating the risk of overfitting attributed to their excessive quantity. Through unsupervised learning techniques, we have succeeded in diminishing the number of independent variables from 311 to 3 via K-means clustering, and to 15 through PCA. These transformed datasets, stemming from unsupervised learning, will subsequently be integrated into diverse models for fitting purposes. Figure 6.4 shows the PCA results.

6.3.2 Support Vector Machine Analysis

Tabulated below are the mean values of RMSE, MAE, R^2, and adjusted R^2, obtained through fivefold cross-validation, for various kernels in the context of SVR. Table 6.1 shows the results of SVR by PCA and K-means clustering.

The presented table reveals that RMSE ranges from 1.12 to 1.86 for models fitted using PCA-transformed data and 1.27 to 1.80 for those derived from K-means clustering. On the whole, the PCA-based approach exhibits a superior fit, characterized by lower RMSE and MAE scores, as well as

TABLE 6.1: Result for Support Vector Machine

	Kernel	RMSE	MAE	R^2	Adjusted R^2
PCA	RBF	1.3471	1.0241	0.5614	0.3830
	Linear	1.1205	0.877	0.6213	0.5667
	Poly	1.3199	0.9964	0.4815	0.4067
	Sigmoid	1.8551	1.2953	0.1646	0.0441
***K*-means Clustering**	RBF	1.2527	0.9638	0.5296	0.5175
	Linear	1.3960	1.0323	0.4183	0.4032
	Poly	1.2181	0.9498	0.5571	0.5457
	Sigmoid	1.7420	1.3955	-0.0205	-0.0469

higher R^2 and adjusted R^2 values. The optimal model emerges as the SVR model with a linear kernel, trained utilizing PCA-transformed data. To enhance this model further, a logical next step involves fine-tuning the hyperparameters associated with the linear kernel in SVR. Conversely, the negative R^2 and adjusted R^2 values linked to the sigmoid kernel in K-means clustering indicate a model that diverges from the data's trends, rendering it less effective compared to a simple intercept term fitting.

6.3.3 Hyperparameter Tuning

Considering that the optimal model is the linear kernel SVR model trained post PCA, our next step involves hyperparameter tuning to optimize parameters for enhanced prediction performance. Based on Table 6.2, for the linear kernel in SVR, a single parameter, the cost parameter C, requires consideration. The default value for C is 1; however, our findings pinpoint the optimal C value at 0.0899. This tuning process yields subtle enhancements, evident through reduced RMSE and MAE values, accompanied by heightened R^2 and adjusted R^2 values. A calculated R^2 of 0.625 and an adjusted R^2 of 0.571 signify that the independent variables elucidate 62.5% of pre-adjustment variation and 57.1% of post-adjustment variation.

TABLE 6.2: Comparison after Hyperparameter Tuning

	C	RMSE	MAE	R^2	Adjusted R^2
Before tuning	1	1.1205	0.877	0.6213	0.5667
After tuning	0.0899	1.1147	0.8639	0.6250	0.5709

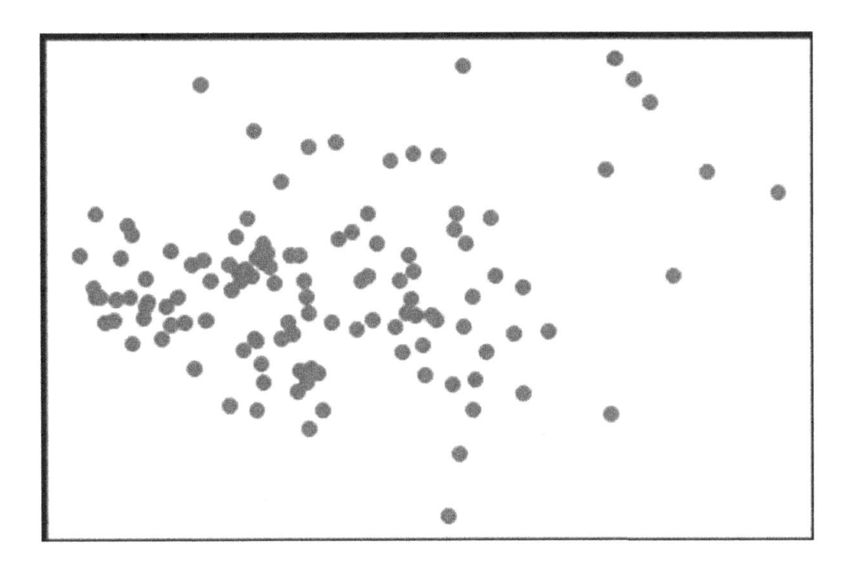

FIGURE 6.5 Residuals vs Fitted Value Plot (SVM).

The equation depicting the SVR model utilizing the linear kernel is provided below:

$$\begin{aligned} y = 3.9382 &+ (-0.5339)W_1 + (0.1891)W_2 + (0.6593)W_3 \\ &+ (-0.1987)W_4 + (0.0315)W_5 + (0.2599)W_6 + 0.2775W_7 \\ &+ (-0.3519)W_8 + (0.0555)W_9 + (0.4578)W_{10} + (-0.1221)W_{11} \\ &+ (0.1042)W_{12} + (0.0416)W_{13} + (0.2611)W_{14} + (-0.2706)W_{15} \end{aligned} \tag{6.12}$$

The residual versus fitted plot provides insight into the prediction pattern. Upon analysis of the plot, no discernible significant pattern emerges. Consequently, we conclude that the prediction lacks adherence to any discernible trend, appearing to exhibit randomness. Figure 6.5 shows residuals versus fitted value plot (SVM).

6.3.4 Multiple Linear Regression

The PCA-transformed data has exhibited improved performance compared to our previous findings. Consequently, this transformed dataset will be employed in fitting the MLR model. A subsequent comparison with the machine learning algorithm's results will be conducted. Prior to examining the outcomes, it is imperative to verify the fulfilment of all linear regression assumptions, thus ensuring the validity of results derived from the MLR analysis.

6.3.4.1 Linearity

Analyzing the plot of actual values versus predicted values reveals a fairly uniform distribution around the diagonal line. This alignment conforms to the constant variance assumption, thereby affirming a linear relationship between the independent and dependent variables. Figure 6.6 shows actual value versus predicted value plot (MLR).

6.3.4.2 Normality

Figures 6.7 and 6.8 show the distribution of residuals (MLR) and Anderson–Darling test for normal distribution. Evidently, the plot illustrates that the residuals conform to a normal distribution. To reinforce this observation,

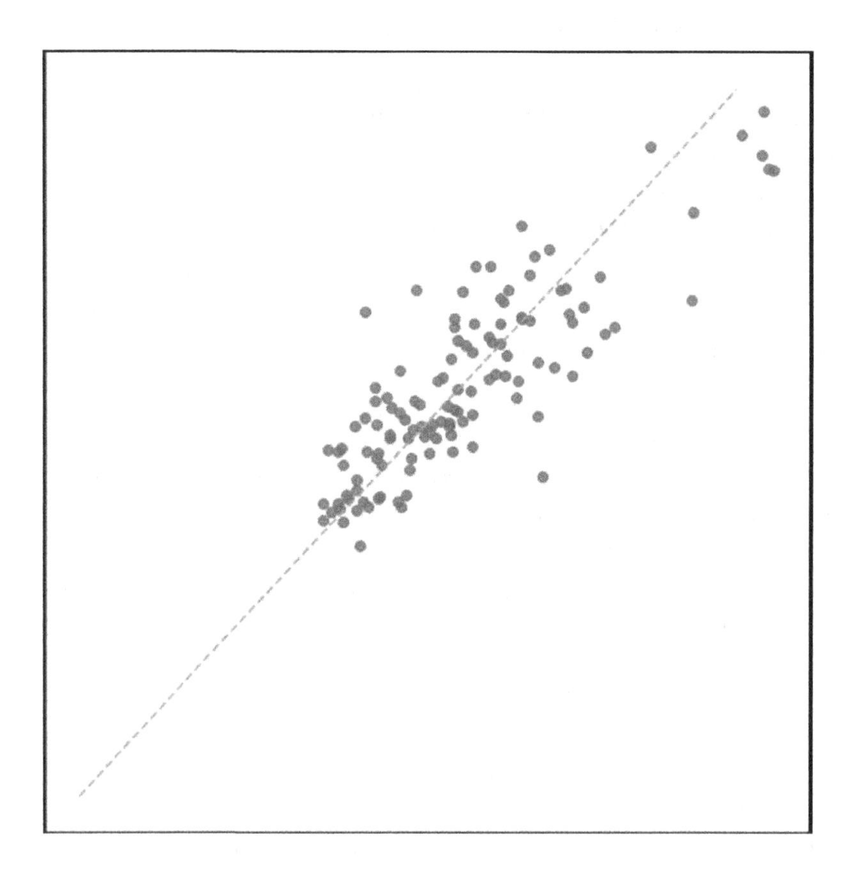

FIGURE 6.6 Actual Value vs Predicted Value Plot (MLR).

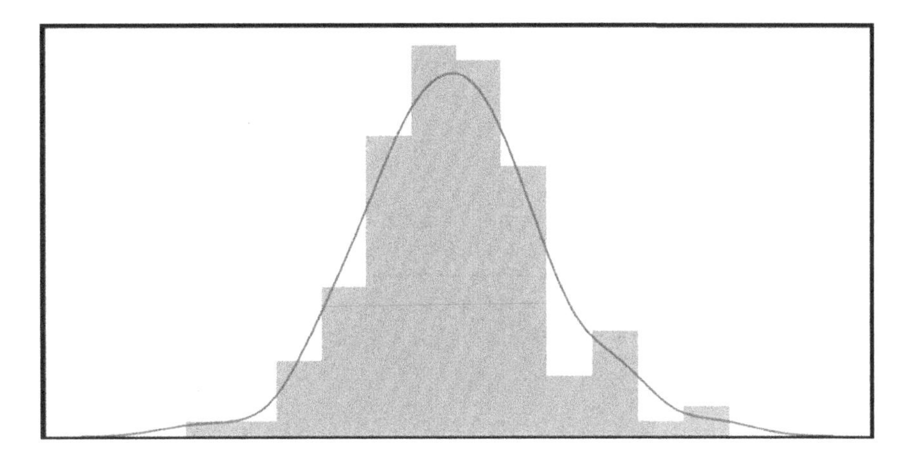

FIGURE 6.7 Distribution of Residuals (MLR).

```
Using the Anderson-Darling test for normal distribution
p-value from the test - below 0.05 generally means non-normal: 0.4171497149701231
Residuals are normally distributed
```

FIGURE 6.8 Anderson-Darling test for normal distribution.

the Anderson–Darling test can be employed. With a significance level set at 0.05, the resulting *p*-value for this hypothesis test is 0.4171, surpassing the 0.05 threshold. Consequently, we accept the null hypothesis, indicating that the residuals indeed exhibit a normal distribution.

6.3.4.3 Autocorrelation

Figure 6.9 shows that the Durbin–Watson statistic, denoted as d, computes to 1.818, positioning itself between the range of 1.5 and 2.5. This placement signifies the absence of substantial autocorrelation within the dataset. As a result, the assumption of non-autocorrelation among error terms is upheld.

6.3.4.4 Homoscedasticity

From Figure 6.10 of residuals plot, the data points exhibit a random scatter across the horizontal zero line, revealing an absence of notable patterns. This arrangement indicates that the residuals maintain a relatively consistent variance. Thus, the assumption of homoscedasticity is satisfied.

```
Performing Durbin-Watson Test
Values of 1.5 < d < 2.5 generally show that there is no autocorrelation in
 the data
0 to 2< is positive autocorrelation
>2 to 4 is negative autocorrelation
-------------------------------------
Durbin-Watson: 1.8181544450135017
Little to no autocorrelation
```

FIGURE 6.9 Durbin–Watson Test.

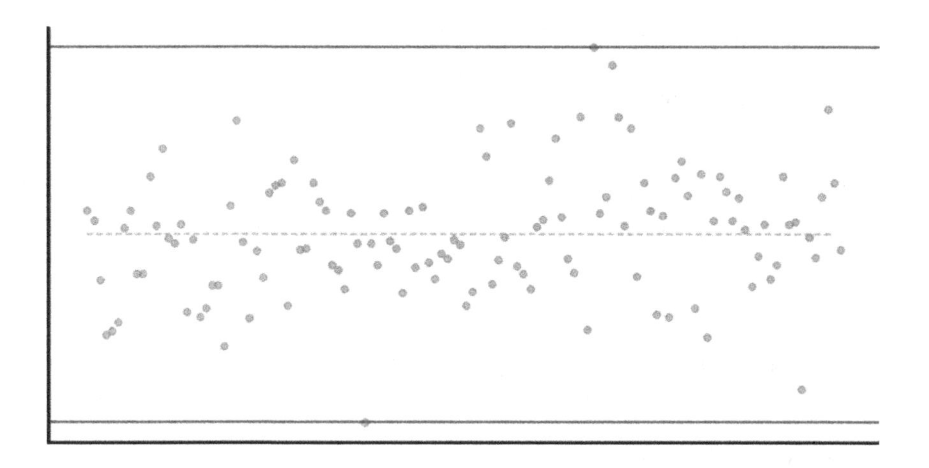

FIGURE 6.10 Residuals Plot.

Nonetheless, it is worth noting that two points align closely with the blue line, potentially indicating outliers. To validate whether these points wield influence, additional diagnostic measures such as Cook's statistics or leverage analysis are necessary. Should these points be deemed influential, their removal from the dataset might be considered.

6.3.4.5 Multicollinearity

The presence of multicollinearity is effectively addressed through PCA, which transforms highly correlated variables into a collection of uncorrelated ones. Consequently, multicollinearity concerns are mitigated following the PCA transformation. At this juncture, all requisite assumptions have been fulfilled, rendering the dataset suitable for fitting into the MLR model.

TABLE 6.3: Comparison of SVM and MLR

	RMSE	MAE	R^2	Adjusted R^2
SVR (Linear)	1.1147	0.8639	0.6250	0.5709
MLR	1.1407	0.8937	0.6018	0.5444

The equation of the MLR is as follows:

$$y = 4.0147 + (-0.5788)W_1 + (0.2195)W_2 + (0.7052)W_3 + (-0.0993)W_4 + (-0.1770)W_5 + (0.3251)W_6 + (0.4706)W_7 + (-0.3127)W_8 + (0.0191)W_9 + (0.7507)W_{10} + (-0.0854)W_{11} + (0.4214)W_{12} + (-0.0338)W_{13} + (0.4908)W_{14} + (-0.5015)W_{15} \tag{6.13}$$

6.3.5 Comparison of SVM and MLR

As evident from Table 6.3, it is readily apparent that the machine learning algorithm, specifically SVR, outperforms the conventional statistical method MLR. SVR exhibits superior performance metrics with lower RMSE and MAE, alongside higher R^2 and adjusted R^2 values, in comparison to the MLR model. Despite the comparable performance of both approaches, it is worth noting that the MLR outcome could be compromised by the presence of two anomalous data points that appear potentially as outliers, consequently undermining result accuracy. In summation, the SVR method stands as the more suitable choice for predicting carcass fatness in contrast to the MLR model.

6.3.6 Overfitting Checking

Based on Table 6.4, the attained adjusted R^2 value stands at 0.6412, signifying its alignment with an acceptable range. An excessively high adjusted R^2 could indicate a potential for overfitting. It is important to clarify that this particular stage is not intended for comprehensive model evaluation, but rather to confirm the absence of overfitting concerns within the fitted model. To comprehensively assess the model's efficacy, a fresh collection

TABLE 6.4: Overfitting Checking

	RMSE	MAE	R^2	Adjusted R^2
Training set (SVM)	1.0348	0.7773	0.6864	0.6412

of previously unseen data can serve as a test dataset. Should the performance of the trained model not meet the desired standards, the possibility of retraining the model may be considered.

6.4 CONCLUSION

In summary, our study underscores the suitability of the SVM regression model as an effective predictive tool for lamb carcass C-site fat depth. Choosing SVR over MLR emerges as the more rational choice, primarily due to SVM's robustness in handling outliers without necessitating the imposition of initial assumptions. This resilience renders SVM a versatile choice for general application, while the utility of a statistical model like MLR can be constrained by its reliance on fulfillment of stringent preliminary assumptions. This distinction may lead to additional time and expenses in scrutinizing data suitability for MLR fitting, potentially mandating further transformations prior to model application. Consequently, our findings emphasize that the SVM regression model is not only more adept at prediction, but also offers a pragmatic solution for real-world datasets where assumptions might not hold uniformly.

ACKNOWLEDGMENT

This study was undertaken through the Advanced Livestock Measurement Technologies Project (ALMTech) and funded by the Department of Agriculture, Rural Research and Development for Profit program and Meat and Livestock Australia.

REFERENCES

1. Cambridge University Press & Assessment. (2022). Carcass. In Cambridge Dictionary.
2. Marimuthu, J., Loudon, K. M. W., & Gardner, G. E. (2021). Prediction of Lamb Carcase C-Site Fat Depth and GR Tissue Depth Using a Non-invasive Portable Microwave System. Meat Science, 181, 108398. https://doi.org/10.1016/j.meatsci.2020.108398.
3. Bzdok, D., Altman, N., & Krzywinski, M. (2018). Statistics Versus Machine Learning. Nature Methods, 15. https://doi.org/10.1038/nmeth.4642.
4. Stewart, M. (2020, July 29). The actual difference between statistics and machine learning. Medium. Retrieved August 10, 2022, from https://towardsdatascience.com/the-actual-difference-between-statistics-and-machine-learning-64b49f07ea3.
5. Rodríguez-Rodríguez, I., Shirvanizadeh, N., Ortiz, A., & Pardo-Quiles, D. J. (2021). Applications of Artificial Intelligence, Machine Learning, Big Data and the Internet of Things to the COVID-19 Pandemic: A Scientometric

Review Using Text Mining. International Journal of Environmental Research and Public Health, 18(16). p. 8578.

6. Dalal, A., Bagherimehrab, M., & Sanders, B. (2021). Quantum-Assisted Support Vector Regression for Detecting Facial Landmarks.
7. Li, E., Yang, F., Ren, M., Zhang, X., Zhou, J., & Khandelwal, M. (2021). Prediction of Blasting Mean Fragment Size Using Support Vector Regression Combined With Five Optimization Algorithms. Journal of Rock Mechanics and Geotechnical Engineering, 13(6), 1380–1397. https://doi.org/10.1016/j.jrmge.2021.07.013
8. Raj, A. (2021). Unlocking the True Power of Support Vector Regression. Medium. https://towardsdatascience.com/unlocking-the-true-power-of-support-vector-regression-847fd123a4a0
9. Holand, S. M. (2019). Principal Components Analysis (PCA).
10. Boser, B. E., Guyon, I. M., & Vapnik, V. N. (1992). A training algorithm for optimal margin classifiers. In Proceedings of the fifth annual workshop on Computational learning theory (pp. 144–152).

CHAPTER 7

Exploring the Power of Convolutional Neural Networks in Face Detection

Muhamad Irsan, Rosilah Hassan,
Mohammad Khatim Hasan, and Meng Chun Lam

7.1 INTRODUCTION

Each person has unique facial features and facial recognition has evolved into a technology that identifies facial patterns. Consequently, this technology can provide insights into the identity of individuals, utilizing various human body features such as voice patterns, fingerprints (Kamelia et al., 2018), eyes (Nojiri et al., 2019), and facial patterns (Verma et al., 2019). Biometric technology, including face recognition, has been extensively researched and developed by experts. It employs a face detection algorithm to differentiate between the facial features of one individual and those of another, leveraging preexisting data stored in databases (Ramadhani et al., 2018). Moreover, smartphones can also detect faces (Widiakumara et al., 2017; Fernando et al., 2019). Face identification pertains to a distinct pattern recognition process that examines the shape of the face to determine its recognizability (Anam, 2018).

Several applications have implemented face detection and facial recognition to cater to various human activities. One such application is utilizing facial parameters for student attendance systems within classrooms (Munawir and Hermasyah, 2020). Facial pattern recognition can validate

DOI: 10.1201/9781003400387-7

students' attendance by detecting smiles through a Wi-Fi network. This verification is based on the student's location and time, using location data determined by matching the service set identifier (SSID) in the database. The framework comprises three interconnected applications. Web, versatile, and administrations are all introduced on a smaller-than-expected computer. The model test achieved an accuracy of 92.6%, while the live test returned 66.7% results. These outcomes indicate a potential overfitting issue in the training model since it performs exceptionally well during model testing but experiences a decline in performance during live testing (Miftakhurrokhmat et al., 2021).

Another study focused on presence-based face recognition and used the eigenface algorithm with principal component analysis (PCA) technique. This approach was tested across different facial expressions, distances, and accessories, yielding a system success rate indicated by a sensitivity value of 73.33%, specificity of 52.17%, and overall accuracy of 86.67%. Specifically, it achieved a 70% success rate in the distance test process and an 85% success rate in detecting people wearing accessories such as headscarves and eyeglasses. Furthermore, it achieved a success rate of 85.33% in identifying different facial expressions (Wiryadinata et al., 2017).

A separate study that employed the Viola-Jones method for detecting the number of people in a room found that non-frontal face positions at distances more than 2–3 m were not detected. However, the detection was successful when the face was in the front position and the distance was less than 2–3 m. The study detected multiple faces within a dataset of 19 images comprising 38 individuals. Of these, 30 people were successfully detected, while 8 were not, resulting in a detection accuracy rate of 79% (Syafitri & Saputra, 2017).

This inquiry aimed to create a facial acknowledgment framework that can tally the number of understudies captured by a camera. The analysts utilized a convolutional neural network (CNN) confront acknowledgment calculation utilizing the OpenCV library and executed the framework on a Raspberry Pi 4 chip gadget prepared with a camera. Raspberry Pi 4 was chosen due to its flexible functionalities, counting remote communication, essential microcontroller operations, Bluetooth module, and the capacity to store data in a database (Zainol et al., 2019).

7.2 RESEARCH METHODS

Several processes and methods were carried out in this study, all illustrated in the Figure 7.1.

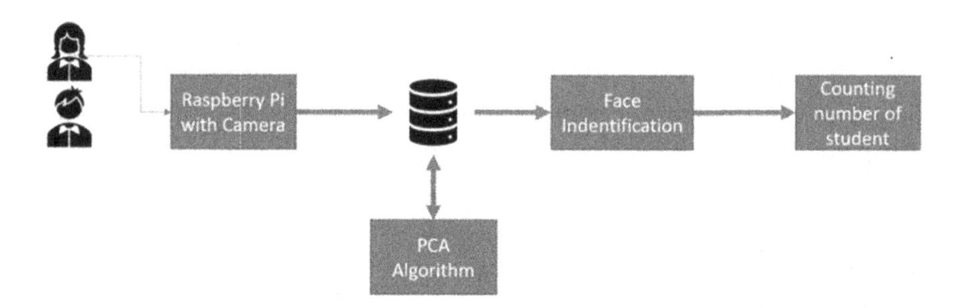

FIGURE 7.1 Flow system.

In this study, researchers first performed observations and collected data on the shape of the student's face. This data collection involved taking pictures of students with their smartphone cameras and collecting additional facial images from online sources such as websites. The collected images were organized and grouped by their respective classes and served as samples for creating datasets for training and testing the system.

Next, the collected data, consisting of the photos of students' faces, were used as input into the system to be used as a dataset. The dataset underwent processing using image processing techniques, explicitly applying the convolutional neural network (CNN) algorithm. This algorithm analyzes and extracts image features to accurately identify student faces.

The output of this process is the identification and recognition of the student faces, where the system can successfully detect and classify the faces based on the trained CNN algorithm.

a. Input image

Student's faces are collected using smartphones or Python programming to create a dataset. Additional faces from the internet are also gathered to augment the dataset. The dataset consists of two types of images: training data and testing data. The first set of images, the training data, is used to train the system. These images are used as input to the system during the training process. A CNN algorithm processes the training data to learn and identify the unique features of the student's face.

The second set of images, the testing data, is used to evaluate the system's performance. These images are separate from the training data and are not used during the training phase. Instead, they assess how well the trained system can recognize and identify the student faces it has not encountered before. By utilizing training and testing data, the system can be trained to identify student faces accurately and then tested to measure its effectiveness on new, unseen images. This approach helps ensure the system performs well beyond the training data in real-world scenarios.

b. Preprocessing image
Before training, images undergo preprocessing to facilitate the CNN algorithm in recognizing and extracting the image's characteristics. Image preprocessing involves multiple stages, including cropping, converting to grayscale, and image augmentation.

Cropping: This stage entails selecting a specific region of interest from an image, such as a student's face while eliminating irrelevant background information. Cropping helps to focus the algorithm's attention on the essential features.

Grayscale conversion: Converting the image to grayscale reduces its complexity by eliminating color information. Grayscale images contain only shades of gray, simplifying the data representation. Grayscaling is often sufficient for facial recognition tasks and reduces computational complexity.

Image augmentation: It is a technique used to enhance the training dataset by applying various transformations to the images. These transformations include rotations, translations, scaling, flipping, and adding noise. By augmenting the dataset, we help the model to become more robust and capable of handling variations in real-world scenarios.

By performing these preprocessing stages, images are prepared in a standardized format that assists the CNN algorithm in recognizing and understanding the distinctive characteristics of the images during the training process.

c. Application of CNN
To identify a student's face using the CNN algorithm, the algorithm goes through two phases: feature learning and classification. CNN algorithm starts by identifying features in the image, and after the model is trained, it uses neurons for the classification process. In CNN models, the input image size is typically $50 \times 50 \times 1$. Several one means the image is grayscale and has one channel. The image first goes through the convolution process, which is convolved with different filters in two layers. Each convolution layer may have different numbers of filters and channel sizes.

After the convolution process, the image enters the pooling stage. Pooling helps to downsample and reduce the spatial dimensions of the feature maps obtained from the convolution layer. This process aids in capturing and emphasizing important features while reducing computational complexity. After pooling, the resulting feature maps are flattened or transformed into a vector form. This process, called the flattening process, rearranges the pooling layer feature maps into a linear order. Once flattened, the features are then passed to a fully connected layer. In this layer, the CNN algorithm performs classification tasks using the learned features to make predictions or decisions regarding the input image. Overall, the CNN model progresses through the stages of feature learning (convolution and pooling) and classification (flattening and fully connected layers) to identify and classify student faces accurately.

7.3 RESULT AND DISCUSSION

The dataset collection process in this study involves obtaining face images by downloading from online sources or manually capturing them using a camera. These images then undergo an editing process that resizes them to 250 x 250 pixels and changes the format to JPG. To facilitate the capture process, a source code program allows the automatic capture of multiple images. The program can capture as many images as needed while determining the size and format of each image. The researchers have collected 13 datasets, each containing various images. In addition, they have introduced variations within each dataset. These variations include images with masks, glasses, and images taken from angles (i.e., right, left, up, and down). The purpose of introducing these variations is to increase the diversity and variability within each dataset, a data preprocessing process using Python programming is then applied to this dataset.

a. Dataset

The already made and collected dataset is called for expansion and preprocessing within the other handle step. However, the dataset must consist of more than 50 data or images for the augmentation process to work. If the dataset contains less than 50 images, the augmentation process will not be performed during preprocessing. This threshold of 50 can be adjusted freely. In this study, researchers aim for many classifications in each dataset, resulting in high variability. Therefore, we use datasets containing more than 50 images.

In addition, only records containing images in ".jpg" format are retrieved; other formats are not considered. Upon accessing the dataset, its images are resized to 50 x 50 pixels. The researchers managed to obtain 13 datasets with a total of 833 images. This indicates that the previously created dataset was successfully transferred to subsequent processes.

b. Application of preprocessing
Preprocessing plays an essential role in preparing images for further processing, such as feature extraction and classification, during the training and testing process. To improve the system's recognition capabilities, preprocessing performs various steps, including B. Extension to grayscale and color conversion. Augmentation is applied to improve the dataset by introducing variations in the images. This includes moving, rotating, and flipping the images. The augmentation process is performed on all images in the dataset, resulting in 17,493 data augmentations.

The color conversion step is also carried out to convert the images from their original format to grayscale. By converting the images to grayscale, the system can easily recognize and process them, simplifying the subsequent stages of feature extraction and classification. These preprocessing techniques ensure that the dataset is enriched with diverse variations and adequately formatted for further analysis and training.

c. Equalizing number of datasets
This phase aims to equalize the data in each dataset and have balanced values. The purpose is to distribute the data more evenly across all datasets. In Figure 7.2, we can see that the Bagas, Gover, and Satria datasets have higher percentages (12%) than the other datasets, with an average percentage of 6%. This discrepancy arises because the Bagas, Gover, and Satria datasets initially contained 100 images each, resulting in more data points than the other datasets. To address this imbalance, a process is conducted to adjust the number of data points in each dataset. This adjustment ensures that all datasets have equal data points, resulting in a more balanced distribution. By equalizing the dataset sizes, the training and testing processes can be more reliable and accurate, as the models will be exposed to similar examples from each class or dataset.

In Figure 7.3, it is observed that the number of datasets has been equalized, resulting in each dataset having the exact value of 7.7%.

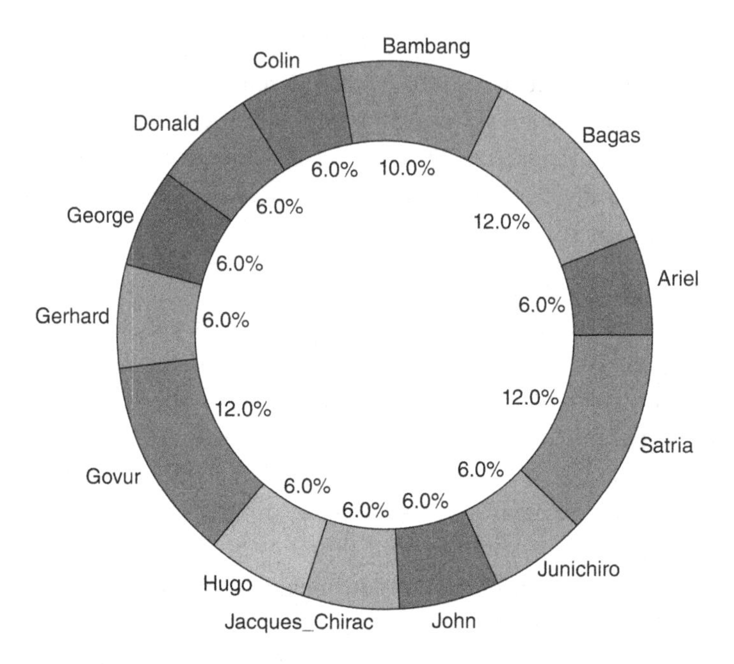

FIGURE 7.2 Before equating.

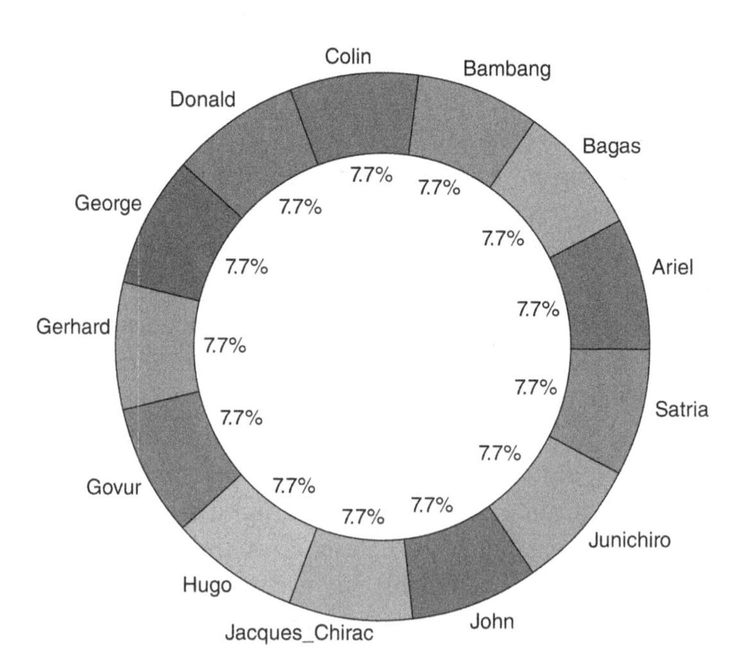

FIGURE 7.3 After equating.

This equalization process ensures that the datasets are balanced and have an equal representation within the overall dataset.

The training and testing processes become fairer and more reliable by equalizing the number of data points across all datasets. It helps prevent any bias or disproportionate influence that may arise due to variations in dataset sizes. With this balanced representation, the models can effectively learn from and generalize across all datasets, leading to more accurate and unbiased results.

d. Distribution of datasets for training and testing
Amid this preparation, the collected dataset that has experienced preprocessing is isolated into two sets: preparing and testing information. To attain this division, the program utilizes the train-test split work, which isolates the clusters and lattices into arbitrary preparing and testing subsets. The train-test split work is commonly utilized to arbitrarily part datasets into preparing and testing information. In this particular handle, the parameter test size is set to 0.15, showing that 15% of the information will be designated for testing purposes, whereas the remaining 85% will be utilized for preparing. In addition, the random state parameter is set to 42. This parameter ensures that the random splitting of the data will be reproducible, meaning that if the process is run multiple times with the same random state value of 42, it will yield the same train-test split. This allows for consistent and comparable results in subsequent experiments or evaluations. By dividing the dataset into training and testing data, it becomes possible to assess the model's performance on unseen data and evaluate its ability to generalize beyond the training set (Table 7.1).

e. Training results
It is great to hear that the training and validation results of the CNN algorithm have been obtained. The researcher employed 13 epochs, each representing a repetition of training on the dataset. The batch size was set to 64, meaning the algorithm processed 64 samples simultaneously. A 15% validation split means that 15% of the training dataset was used as another validation set to evaluate the model's

TABLE 7.1: Distribution of Train and Test

	X	Y
Train	(11050, 50, 50)	(11050, 13)
Test	(1950, 50, 50)	(1950, 13)

performance during training. The results indicate an accuracy of 99% and a loss of 0.0074 on the training set. This means the model achieved high accuracy and a low loss value on the training data. In addition, a validation accuracy of 98% and a validation loss of 0.0735 were obtained on the validation set, demonstrating the model's ability to generalize well on unseen data. The epoch time of 40 minutes represents the duration it took to complete the training process for all 13 epochs.

The results show that the accuracy value is higher than the validation accuracy value, which suggests that the model performed better on the training set than on the validation set. Nevertheless, the overall training process appears to have run well. We can refer to the training results graph provided in the form of figure to comprehensively understand the entire epoch process. Figure 7.4 likely depicts the trends of accuracy and loss during the training process.

f. Testing result

Once the preparation is complete, the next step is the testing stage, where we assess the model's execution on modern information. This unused information is gotten by part of the initial dataset into preparing and testing information. An add-up to 1,950 test information focuses is utilized for the testing process. During the testing stage, the model's expectations are compared against the real course names of the test information. The disarray framework is utilized to summarize and analyze the model's execution. It comprises four fundamental values: genuine positives (accurately classified positive tests), genuine negatives (accurately classified negative tests), wrong positives (erroneously classified positive tests), and wrong negatives (inaccurately classified negative tests). By examining the values in the confusion matrix, various metrics such as accuracy, precision, recall, and F1-score can be calculated to assess the model's effectiveness on the test data. To generate the confusion matrix, it is necessary to have the model's predictions and the corresponding actual class labels for the test data. The matrix can be constructed with this information, providing insights into the model's correct and incorrect classifications.

Figure 7.5 displays the prediction results of the model for the testing data, indicating favorable outcomes. Each accurate label is accompanied by an explanation of the prediction results as follows:

- Ariel: The model correctly classified 158 instances as Ariel, with one instance misclassified as Gerhard, one as Gover, and one as Hugo.

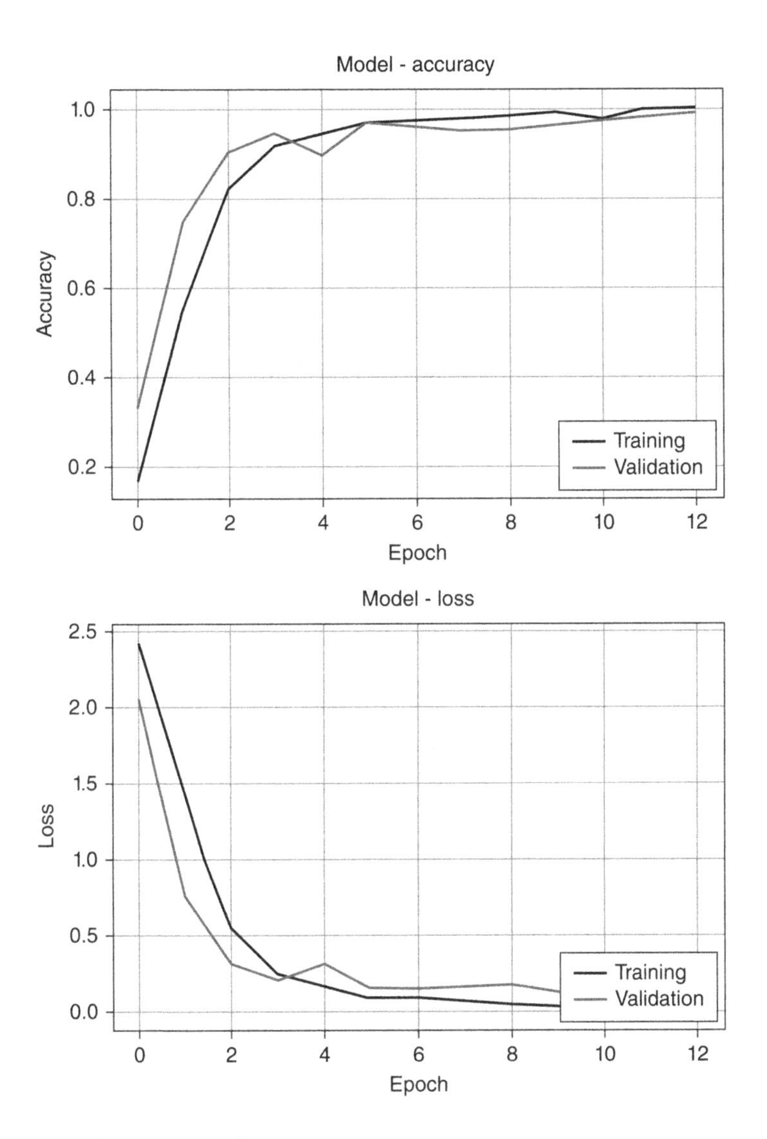

FIGURE 7.4 Training result.

- Bagas: The model accurately classified 146 instances as Bagas, with one instance misclassified as Bambang, one as Gover, and one as Hugo.
- Bambang: The model correctly classified 150 instances as Bambang, with one instance misclassified as Govur.
- Colin: The model accurately classified 148 instances as Colin, with one misclassified as Jacques Chirac.

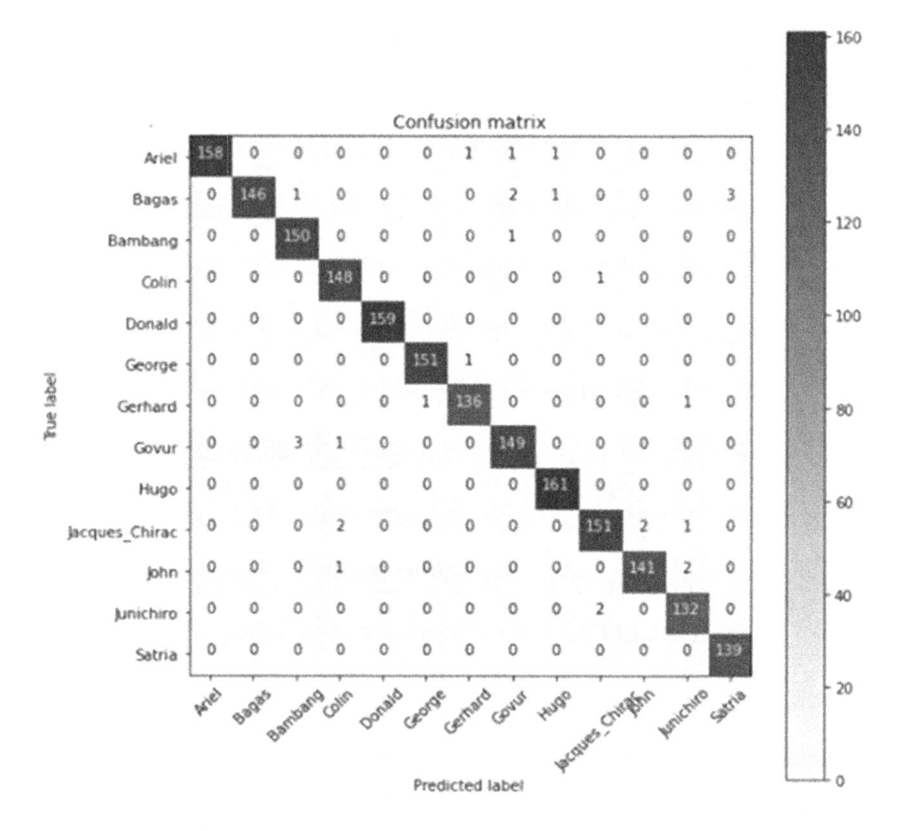

FIGURE 7.5 Confusion matrix.

- Donald: The model correctly classified 159 instances as Donald without any misclassifications.
- George: The model accurately classified 151 instances as George, with one misclassified as Gerhard.
- Gerhard: The model correctly classified 136 instances as Gerhard, with one misclassified as George and one as Junichiro.
- Govur: The model accurately classified 149 instances as Govur, with three misclassified as Bambang and one as Colin.
- Hugo: The model correctly classified 161 instances as Hugo without any misclassifications.
- Jacques Chirac: The model accurately classified 151 instances as Jacques Chirac, with two instances misclassified as Colin, two as John, and one as Junichiro.

- John: The model correctly classified 141 instances as John, with one misclassified as Colin and two as Junichiro.
- Junichiro: The model accurately classified 132 instances as Junichiro, with two misclassified as Jacques Chirac.
- Satria: The model correctly classified 139 instances as Satria without any misclassifications.

The overall accuracy score determined by the CNN model using a test sample of 1,950 data points was 98%. These results show that the model performed well and accurately classified the test data.

g. Work system

A block diagram is a graphical representation of a system or process, illustrating its components and their interconnections. It typically consists of blocks or boxes representing individual components, along with arrows or lines indicating the flow of information or signals between the components.

In the context of our study, the block diagram Figure 7.6, likely illustrates the various components of the system implemented using a Raspberry Pi. It may include blocks representing the Raspberry Pi, the camera module, the monitor, and other relevant components. The arrows or lines in the diagram would indicate the flow of data or information between these components. The block diagram serves as a visual representation of the system's architecture, providing an overview of how the different components interact and contribute to the overall functionality of the face classification system.

Figure 7.6 represents a series of tools or components involved in the system implementation. Here is a breakdown of the key elements:

1. Microprocessor (Raspberry Pi): The microprocessor, specifically the Raspberry Pi, serves as the central component or "brain" of the system. It manages and processes the data received from the camera module.
2. Camera module: The system incorporates a camera module compatible with the Raspberry Pi. This camera module captures the input, which is the object (faces) to be detected.
3. Microprocessor processes: The microprocessor receives the input from the camera module and performs the necessary processing.

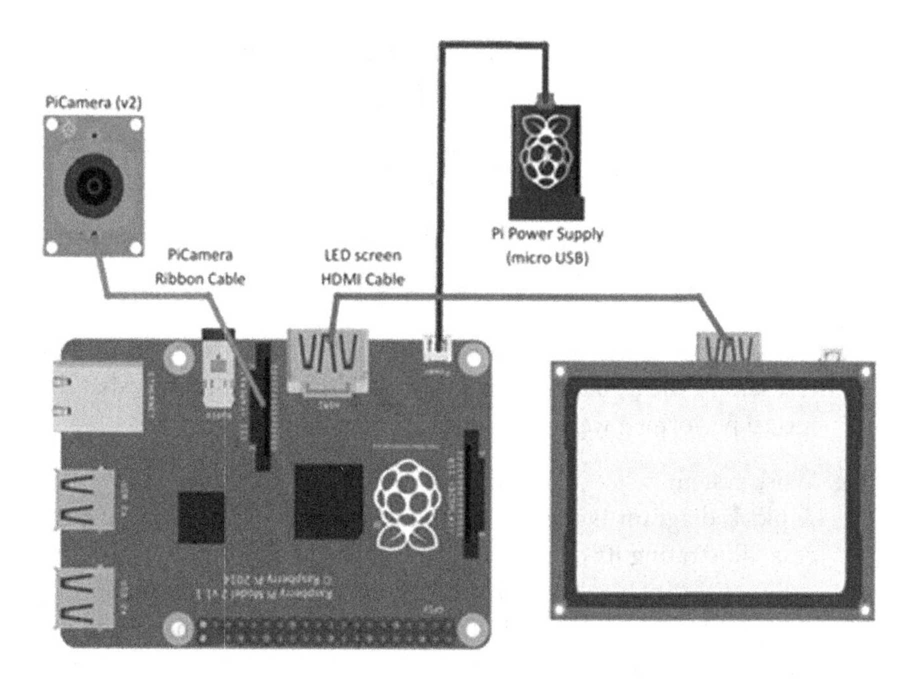

FIGURE 7.6 Block diagram.

This includes running the Python program, which utilizes the previously generated model file (in .pb format) for face detection.

4. Model file: The model file, created through Python programming, contains the trained model for face detection. This file is loaded into the Raspberry Pi to be utilized during the processing stage.
5. Package or library: The necessary packages or libraries required for the system, which have been developed and customized for the Raspberry Pi, are installed. These packages enable the execution of the program and facilitate the face-detection process.
6. Output display: The LCD monitor is used to display the output of the face detection process. The results obtained from the microprocessor's processing are presented on the monitor screen.

The system performs real-time face detection using the camera module by combining these components and running the program on the Raspberry Pi. The results are then displayed on the connected LCD monitor.

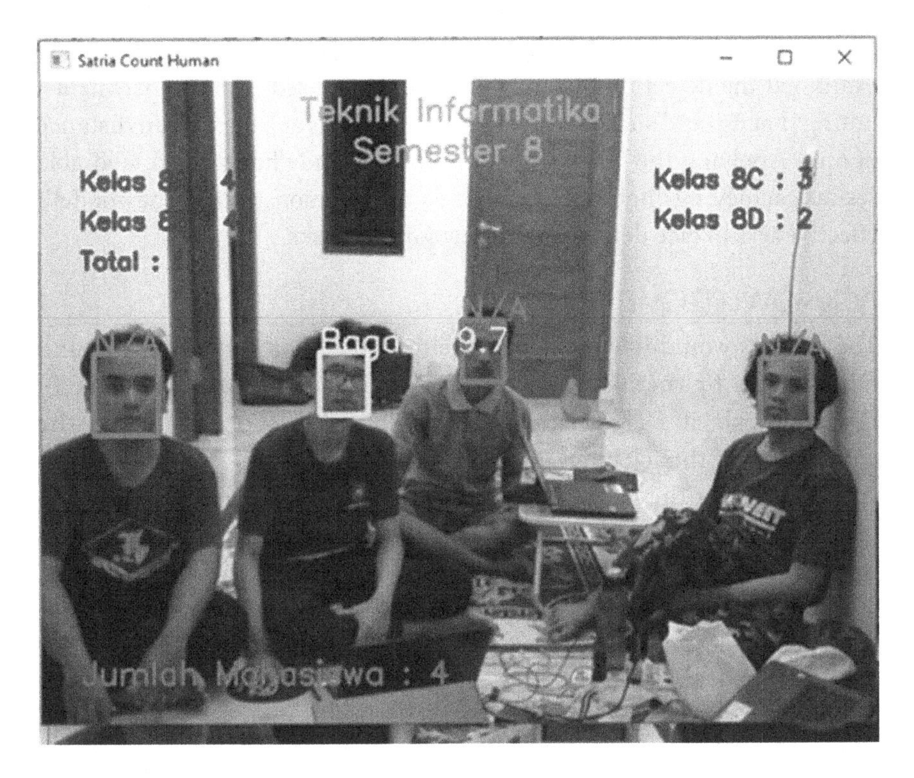

FIGURE 7.7 System test result.

However, it is a diagram based on the information provided. Figure 7.7 shows the result of counting people in a room using the developed system. The experiment required testing 4 of her at different distances.1–2 m, 2–4 m, 4–5 m, and 5–6 m.

7.4 CONCLUSION

Based on the examination conducted in this experiment, it can be concluded that the CNN demonstration executed within the investigation has accomplished tremendous and precise outputs. This show takes a 50 × 50 picture input and applies 6 convolutions with 32, 64, and 128 channel sizes. A 3 × 3 bit move step of esteem one is utilized in combination with the ReLU enactment work. Max pooling with a 2 × 2 kernel and walk two is additionally utilized. The model's training process involved 13 epochs with a batch size of 64. The training dataset comprised 11,050 samples, while the testing dataset contained 1,950 samples. The achieved training accuracy was 99%, indicating a high level of accuracy in classifying the training data. The testing accuracy value reached 98%, demonstrating the model's ability to generalize well to unseen data. Furthermore, the study examined the detection distance of the camera used in the experiments, noting that it could successfully detect individuals at a maximum distance of 6 m. Applying the CNN algorithm in this study has yielded favorable results, achieving high accuracy levels and demonstrating the model's effectiveness in face detection and recognition tasks.

ACKNOWLEDGMENT

The authors would like to specify their appreciation for the bolster and offices given by the Network and Communication Technology (NCT) Research Lab at FTSM, UKM, which incredibly contributed to the completion of this chapter. The authors would like to extend our deep appreciation for the permission and support from the copyright owners of the images used in this document, and expand their appreciation to the editor and the commentators for their important input, comments, and proposals. Their input played a vital part in improving the quality and effectiveness of the chapter. This work was funded by UKM research grant DCP-2017-020/2.

REFERENCES

Anam, M. K. (2018). Metode Eigenface/Principal Component Analysis (PCA) Untuk Identifikasi Wajah Manusia. *JUTIS*, *6*(2), 82–88.

Fernando, E., Andwiyan, D., Fitria Murad, D., Touriano, D., & Irsan, M. (2019). Face Recognition System Using Deep Neural Networks With Convolutional Neural Networks. *Journal of Physics: Conference Series*, *1235*(1). https://doi.org/10.1088/1742-6596/1235/1/012004

Kamelia, L., Hamidi, E. A. D., Darmalaksana, W., & Nugraha, A. (2018). Real-Time Online Attendance System Based on Fingerprint and GPS in the Smartphone. *Proceeding of 2018 4th International Conference on Wireless and Telematics, ICWT 2018*. https://doi.org/10.1109/ICWT.2018.8527837.

Miftakhurrokhmat, Rajagede, R. A., & Rahmadi, R. (2021). Presensi Kelas Berbasis Pola Wajah, Senyum Dan Wi-Fi Terdekat Dengan Deep Learning. *Jurnal RESTI (Rekayasa Sistem Dan Teknologi Informasi)*, *5*(1), 31–38. https://doi.org/10.29207/resti.v5i1.2575

Munawir, Fitria, L., & Hermasyah, M. (2020). Implementasi Face Recognition Pada Absensi Kehadiran Mahasiswa Menggunakan Metode Haar Cascade Classifier. *InfoTekJar : Jurnal Nasional Informatika Dan Teknologi Jaringan*, *5*(1), 40–43.

Nojiri, N., Kong, X., Meng, L., & Shimakawa, H. (2019). Discussion on Machine Learning and Deep Learning-Based Makeup Considered Eye Status Recognition for Driver Drowsiness. *Procedia Computer Science*, *147*, 264–270. https://doi.org/10.1016/j.procs.2019.01.252

Ramadhani, A. L., Musa, P., & Wibowo, E. P. (2018). Human face recognition application using PCA and eigenface approach. *Proceedings of the 2nd International Conference on Informatics and Computing, ICIC 2017*. https://doi.org/10.1109/IAC.2017.8280652.

Syafitri, N., & Saputra, A. (2017). Prototype Pendeteksi Jumlah Orang Dalam Ruangan. *IT Journal Research and Development, 1*(2), 36–48. https://doi.org/10.25299/itjrd.2017.vol1(2).678.

Verma, A., Singh, P., & Rani Alex, J. S. (2019). Modified Convolutional Neural Network Architecture Analysis for Facial Emotion Recognition. *International Conference on Systems, Signals, and Image Processing, 2019-June*, 169–173. https://doi.org/10.1109/IWSSIP.2019.8787215.

Widiakumara, I. K. S., Putra, I. K. G. D., & Wibawa, K. S. (2017). Aplikasi Identifikasi Wajah Berbasis Android. *Lontar Komputer : Jurnal Ilmiah Teknologi Informasi, 8*(3), 200. https://doi.org/10.24843/lkjiti.2017.v08.i03.p06.

Wiryadinata, R., Istiyah, U., Fahrizal, R., Priswanto, P., & Wardoyo, S. (2017). Sistem Presensi Menggunakan Algoritme Eigenface Dengan Deteksi Aksesoris Dan Ekspresi Wajah. *Jurnal Nasional Teknik Elektro Dan Teknologi Informasi (JNTETI), 6*(2), 222–229. https://journal.ugm.ac.id/v3/JNTETI/article/view/2859.

Zainol, M. F., Mohamed Farook, R. S., Hassan, R., Abdul Halim, A. H., Abdul Rejab, M. R., & Husin, Z. (2019). A New IoT Patient Monitoring System for Hemodialysis Treatment. *2019 IEEE Conference on Open Systems, ICOS 2019*. https://doi.org/10.1109/ICOS47562.2019.8975703.

CHAPTER 8

Protecting Higher Learning Institutions from Phishing Attacks: A Staff Awareness Program

Norliza Katuk, Adi Badiozaman Ruhani, Muzdalini Malik, Abdul Kadir Mahamood, and Mohd Shamshul Anuar Omar

8.1 INTRODUCTION

The rapid digitalization of various aspects of modern life, accelerated by COVID-19 restrictions, has led to a corresponding increase in cyberattacks worldwide (Scherb et al., 2023). These malicious activities range from phishing scams and ransomware attacks to data breaches and identity theft, posing significant threats to businesses, governments, and individuals. A particularly alarming trend is the continuous growth of phishing attacks, which have experienced a staggering rise in recent years. According to the Anti-phishing Working Group (2022), year 2022 was a record-breaking year for phishing, with more than 4.7 million attacks logged. Since the beginning of 2019, these attacks have increased by over 150% per year, highlighting the urgency of addressing this cybersecurity issue. In the final quarter of 2022 alone, the Anti-phishing Working Group observed 1,350,037 phishing attacks, further emphasizing the scale of this ongoing problem (Anti-phishing Working Group, 2022). A closer examination of the data reveals that the financial sector, including banks,

DOI: 10.1201/9781003400387-8

has borne the brunt of these attacks, accounting for 27.7% of all phishing incidents (Anti-phishing Working Group, 2022). This figure represents a notable increase from 23.2% in the third quarter of 2022 and underscores the vulnerability of this critical industry.

Phishing attacks have emerged as one of the most significant threats confronting businesses and the general public in recent years, posing severe challenges to the security of sensitive information and systems (Alwanain, 2019). These insidious attacks are particularly concerning due to their ubiquity, as they can be executed anywhere in the world and target any device or user (Manoharan et al., 2022). As a result, organizations and individuals must be increasingly vigilant and prepared to counteract this pervasive and ever-evolving menace. Alarmingly, the education sector has become the prime target for cybercriminals, experiencing an unprecedented surge in cyberattacks. According to SentinelOne (2022), educational institutions faced an average of 2,297 attacks per week during the first half of 2022, marking a 44% increase compared to 2021. This trend is evident not only in the United States, where data indicates a consistent rise in the volume of monthly cyberattacks since 2021, but also in the United Kingdom. Government statistics reveal that a staggering 62% of higher education institutions in the United Kingdom reported encountering breaches or attacks at least once a week over the previous 12 months (SentinelOne, 2022). Given the severity and prevalence of phishing attacks, it is crucial for organizations, particularly those in the education sector, to develop and implement comprehensive action plans to enhance cybersecurity awareness and safeguard sensitive information (Alharbi & Tassaddiq, 2021). Institutions can empower their staff and students with the knowledge and tools necessary to recognize and respond effectively to phishing attempts by prioritizing education and training, thereby minimizing the likelihood of successful attacks.

Organizations must also invest in robust security measures and protocols designed to provide multilayered protection against phishing and other cybercrimes and raise awareness. Educational institutions can significantly reduce the risks associated with phishing attacks and safeguard their valuable assets by proactively addressing vulnerabilities and staying abreast of emerging threats, including personal and financial data, intellectual property, and institutional reputations. The rapidly evolving landscape of phishing attacks presents a formidable challenge to organizations worldwide, with the education sector being particularly vulnerable to this pervasive threat. Institutions must prioritize cybersecurity awareness, implement robust security measures, and foster a culture of vigilance among their staff and students to counteract this growing menace. Organizations can effectively mitigate the risks associated with phishing attacks and ensure the continued protection of their valuable assets and information by adopting a proactive and comprehensive approach to cybersecurity.

8.2 CURRENT STATE OF PHISHING IN HIGHER LEARNING INSTITUTIONS

In today's interconnected world, email has become a fundamental communication tool for organizations across various industries, including higher learning institutions. This reliance on email has inadvertently given rise to the proliferation of phishing attacks, which have become a significant threat to the security of these organizations (Pinto et al., 2022). The frequency and sophistication of phishing attempts have increased over the years, with a staggering 83% of institutions reporting encounters with phishing attacks in 2021 (Okokpujie et al., 2023). One of the primary challenges in combating phishing attacks is the constantly evolving nature of these threats. Recent research indicates that conventional awareness training, such as phishing campaigns, tends to yield only short-term benefits without significantly enhancing employees' long-term resistance to phishing emails (Scherb et al., 2023). This limited effectiveness can be attributed to the increasing sophistication of phishing emails, making it more challenging to identify and differentiate from legitimate correspondence. Furthermore, employees' attitudes and workloads can also hinder the effectiveness of anti-phishing measures. Some staff members may not fully appreciate the severity of phishing threats or maybe too preoccupied with their daily tasks to scrutinize every email they receive thoroughly (Scherb et al., 2023). This combination of sophisticated attacks and potential employee complacency underscores the need for higher learning institutions to adopt innovative and comprehensive strategies to address the growing risks associated with phishing.

Cybercriminals constantly evolve their methods and strategies to exploit vulnerabilities in digital systems, posing significant challenges to individuals and organizations. Alharbi and Tassaddiq (2021) explained that these malicious actors do not always rely on the same attack vectors; instead, they adapt and shift between various approaches, such as email phishing and network traffic manipulation, to achieve their deceptive goals. A deep understanding of cybercriminals' common attack vectors and techniques is crucial for developing effective security measures and safeguarding sensitive information. One of the most prevalent techniques that cybercriminals use is phishing, a form of social engineering that exploits human psychology and trust to deceive victims into providing sensitive information or performing actions that compromise security. Aljeaid et al. (2020) explain that phishing is not limited to email-based schemes; instead, various methods can be adopted to launch such attacks, including SMS, websites, and phone calls. Cybercriminals can increase the likelihood of successfully targeting unsuspecting victims by leveraging these diverse channels.

For example, SMS phishing, or smishing, is a common tactic wherein cybercriminals send fraudulent text messages that appear to originate from trusted sources, such as financial institutions or government agencies. These messages often contain urgent requests or alarming information designed to elicit an immediate response from the recipient, who may be tricked into clicking on malicious links or divulging sensitive information. Similarly, website phishing involves the creation of fake websites that closely resemble legitimate ones, luring victims into entering their login credentials or other personal data, which cybercriminals can then exploit. Telephone-based phishing attacks, or vishing, are another common method cybercriminals employ. In these scenarios, attackers impersonate representatives of reputable organizations, such as banks or government agencies, to convince victims to divulge sensitive information or perform actions that compromise security. Vishing attacks capitalize on the trust individuals typically place in voice-based communication, making it an effective means of deception.

In addition to these phishing techniques, cybercriminals exploit network vulnerabilities to gain unauthorized access to systems and information. Alharbi and Tassaddiq (2021) emphasized that attackers often shift between various strategies, including manipulating network traffic, to deceive their targets. Cybercriminals can intercept sensitive data, inject malicious code, or manipulate system behavior to their advantage by intercepting or redirecting network communication. Aljeaid et al. (2020) discuss various phishing techniques employed by cybercriminals, including spear phishing, SMS phishing (also known as smishing), Voice-over-Internet Protocol (VoIP)-based vishing scams, whaling, clone phishing, social networking-based phishing, and watering hole attacks. Each method represents a distinct approach to deceive and exploit victims, illustrating the diverse tactics malicious actors utilize in the digital realm (Table 8.1).

TABLE 8.1 Types of Phishing

Phishing Type	Description
Spear phishing	A targeted form of phishing that aims explicitly at a particular individual or organization, often using personalized details to appear more convincing
SMS phishing (Smishing)	A phishing technique that uses fraudulent text messages appearing to come from trusted sources, often containing urgent requests or alarming information to trick victims into clicking on malicious links or providing sensitive information
Vishing scams (VoIP-based)	Phishing attacks that use Voice-over-Internet Protocol technology to impersonate representatives of reputable organizations via phone calls, exploiting trust in voice-based communication to deceive victims into sharing sensitive data or performing compromising actions
Whaling	A form of phishing that targets explicitly high-level executives or key decision-makers within an organization, using tailored tactics to exploit these individuals' unique responsibilities and access privileges
Clone phishing	A type of phishing attack that involves creating replicas of legitimate messages or websites, with subtle alterations to trick victims into providing their credentials or personal information
Social networking-based	Phishing attacks that leverage popular social networking platforms to deceive users, often through fake profiles, direct messages, or posts containing malicious links or requests for sensitive information
Watering hole attack	In a targeted attack, cybercriminals compromise a legitimate website that their intended victims frequently visit and then use it to launch phishing campaigns or other malicious activities against the targeted group

Source: Aljeaid et al. (2020).

Phishing attacks exploit human psychology and trust, luring unsuspecting victims into providing sensitive information or granting unauthorized access to their systems. At its core, a phishing attack typically involves a series of carefully orchestrated steps designed to maximize the attacker's chances of success while minimizing their exposure to detection. The first step in a phishing attack is the preparation stage, during which the attacker meticulously crafts a convincing email that impersonates a reputable organization or service provider. It often involves mimicking the targeted entity's visual elements and tone and crafting a subject line and content that triggers a sense of urgency or curiosity. The attacker aims to prompt the recipient into taking immediate action, thereby increasing their susceptibility to the scam. The attacker proceeds to the distribution stage once the phishing email has been assembled. It involves sending malicious messages to many potential victims, whose contact information is often obtained through various illicit means such as data breaches, social

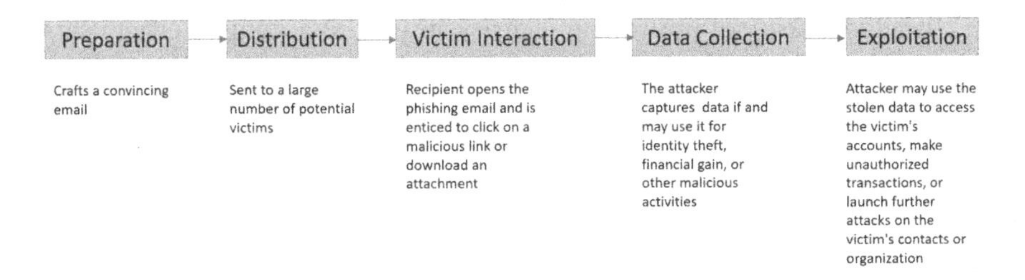

FIGURE 8.1 The process of a phishing attack.

media, or the dark web. The sheer volume of emails distributed increases the likelihood of the attacker ensnaring at least a few victims.

Upon receiving the phishing email, the recipient may be enticed to click on a malicious link or download an attachment. This interaction, which constitutes the third step of the phishing attack, often leads the victim to a fake login page or installs malware on their device. The attacker aims to deceive recipients into providing credentials or personal information by creating a legitimate and trustworthy environment. Once the victim enters their information on the fake login page, the attacker captures this data and moves on to the exploitation stage. The attacker can engage in various malicious activities, from identity theft to unauthorized financial transactions with stolen credentials or personal information. Sometimes, the attacker may use the acquired data to launch further attacks on the victim's contacts or organization, perpetuating the deception and exploitation cycle. Figure 8.1 shows the process of a phishing attack.

8.3 IMPLEMENTING A PHISHING AWARENESS PROGRAM

Numerous efforts have been made to curb the incidence of phishing attacks, including developing anti-phishing toolbars that serve as web browser plug-ins and warn users when they access suspected phishing sites (Alwanain, 2019). However, as cyber threats become increasingly sophisticated, exploring innovative approaches to bolster awareness and understanding of these risks is crucial. One such approach is using serious games, as proposed by Scherb et al. (2023), instead of traditional awareness campaigns. These games have demonstrated a positive short-term effect on increasing cybersecurity awareness, highlighting their potential as an effective tool for educating technical and non-technical employees about common cyber threats. According to Scherb et al. (2023), most attacks could have been avoided if non-technical employees had received proper security awareness education.

The significance of improving security awareness among internet users cannot be overstated, especially in the context of the tremendous growth of online services (Alwanain, 2019). Individuals can make informed choices and adopt safe online behaviors by enhancing their understanding of common cyber threats, thereby reducing their susceptibility to phishing attacks and other forms of cybercrime. As Pinto et al. (2022) point out, cybersecurity awareness and training play a fundamental role in preventing and mitigating a wide range of cyberattacks, emphasizing the need for organizations to prioritize educating their technical and non-technical human resources. An example of the effectiveness of serious games in raising cybersecurity awareness is their ability to engage users in an interactive, immersive learning experience (Scherb et al., 2023). These games can help users develop practical skills and strategies for identifying and responding to phishing attacks by simulating real-world scenarios and challenges, fostering a more proactive approach to cybersecurity.

In the ongoing battle against phishing attacks, exploring diverse strategies to raise awareness among various types of users is essential. While machine learning techniques have emerged as a powerful tool, it is essential not to overlook the value of more traditional approaches, such as user training and education (Baadel et al., 2021). Organizations can add an extra layer of defense against phishing attacks by incorporating these conventional methods, complementing the capabilities of more advanced technologies and ensuring a more comprehensive and robust cybersecurity posture. Activities such as cybersecurity awareness campaigns encompass a variety of elements, including training programs, informational flyers, and webinars, as noted by Georgiadou et al. (2021). These diverse approaches aim to educate users and raise awareness about cybersecurity threats and best practices. Alhashmi et al. (2021) categorized cybersecurity awareness delivery methods into three primary types: face-to-face classes, self-directed classes, and embedded classes, as illustrated in Figure 8.2. Each method offers a unique approach to conveying cybersecurity knowledge and fostering user awareness.

Organizations must develop practical phishing educational tools and strategies to stay one step ahead of these threats as phishing attacks evolve and become more sophisticated. Researchers have been hard at work devising various techniques to alert users to potentially risky email content and links, helping them avoid falling victim to these attacks (Shepherd & Szymkowiak, 2023). However, relying solely on passive awareness methods, such as emails, newsletters, and SMS notifications, may not be

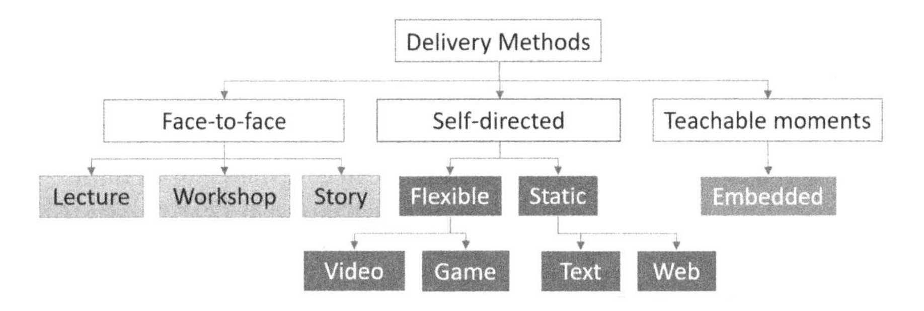

FIGURE 8.2 The process of a phishing attack (Alhashmi et al., 2021).

sufficient in addressing the growing complexity of phishing attacks. As Aljeaid et al. (2020) suggested, integrated, proactive training programs targeted at different age groups are necessary. These programs could be taught in schools, universities, and organizations, enabling a wide range of individuals to develop the skills and knowledge required to identify and avoid phishing threats. These proactive training programs can help ensure that individuals at various stages of their lives are equipped with the tools and strategies they need to protect themselves from phishing attacks. For example, younger users, who may be more susceptible to certain types of phishing scams due to their inexperience with technology, could benefit from age-appropriate training materials and activities that teach them about the risks of sharing personal information online. Meanwhile, older users may require different types of training, focusing on the unique challenges they face in the digital age, such as recognizing fraudulent emails disguised as legitimate communications from banks, retailers, or other trusted entities.

One of the key factors in motivating individuals to take phishing threats seriously and take the necessary precautions is an appreciation of the potential consequences of falling victim to such attacks (Scherb et al., 2023). Illustrating the devastating effects of phishing on both personal and organizational levels, such as identity theft, financial loss, and reputational damage, can help users understand the importance of remaining vigilant and taking proactive steps to protect their sensitive information. Organizations could consider incorporating real-world examples and case studies into their phishing education efforts. Organizations can provide users with valuable insights and lessons they can apply in their own lives by sharing stories of actual phishing attacks and their consequences, as well as the strategies employed to combat them. Furthermore, interactive training exercises, such as mock phishing simulations, can help users develop practical skills in identifying and responding to phishing threats, reinforcing the lessons learned through other educational resources.

One of the critical aspects of implementing a phishing awareness program is the evaluation of its effectiveness. It can be achieved by examining different threat techniques and scenarios, as Nifakos et al. (2021) suggested. Organizations can gain valuable insights into the strengths and weaknesses of their training and awareness initiatives by investigating a wide range of potential risks and challenges, enabling them to make informed decisions about optimizing these programs for maximum impact. For instance, a company could conduct a series of simulated phishing attacks, each employing a different technique or targeting a specific group of users, to assess the preparedness of its employees in the face of these threats. The organization can identify areas where additional training or resources may be needed, as well as any patterns or trends that suggest particular vulnerabilities or areas of concern by analyzing the results of these simulations. Moreover, the evaluation process can also serve as an opportunity to gather user feedback regarding their experiences with the training and awareness campaigns. It can help organizations better understand their employees' needs and preferences and tailor their programs accordingly. For example, some users prefer hands-on, interactive training sessions, while others find written materials or online resources more effective. Organizations can ensure that their phishing awareness programs are adequate for their diverse users by considering these individual preferences.

8.4 CASE STUDY: EVALUATING A PHISHING AWARENESS PROGRAM

In November 2021, a pilot phishing awareness program was conducted at a government-funded higher learning institution in Malaysia, involving 24 administrative staff members out of approximately 500 total staff. The distribution of participants according to their positions showed that 12.50% of the respondents were management staff, with an equal percentage representing administrative (support) staff. Regarding gender, 13.54% of the respondents were male, while 11.46% were female. Most participants (23.96%) were Malay, with only one Chinese respondent (4%) in the group. The educational backgrounds of the respondents varied, with 34% holding Master's degrees, 4% possessing Bachelor's degrees, 33% being Diploma holders, and 7.29% having certificates. The program aimed to investigate the effectiveness of awareness materials and strategies for combating phishing. The goal was to determine how well these resources and approaches worked to replicate them in a broader context or different setting.

8.4.1 Overview of The Program and Its Implementation

During the session, the respondents were gathered in a virtual meeting room where four facilitators from the IT department were present. The pilot phishing awareness program used (1) pre-test and post-test questions, (2) posters, (3) infographics, and (4) video presentations. At the beginning of the session, the respondents were provided with a pre-test question. They were then presented with Poster 1 to view, followed by perception questions in a Microsoft Form. It was followed by the presentation of Poster 2 and its corresponding perception form. Throughout the session, participants were given the freedom to ask questions at any time. Following that, the participants were presented with Infographic 1 and Infographic 2, along with their respective perception questions. Lastly, a video presentation was shown, and the session concluded with a post-test. It's worth noting that all program content was created in English, while Malay served as the primary language for communication and instructions. The participants were specifically chosen from the top management and administrative employees. Figure 8.3 shows the process.

For the pre- and post-test questionnaires, we chose questions from the phishing quiz (https://phishingquiz.withgoogle.com/) that focused on detecting and identifying phishing threats. Both the pre-test and post-test consisted of eight questions, which were used in the pilot tests. The participants utilized Microsoft Forms to provide their answers while watching videos presented during the test. After displaying posters, infographics, and videos on phishing security awareness, the participants were requested to complete a questionnaire. These questionnaires aimed to assess the participant's perception of the contents and materials and determine how they could enhance their understanding of phishing awareness.

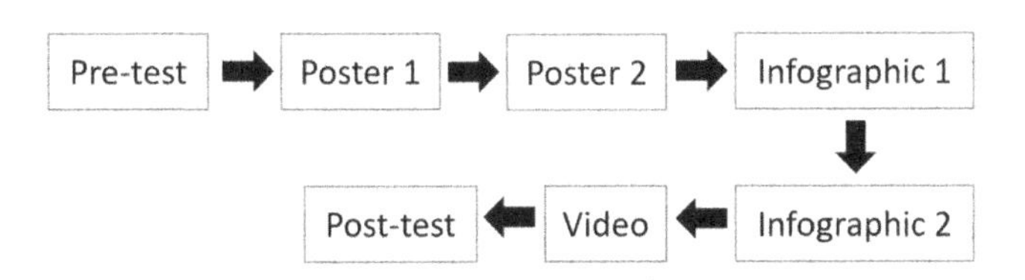

FIGURE 8.3 The flow of delivering material of the program.

The e-poster presentation is a commonly used approach for conveying information, emphasizing visual appeal significantly. Such posters' design must be adaptable to ensure adequate knowledge transmission. We searched for multiple sources to find suitable posters, yielding more than two options. Two posters were chosen from this selection for their clarity and succinctness in conveying the intended content. The first e-poster obtained from InspiredLearning.com (https://inspiredelearning.com/resource/cybersecurity-posters/) reminds participants to think before clicking. The second e-poster was created by Standford University (https://uit.stanford.edu/service/phishingawareness/infographic).

Infographics are effective tools for presenting information or knowledge in a visually appealing and colorful manner that can be easily understood at a glance. Two of the most suitable infographics have been selected to help the target group identify phishing emails or links. Infographic 1 was adapted from InspiredLearning (https://inspiredelearning.com/resource/how-to-identify-a-phishing-email/). It explains the meaning of phishing and guides in detecting a phishing email. These infographics effectively illustrate the structure and components of phishing emails, making the information more engaging, attention-grabbing, and comprehensible to the audience.

The second infographic (https://inspiredelearning.com/resource/social-media-phishing-infographic/) provides valuable information and knowledge about social phishing. This particular infographic was chosen due to its well-organized arrangement of graphics and easily comprehensible content. The infographic delves deeper into the following aspects related to social media:

i. Definition of social media phishing

ii. Reasons why hackers are drawn to social media

iii. Social phishing statistics

iv. Types of social phishing scams

v. Tips to avoid falling victim to social media phishing scams

The infographic aims to enhance the audience's understanding of social media phishing and equip them with the necessary knowledge to protect themselves by presenting this information visually appealing and structured.

The video serves as a medium that provides users with both visual and auditory learning experiences. Two videos from the international cyber risk and security quotient YouTube channels were selected to enhance phishing awareness. These videos played a significant role in the security awareness program. The first video, lasting 3 minutes, focused on spear phishing, phishing, and smishing. It encompassed definitions of malware, instructions for deleting phishing emails, and guidance on reporting phishing attempts. The second captivating video, titled "Phishing: A Game of Deception," delved further into phishing attacks, such as those carried out via email, text and instant messaging, and phone calls. The video also provided insights into how individuals can proactively mitigate phishing attacks. Participants could gain valuable knowledge and understanding of phishing threats and effective countermeasures by incorporating these videos into the program.

8.4.2 Measuring the Program's Effectiveness

The program's purpose is to evaluate its effectiveness by identifying any shortcomings in the techniques used to deliver the necessary materials and the content itself. The program assessment can be conducted using qualitative, quantitative, or a combination of both methods. However, the evaluation process described here primarily focuses on quantitative methods. It is recommended to prioritize quantifiable and repeatable findings when selecting appropriate metrics for the evaluation, as outlined in relevant recommendations. Quantitative methodologies aim to provide evaluation results objectively while establishing benchmarks for future assessments. One method that can be employed to understand better the program's effectiveness is using metrics such as key performance indicators. The evaluation employed various methods for this study, including testing, participation observation, data collection, and performance analysis. Table 8.2 presents the relevant components associated with each of these methods.

TABLE 8.2 Methods and Elements for Evaluation

Method	Element
Test	Pre- and post-test
Participation	Attendance, completion
Data collection	Questionnaire, feedback form
Performance	Subject (participant) and program

Two key factors are considered when evaluating the effectiveness of an awareness campaign: whether the information has reached the intended target and whether it has impacted the target audience (Rantos et al., 2012). During the briefing session, the awareness team anticipated that not all participants would attend, resulting in some individuals not receiving the awareness materials and proper explanations. The efficacy of resources like e-posters and infographics depends on how they are delivered. Despite extensive distribution, certain participants may go unnoticed or ignore these materials. Attendance data provides insight into the number of people who received the information, while polls and questionnaires gauge the impact of the session on participants. However, it is essential to note that these metrics do not necessarily indicate whether the acquired knowledge will impact their regular routines. These are the quantitative methods employed in the pilot test, as presented in Table 8.3.

Questionnaire-based or audience satisfaction surveys were utilized to evaluate the program's effectiveness. These surveys are conducted after the program's completion to identify any organizational flaws. The survey findings offer valuable information regarding the program's success, exposing shortcomings and highlighting areas that need improvement for future iterations. Statistical approaches are applied to comprehensively understand the overall assessment (Rantos et al., 2012). Participants' comments and recommendations can unveil significant flaws in the program's implementation and distribution methods, making them a practical qualitative approach. It is possible to gauge the level of interest and acceptance of the security awareness culture by analyzing attendee engagement during the sessions, such as asking questions, providing answers, and offering instructor feedback. In addition, ideas expressed by employees through feedback forms can be evaluated, providing insights into the quality of proposals and indicating the level of interest generated.

The hypothesis for this pilot test posits that the mean score of the post-test will be higher than that of the pre-test, and the standard deviation of

TABLE 8.3 Quantitative Methods

Summation	Summation of Participants, Test Score, and Rating Scale of Materials and Contents
Mean	Average scores and rating scale of materials and contents
Standard deviation	The measure of the dispersion of data values from the score means
Percentage	Percentage of attendees and demographic data, correct answers and respondents

the post-test will be lower than that of the pre-test. The phishing quiz was administered as a pre-test and post-test to assess the levels of awareness and understanding before and after the program. The results indicated a mean score of 4.92 for the pre-test and 4.54 for the post-test. It suggests a decrease in the score of 0.38 from the pre-test to the post-test. The standard deviation for the pre-test was 1.32, whereas it decreased to 1.18 for the post-test. Based on the mean scores, these findings are surprising because the program aimed to enhance understanding of phishing, yet the results indicate a lower score in the post-test. Regarding the distribution of scores, the pre-test scores were more widely spread than the post-test. There could be several reasons, such as (1) the respondents may have guessed while answering the quiz, and (2) they may lack the necessary knowledge or experience to evaluate technical problems.

However, the respondents became more aware and understanding after completing the program. They developed a sense of caution and critical thinking in their actions. The narrowing range of standard deviation in the post-test supports this assumption. Twenty-one respondents (or 87.5%) rated the program content and materials with a score of 8 or higher on a scale of 1 to 10, reinforcing this assumption. Consequently, there is no dispute regarding the content and materials' effectiveness. Furthermore, the educational level of the respondents also influenced the outcomes. In this case, eight responders (33.3%) with a master's degree demonstrated good results in the post-test. There is a relationship between education levels and test results in which higher education levels correspond to higher competency levels.

Nevertheless, 100% of the participants strongly agreed or agreed that the program was successful and valuable overall, according to the survey findings conducted at the end of the session to evaluate the effectiveness of the course. Furthermore, they strongly agreed or agreed with the following statements:

i. The course objectives were achieved.

ii. The course content and assigned training/workshops were appropriate.

iii. The delivery of the course was good and effective.

iv. The teaching aids were used effectively.

v. The course venue provided a conducive atmosphere.

vi. The program planning and implementation were carried out smoothly.

vii. The allocated time for each module was appropriate.

viii. Their knowledge and understanding increased compared to before.

ix. They felt more confident in applying what they had learned compared to before.

These survey responses indicate a high level of satisfaction among the participants regarding the course's success, usefulness, content, delivery, and overall impact on their knowledge and confidence.

8.4.3 Lessons Learned and Areas for Improvement

The findings from this pilot test demonstrate that various factors, such as education level, user competency, computer literacy, and personality attributes, can significantly influence the outcomes attained in the phishing awareness program (Asfoor & Rahim, 2018). Due to the limited duration of the program, the number of participants involved in this phishing awareness initiative was relatively small, with only 24 individuals. Consequently, the obtained results may not fully meet the intended objectives. However, conducting a larger-scale study with a more extensive subject sampling is necessary to ensure broader validity. Based on the findings of this pilot test, it can be assumed that some participants did not pay full attention and failed to thoroughly read all the information presented during the security awareness program. As a result, the program's actual goals were not effectively achieved. Many participants mistakenly believed that their IT security devices and information security solutions were sufficient to prevent dangerous emails from reaching their inboxes. In reality, the opposite occurred, emphasizing the need for all parties to exercise caution and take necessary precautions when handling their ICT assets.

Several suggestions can be considered to improve future iterations of the program. These include incorporating simulations or phishing exercises as preparation for the test, grouping participants based on their educational background or organizational roles, tailoring materials and content for each group, and closely monitoring participant engagement during each session or module. Repetition may be necessary to achieve better results and meet the objectives of the awareness program. Conducting these awareness initiatives more frequently and in diverse formats can be beneficial.

Participants may also benefit from repeated exposure before their behaviors regarding cybersecurity are effectively changed. For instance, they consistently disseminate information on cyber security awareness through posters and emails to all stakeholders. Changing security behavior among higher education employees is a challenging endeavor that demands serious attention. Continuous improvement is the key to a successful awareness campaign, and this can only be determined by employing the same effectiveness measuring techniques. It is recommended that such awareness sessions be conducted monthly to enhance the knowledge and safety of staff and students.

In the realm of cybersecurity, establishing a culture of security awareness is of paramount importance. It is crucial to have well-structured cybersecurity awareness and training programs that equip individuals with fundamental cybersecurity knowledge (Alharbi & Tassaddiq, 2021). These programs should educate people on the risks and threats present in the digital landscape and provide them with the necessary skills to protect themselves and their organizations. Regardless of educational background, qualifications, occupation, age, or gender, end users are susceptible to phishing attacks if they lack sufficient security knowledge and awareness (Aljeaid et al., 2020). The effectiveness of security measures depends on the individuals' understanding and vigilance in identifying and mitigating potential risks. Hence, security awareness must be ingrained in the mindset of individuals from all walks of life. Instilling security awareness as a cultural norm from a young age is essential. We can promote cyber awareness and cultivate sustainable, safe cyber behavior by integrating cybersecurity education into early childhood learning (Aljeaid et al., 2020). Furthermore, individuals will be better prepared to navigate the digital landscape cautiously and make informed decisions regarding their online activities by embedding security practices into the fabric of society.

Creating a culture of security awareness requires a comprehensive approach encompassing various aspects. One crucial component is education and training programs that cater to individuals of all ages and backgrounds. These programs should deliver information in a clear and accessible manner, providing practical knowledge and actionable steps to protect against cyber threats. Furthermore, organizations should promote a security-conscious environment by incorporating security practices into their policies and procedures. Employees will be more inclined to prioritize security in their day-to-day activities by fostering a culture that values cybersecurity and encourages proactive measures. Regular training sessions, workshops, and simulations can reinforce security awareness and keep individuals updated with the latest threats and best practices. Technology plays a significant role in facilitating security awareness. Innovative tools such as interactive e-learning platforms, simulations, and gamified training modules can enhance engagement and knowledge retention. These technologies provide hands-on experiences and real-world scenarios that enable individuals to practice their cybersecurity skills in a safe environment.

8.5 CONCLUSION

In conclusion, this chapter has highlighted the importance of addressing phishing attacks in higher education and the potential for effective phishing awareness programs to reduce risk. Phishing attacks continue to be a significant threat to individuals and institutions, targeting sensitive information and compromising the security of systems. Higher education institutions, prime targets due to their vast networks and valuable data, must prioritize phishing awareness and proactively mitigate the risk. Phishing awareness programs equip individuals with the knowledge and skills necessary to identify and respond to phishing attempts effectively. Institutions can significantly reduce the success rate of such attacks by educating students, faculty, and staff about the tactics employed by attackers and providing them with practical guidance on spotting and reporting phishing emails or links. These programs play a vital role in building a culture of security awareness, where individuals are vigilant, informed, and actively contribute to protecting their institution's digital assets. Furthermore, institutions have a critical role in fostering a culture of security awareness within their campuses. Institutions can set the tone for a secure environment by prioritizing cybersecurity and integrating it into their policies, procedures, and training initiatives. It involves creating a security-conscious culture that emphasizes protecting sensitive data, encourages individuals to report suspicious activities, and provides ongoing support and resources for maintaining security awareness. Furthermore, collaboration between departments, such as IT, security, and education, is crucial in developing comprehensive and effective phishing awareness programs.

Although significant progress has been made in phishing awareness programs in higher education, there is still room for improvement and future work. First and foremost, evaluating these programs is essential to measure their effectiveness and identify areas for enhancement. Institutions should conduct regular assessments, collect participant feedback, and analyze metrics to gauge the impact of their initiatives. This evaluation process can inform future program development and ensure resources are allocated to the most effective strategies. In addition, as phishing attacks continue to evolve, phishing awareness programs must adapt and stay current with the latest tactics employed by attackers. Ongoing training and education should be provided to address emerging threats and equip individuals with the knowledge to recognize new phishing techniques. Institutions can also leverage technology solutions, such as simulated phishing exercises and interactive training platforms, to enhance engagement and provide hands-on experiences that reinforce learning. Furthermore, collaboration and information sharing among higher education institutions are crucial in combating phishing attacks. Institutions can collectively strengthen their security posture and contribute to a safer overall ecosystem by sharing best practices, success stories, and lessons learned. Collaborative efforts can involve sharing threat intelligence, participating in joint training initiatives, and establishing communities of practice to promote continuous learning and improvement.

REFERENCES

Alharbi, T., & Tassaddiq, A. (2021). Assessment of cybersecurity awareness among students of Majmaah University. Big Data and Cognitive Computing, 5(2), Article 23. https://doi.org/10.3390/bdcc5020023

Alhashmi, A. A., Darem, A., & Abawajy, J. H. (2021). Taxonomy of cybersecurity awareness delivery methods: A countermeasure for phishing threats. International Journal of Advanced Computer Science and Applications, 12(10), 29–35. https://doi.org/10.14569/IJACSA.2021.0121004

Aljeaid, D., Alzhrani, A., Alrougi, M., & Almalki, O. (2020). Assessment of end-user susceptibility to cybersecurity threats in Saudi Arabia by simulating phishing attacks. Information (Switzerland), 11(12), 1–19, Article 547. https://doi.org/10.3390/info11120547

Alwanain, M. I. (2019). An evaluation of user awareness for the detection of phishing emails. International Journal of Advanced Computer Science and Applications, 10(10), 323–328. https://doi.org/10.14569/ijacsa.2019.0101046

Anti-phishing Working Group. (2022). Phishing Activity Trends Reports: 4th Quarter 2022. https://apwg.org/trendsreports/

Asfoor, A. H., & Rahim, F. A. (2018). The potential factors influencing information security awareness on phishing attacks from various industries: A systematic literature review (SLR). International Journal of Engineering and Technology (UAE), 7(4.29 Special Issue 29), 25–30.

Baadel, S., Thabtah, F., & Lu, J. (2021). Cybersecurity awareness: A critical analysis of education and law enforcement methods. Informatica (Slovenia), 45(3), 335–345. https://doi.org/10.31449/INF.V45I3.3328.

Georgiadou, A., Michalitsi-Psarrou, A., Gioulekas, F., Stamatiadis, E., Tzikas, A., Gounaris, K., & Askounis, D. (2021). Hospitals' cybersecurity culture during the COVID-19 crisis. Healthcare (Switzerland), 9(10), Article 1335. https://doi.org/10.3390/healthcare9101335

Manoharan, S., Katuk, N., Hassan, S., & Ahmad, R. (2022). To click or not to click the link: The factors influencing internet banking users' intention in responding to phishing emails. Information and Computer Security, 30(1), 37–62. https://doi.org/10.1108/ICS-04-2021-0046

Nifakos, S., Chandramouli, K., Nikolaou, C. K., Papachristou, P., Koch, S., Panaousis, E., & Bonacina, S. (2021). Influence of human factors on cyber security within healthcare organisations: A systematic review [review]. Sensors, 21(15), Article 5119. https://doi.org/10.3390/s21155119

Okokpujie, K., Kennedy, C. G., Nnodu, K., & Noma-Osagha, E. (2023). Cybersecurity awareness: Investigating Students' susceptibility to phishing attacks for sustainable safe email usage in academic environment (A case study of a Nigerian leading university). International Journal of Sustainable Development and Planning, 18(1), 255–263. https://doi.org/10.18280/ijsdp.180127

Pinto, L., Brito, C., Marinho, V., & Pinto, P. (2022). Assessing the relevance of cybersecurity training and policies to prevent and mitigate the impact of phishing attacks. Journal of Internet Services and Information Security, 12(4), 23–38. https://doi.org/10.58346/JISIS.2022.I4.002

Rantos, K., Fysarakis, K., & Manifavas, C. (2012). How effective is your security awareness program? An evaluation methodology. Information Security Journal: A Global Perspective, 21(6), 328–345.

Scherb, C., Bryan, L., Grimberg, F., Grieder, H., & Maurer, M. (2023). A cyber-attack simulation for teaching cybersecurity. EPiC Series in Computing. https://doi.org/10.29007/dkdw

SentinelOne. (2022). Cyber Risks in the Education Sector | Why Cybersecurity Needs to Be Top of the Class. SentinelOne. https://www.sentinelone.com/blog/cyber-risks-in-the-education-sector-why-cybersecurity-needs-to-be-top-of-the-class/

Shepherd, L. A., & Szymkowiak, A. (2023). Investigating Phishing Awareness Using Virtual Agents and Eye Movements. Eye Tracking Research and Applications Symposium (ETRA). https://doi.org/10.1145/3588015.3590113

CHAPTER 9

Intelligence Random Forest Application in Developing Regression Model from Lamb Carcass C-Site Fat Depth Data

Sin Jie, Elayaraja Aruchunan, Nur Anisah Mohamed A. Rahman, Majid Khan Majahar Ali, Suhana Mohezar Ali, Muhammad Ashraf Khalid, Jayaseelan Marimuthu, Graham Edwin Gardner, and Samsul Ariffin Abdul Karim

9.1 INTRODUCTION

The correlation between carcass market value and fat depth is typically negative [1]. Higher fat depth results in reduced saleable meat yield and increased trimming costs to meet market sstandards. Enhancing the efficiency and profitability of the lamb supply chain necessitates a noninvasive, cost-effective, and efficient method for assessing lamb carcass fatness. Figure 9.1 shows illustration of C-site on lamb carcass.

DOI: 10.1201/9781003400387-9

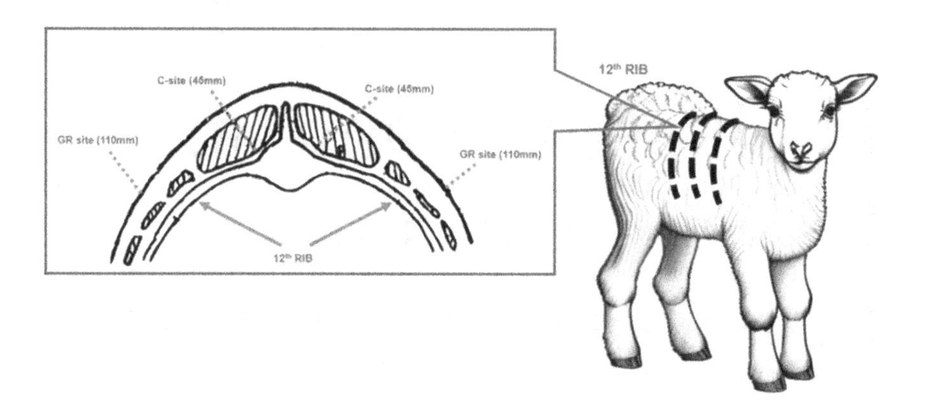

FIGURE 9.1 The Lamb carcass.

Traditional statistical methods and machine learning algorithms share a common objective of foreseeing unforeseen outcomes or future behaviors based on existing data [2]. Machine learning algorithms have historically leaned heavily on computational power [3], whereas traditional statistical methods initially evolved in settings where computational capacity was lacking [4]. Consequently, conventional statistical methods hinge on limited data samples and numerous assumptions about data and distributions. Conversely, machine learning algorithms tend to embrace fewer assumptions about the issue at hand, adopting more flexible methodologies and strategies to discover solutions, often leveraging heuristics [5].

The random forest algorithm is a machine learning tool utilized for both classification and regression purposes. Breiman (2001) introduced random forest, amalgamating his bagging sampling technique with the random feature selection to create forests of decision trees [6, 7]. Throughout the past decade, random forest has found applications across various domains, with new utilities continually emerging. For example, in the realm of engineering, random forest regression (RFR) was introduced to predict the capacity of lithium ion batteries. This method learns the battery capacity's reliance on features extracted from charging voltage and capacity measurements [8]. In social sciences, researchers proposed the use of RFR to estimate poverty levels in Bangladesh. The model, trained on Bangladeshi data, was subsequently applied to both Bangladesh and Nepal, demonstrating robust predictive capabilities for poverty estimation [9]. Within the field of medical science, Jing Zhao (2019) formulated a random forest regression model that precisely predicted the estimated glomerular filtration rate, aiding in the identification of chronic kidney disease [10].

In contrast, multiple linear regression (MLR) stands as a renowned traditional statistical technique. It is a valuable instrument for uncovering connections between a dependent variable and multiple independent variables [11]. MLR's strong interpretability has led to its widespread use across diverse research disciplines. For instance, in the sphere of waste management research, an MLR model was devised to prognosticate the rate of municipal solid waste generation in Mexico [12]. In manufacturing, an MLR model was presented to approximate hourly photovoltaic production, factoring in the performance ratio [13]. In the domain of education, MLR coupled with principal component analysis (PCA) has been harnessed to foresee students' academic performance [14].

A comprehensive comparison was carried out between multiple machine learning algorithms and MLR. To model indoor PM2.5 concentration, both MLR and RFR were employed. The findings revealed that RFR exhibited superior performance in modeling indoor pollution [15]. Moreover, both MLR and RFR were employed to gauge soil infiltration rates, with RFR showing promise in predicting the spatial distribution of infiltration rates [16]. In addition, property prices in Slovenia were projected using MLR and RFR, with RFR outperforming MLR in price prediction [17].

Australia has pioneered a noninvasive method to measure fat depth using an ultrawide-band microwave system (MiS) that operates on energy-efficient, nonionizing electromagnetic radiation [18]. Murdoch University has developed a cost-effective mobile MiS prototype, equipped with sample broadband antennas, to forecast carcass fat depth. Following data collection, two prediction model strategies were assessed. The first involved the use of partial least squares regression (PLSR), while the second integrated an ensemble stacking method, combining random forest and support vector machine (SVM). In the paper [19], concluded that the ensemble stacking method outperformed PLSR [18].

The benefits of the MiS technology include its safety, as it poses no health risks, and its user-friendliness, requiring minimal operator training beyond proper antenna positioning at the designated measurement spot. Figure 9.2 shows measuring carcass C-site fat depth using microwave device.

The primary objective of this research is to construct an effective model for predicting the fat depth of lamb carcasses at the C-site. In addition, we aim to compare the predictive accuracy of two models: RFR model and the MLR model, using the dataset for lamb carcass C-site fat depth.

FIGURE 9.2 Measuring Carcase C-site Fat Depth using Microwave Device.

This study is centered on crafting a predictive model that can estimate the fat depth of lamb carcasses specifically at the C-site. Furthermore, in order to identify the superior model, this investigation will assess the performance of both RFR and MLR models in forecasting the fat depth of lamb carcasses at the C-site. The dataset for this research was generously provided by Jayaseelan Marimuthu and Graham Edwin Gardner from Murdoch University. The findings of this study hold potential benefits for the meat industry, as it will offer insights into the effectiveness of both RFR and MLR models in predicting lamb carcass fat depth at the C-site. In addition, this research could serve as a valuable resource for future scholars, providing them with relevant information that could prove useful in their own studies. Moreover, some of the questions that future researchers might have could potentially find answers within the findings of this investigation.

9.2 MATERIALS AND METHODS

9.2.1 Data

The dataset was provided by Jayaseelan Marimuthu and Graham Edwin Gardner from Murdoch University. The data consists of 311 independent variables, which are frequency domain microwave signal, and one dependent variable, which is the fat depth of lamb carcass at C-site in millimeters (mm). The flowchart of research was executed in accordance with the layout depicted in Figure 9.3.

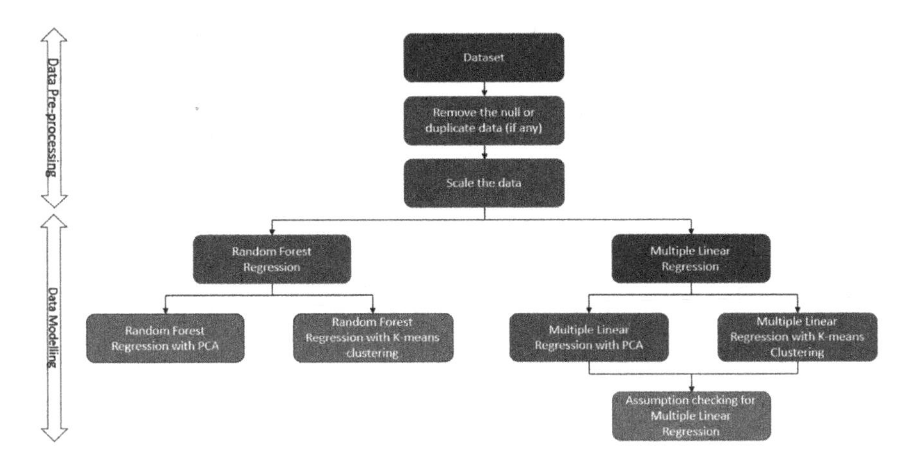

FIGURE 9.3 Flowchart of research.

9.2.2 Random Forest Regression

RFR is an ensemble learning algorithm based on a large number of decision trees [3]. To create a single decision tree, it employs the bootstrap method to extract randomized samples from original samples. A random feature subspace is utilized to select a sorting point at each node of the decision tree. Finally, these decision trees are integrated to get the final forecast result using a majority vote. Figure 9.4 shows an illustration of RFR algorithm.

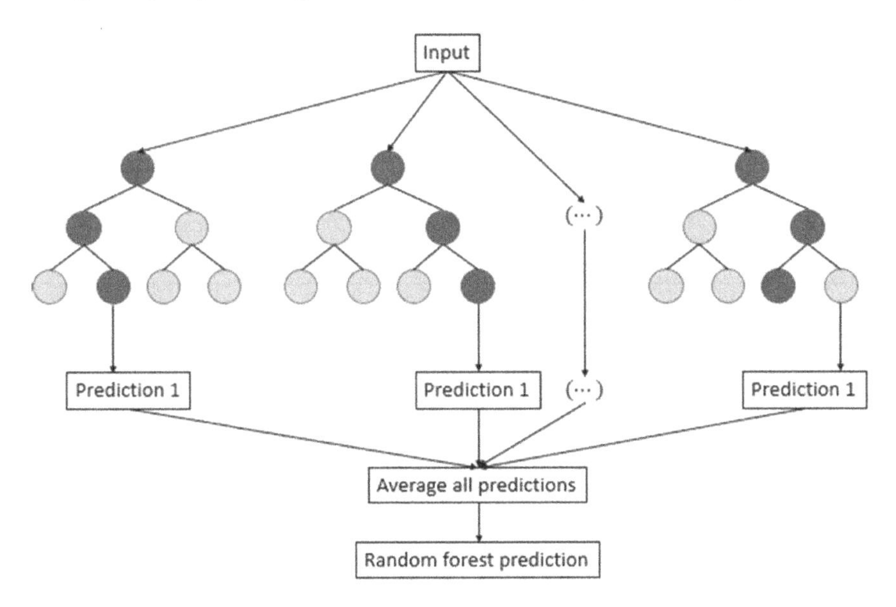

FIGURE 9.4 Illustration of Random Forest Regression.

9.2.3 Multiple Linear Regression

MLR is a traditional statistical method that has been widely used in identifying the linear association between the independent variables x and dependent variable y [11]:

$$y = \beta_0 + \beta_1 x_1 + \beta_2 x_2 + \cdots + \beta_k x_k + \varepsilon \tag{9.1}$$

where $j = 1, 2, \ldots, k$.

The five assumptions of MLR are outlined in Table 9.1.

9.2.4 Principal Component Analysis

As the number of dimensions increases, the computational burden increases, the risk of overfitting increases and multicollinearity issues arise. Thus, to lower the dimensionality, PCA will com press the variables from the original large dataset into new variables, named as principal components. First, PCA solves the eigenvalue and eigenvector issue. Second, PCA compresses the information into the first principal components as much as possible. Third, PCA compressed the maximum possible information into the second principal component from the remaining variables in the original dataset. The process is repeated until all the original variables are converted into principal components. The advantage of PCA is less accuracy reduction after the dimensionality reduction, but the interpretability of principal components is low [20].

9.2.5 K-Fold Cross-Validation

In machine learning, the dataset usually will be split into two sets: a train set for model training and a test set for model validation. However, for the

TABLE 9.1: Assumptions of Multiple Linear Regression

Assumption	Assumption Met If
Linearity	The points are randomly scattered around or lie on the diagonal line in the predicted value versus actual value plot
Normality	The histogram shows the normal distribution and the p-value from the Anderson–Darling test for normality is more significant than 0.05
Homoscedasticity	The residual plot against the independent variable should show a random pattern (equally scattered around zero and not forming a U-shaped curve)
Multicollinearity	The variables should not be highly correlated to each other. The variance inflation factor (VIF) of each variable should be lower than 10
Independence	The value obtained is between 1.5 to 2.5 in Durbin–Watson test

smaller dataset, k-fold cross-validation is a popular method to evaluate the trained model. Instead of splitting the data into two sets, k-fold cross-validation randomly splits the dataset into k groups uniformly. The first group will serve as a validation set, then compute the mean square error, MSE_1. Next, the second group will serve as a validation set, then compute MSE_2. This process will be repeated for k times to obtain MSE_1, MSE_2, …, MSE_k. Averaging these values yields the k-fold cross-validation estimate [21].

9.2.6 Mean Square Error

$$MSE = \frac{\sum_{i=1}^{N}\left(x_i - \hat{x}_i\right)^2}{N} \tag{9.2}$$

where

N = Sample size,
x_i = actual observations, and
$\hat{x}_i$ = estimated observations

Mean square error (MSE) is the average of the squared difference between the original values and predicted values. It measures the variance of the residuals.

9.2.7 Root Mean Square Error

$$RMSE = \sqrt{\frac{\sum_{i=1}^{N}\left(x_i - \hat{x}_i\right)^2}{N}} \tag{9.3}$$

where

N = Sample size,
x_i = actual observation, and
$\hat{x}_i$ = estimated observation.

Hence it is common to use root mean square error (RMSE) in evaluating the model accuracy where the lower the RMSE, the closer the estimated value as compared to the actual value. A model that minimizes RMSE will result in a higher accuracy of the predicted value.

9.2.8 Mean Absolute Error

$$MAE = \frac{\sum_{i=1}^{N}\left|\hat{y}_i - y_i\right|}{N} \tag{9.4}$$

where

N = Sample size,
x_i = actual observation, and
$\hat{x}_i$ = estimated observation.

Mean absolute error (MAE) is a measure of errors between paired observations. It can be seen as an arithmetic average of absolute errors. MAE is widely used as it is easy in terms of understanding and computation.

9.2.9 R-Squared (R^2)

It is the variation proportion that is explained by the predictor variables. In other words, it also explains how well the data fits the model. The better model will have higher R-squared.

$$R^2 = 1 - \frac{\Sigma\left(y_i - \hat{y}_i\right)^2}{\Sigma\left(y_i - \bar{y}\right)^2} \tag{9.5}$$

where

y_i = actual observation,
$\hat{y}_i$ = estimated observation, and
$\bar{y}$ = average actual observation.

The coefficient of determination, or R-squared, is the proportion of the variation in the dependent variable that is predictable from the independent variables. An R-squared of 1 tells us that all of the variations from the dependent variable can be explained by the features.

9.2.10 Adjusted R-Squared

It adjusts the root squared when there are too many variables in a model like n is the number of observations, R^2 is the R-squared, and k is the number of independent variables. It is described as follows:

$$R_{adj}^2 = 1 - \left[\frac{\left(1 - R^2\right)(n-1)}{n-k-1}\right]. \tag{9.6}$$

9.3 RESULTS AND DISCUSSION

9.3.1 Random Forest Regression and Multiple Linear Regression

Table 9.2 presents a comparison between RFR and MLR, excluding unsupervised learning methods. The aim was to assess predictive accuracy.

TABLE 9.2: Comparison of Random Forest Regression and Multiple Linear Regression without Unsupervised Learning Methods

Model	MSE	RMSE	*R*-squared	Adjusted *R*-squared	MAE
Random forest regression	1.37	1.16	0.59	1.25	0.94
Multiple linear regression	2.22	1.47	0.33	1.42	1.18

From the table, it is evident that RFR and MLR yielded adjusted R-squared values of 1.25 and 1.42, respectively. Typically, the adjusted R-squared value should be less than 1, representing the proportion of dependent variable variance explained by the model. The anomalous values observed here, exceeding 1, hold no statistical significance. These peculiar results might be attributed to the dataset's small sample size and high count of independent variables.

To address this, PCA and K-means [22] clustering were subsequently implemented to mitigate dimensionality issues.

9.3.2 Results of Principal Component Analysis

Referring to Figure 9.5, PCA has effectively decreased the variable count to 15, encompassing over 95% of the total explained variance. Post PCA, the number of independent variables was reduced from 311 to 15. Subsequently, the newly derived 15 principal components from the PCA will be integrated into both the RFR and MLR models.

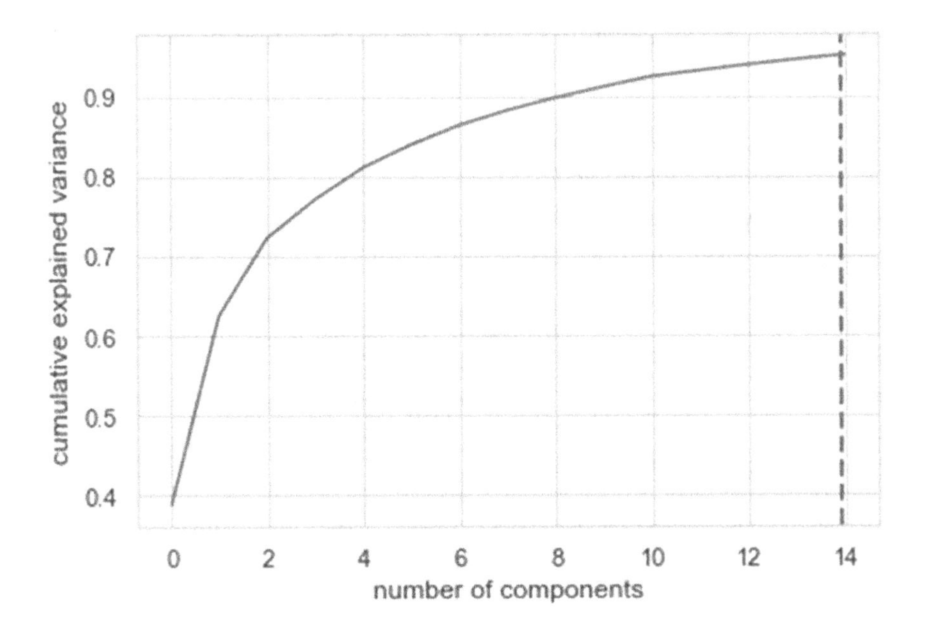

FIGURE 9.5 The Principal Component Analysis.

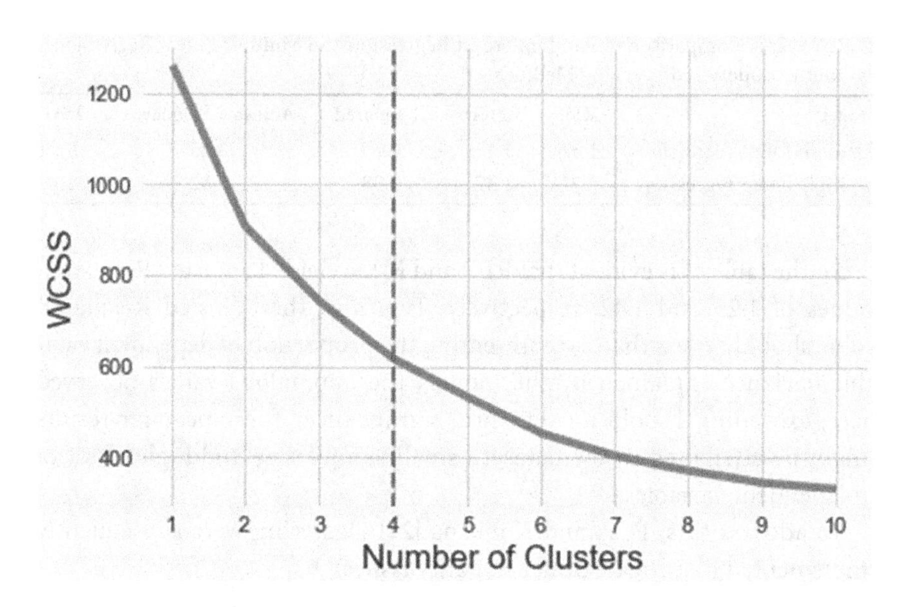

FIGURE 9.6 Elbow Method Analysis.

9.3.3 Results of *K*-Means Clustering

Referencing Figure 9.6, which depicts the Elbow method, we determined that the optimal number of clusters is $K = 4$. Through this method, all 311 variables were categorized into 4 clusters, with sizes 50, 101, 86, and 74, respectively. These four newly formed clusters resulting from the *K*-means clustering will subsequently be employed in the fitting process of both RFR and MLR models.

9.3.4 Comparison between RFR and MLR with PCA and *K*-Means Clustering

Table 9.3 reveals that among these four models, MLR with *K*-means clustering showcases the most favorable performance. This is evident across

TABLE 9.3: Comparison between Random Forest Regression and Multiple Linear Regression with PCA and *K*-Means Clustering

Model	MSE	RMSE	R-Squared	Adjusted R-Squared	MAE
Random forest regression with PCA	1.58	1.24	0.54	0.48	0.99
Random forest regression with K-means	1.33	1.13	0.60	0.58	0.89
Multiple linear regression with PCA	1.36	1.15	0.60	0.53	0.90
Multiple linear regression with K-means	1.15	1.06	0.65	0.64	0.84

all evaluation metrics, as evidenced by the MSE and RMSE values of 1.15 and 1.06, respectively. The minimized MSE and RMSE values signify that the data in MLR with *K*-means clustering closely aligns with the optimal fitting line. Furthermore, the MAE of MLR with *K*-means clustering is at its lowest, measuring 0.84. This outcome implies that this particular model exhibits the smallest average residual magnitude. Moreover, MLR with *K*-means clustering yields the highest *R*-squared and adjusted *R*-squared values, reaching 0.65 and 0.64, respectively. These outcomes underscore that the independent variables can elucidate 65% of the dependent variable's variance before adjustment and 64% post-adjustment. The corresponding equation for MLR with *K*-means clustering is as follows:

$$y = 9.0241 - 2.5703x_1 + 3.8183x_2 - 0.5710x_3 - 9.3509x_4 \quad (9.7)$$

Conversely, the model that achieved the second-best results is RFR with *K*-means clustering. In this case, the MSE and RMSE values stand at 1.33 and 1.13, respectively. Furthermore, the MAE for this model amounts to 0.84.

In addition, RFR with *K*-means clustering yield the second-highest *R* -squared and adjusted *R*-squared value, which are 0.60 and 0.58, respectively. This indicates that the independent variables could explain 60% of the dependent variable before adjusted and 58% of the dependent variable after adjusted. The visualization of the first decision tree of RFR with *K*-means clustering as shown in Figure 9.7.

As RFR with *K*-means clustering consists of 100 decision trees which lead to a highly complex illustration, thus, only the first decision tree is shown in Figure 9.7.

9.3.5 Assumption Checking of Multiple Linear Regression

9.3.5.1 Linearity

Assumption 1: Linear relationship between the target and the feature checking with a scatter plot of actual versus predicted. Predictions should follow the diagonal line.

In Figure 9.8, which displays the plot of actual values against predicted values, it can be observed that the data points are uniformly scattered along the diagonal line. As a result, we can deduce that the assumption of linearity has been met.

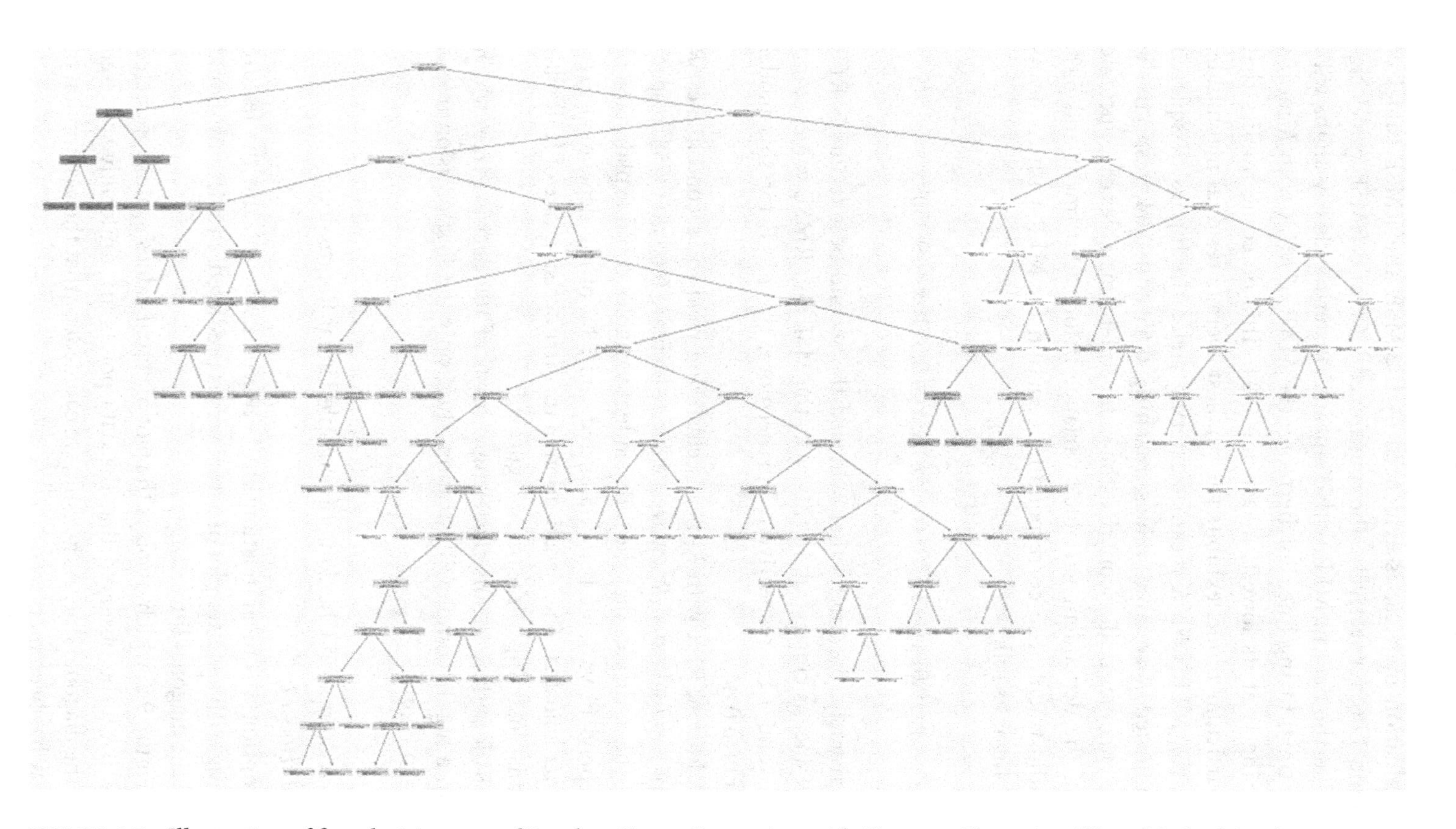

FIGURE 9.7 Illustration of first decision tree of Random Forest Regression with *K*-means Clustering Elbow Method Analysis.

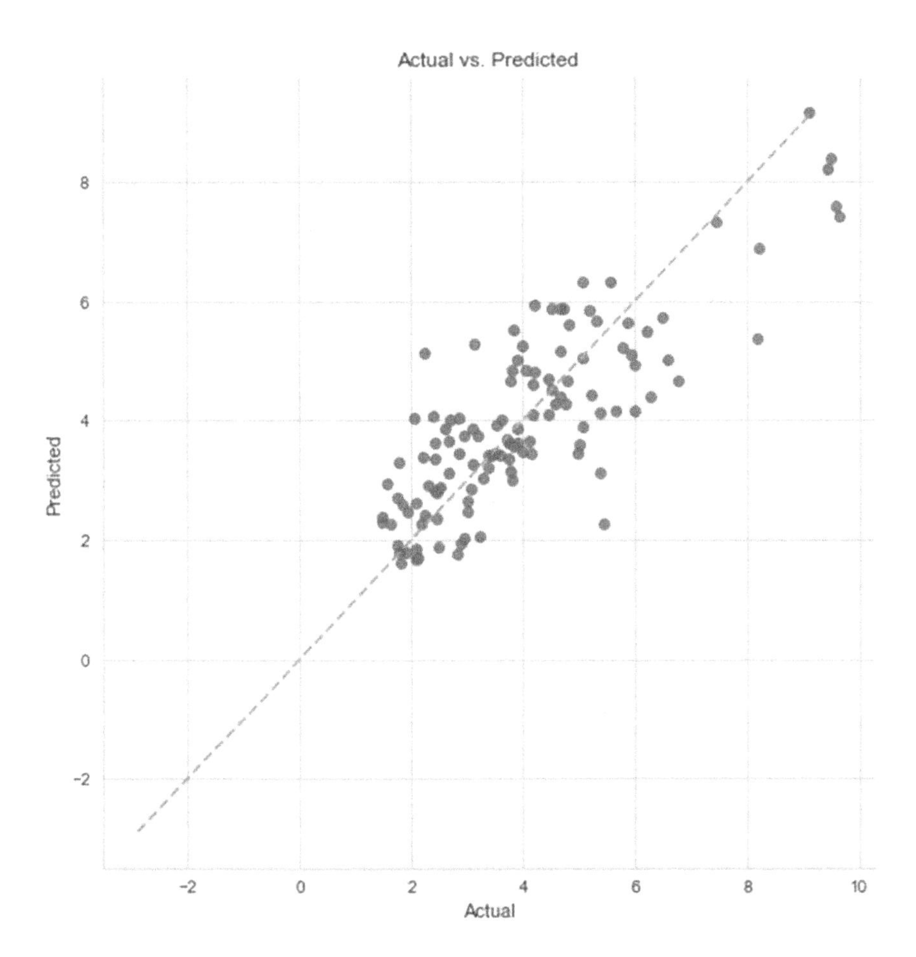

FIGURE 9.8 Linearity assumption checking of MLR.

9.3.5.2 Normality

Assumption 2: The error terms are normally distributed.

Employing the Anderson–Darling test to assess normal distribution, the *p*-value from the test is exhibited in Figure 9.9. Generally, a *p*-value below 0.05 suggests non-normality. In this case, the *p*-value is 0.3668267, indicating that the residuals follow a normal distribution.

9.3.5.3 Homoscedasticity

Assumption 3: Homoscedasticity of error terms.

Residuals should have relative constant variance.

From the residuals plot in Figure 9.10, it observed that the residuals are randomly distributed around zero, thus, we conclude that residuals have

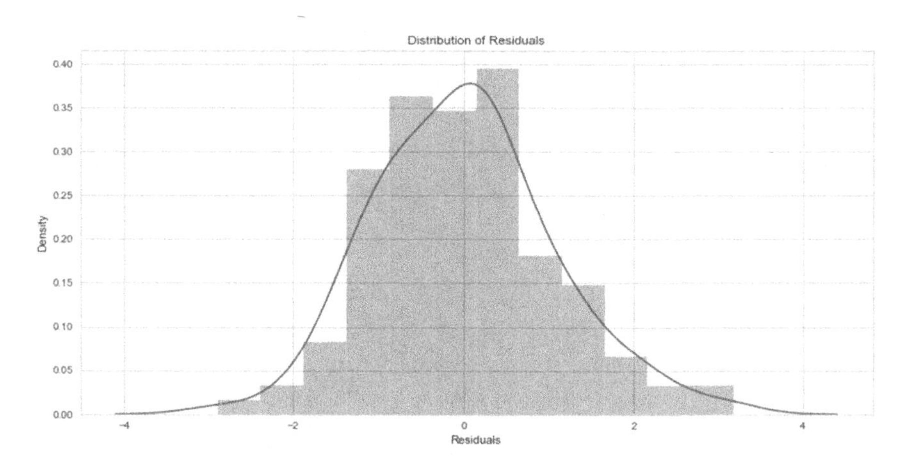

FIGURE 9.9 Normality assumption checking of MLR.

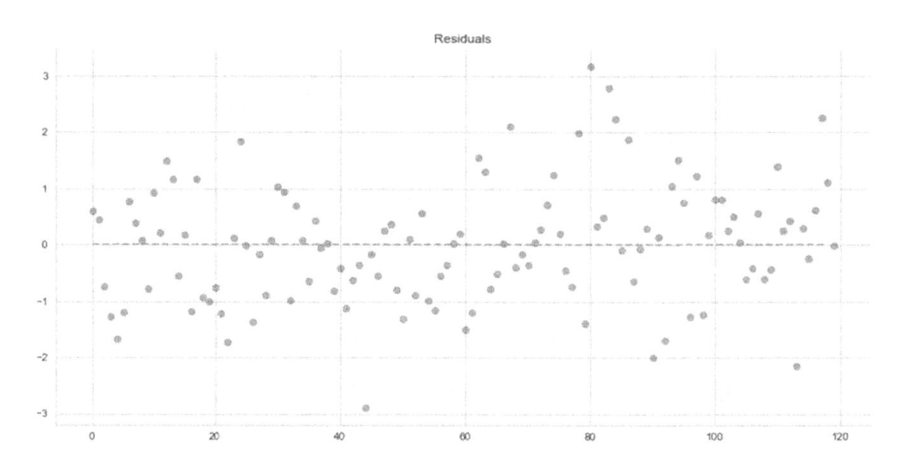

FIGURE 9.10 Homoscedasticity assumption checking of MLR.

constant variance. We could observe that there is an outlier (within the green circle) that is larger than 3. Further diagnostic should be carried out to identify whether the outlier is influential.

9.3.5.4 Multicollinearity

Assumption 4: Multicollinearity.

From Table 9.4, it can be observed that multicollinearity issue exists. Cluster 1, 2, and 3 have the high value of variance inflation factor (VIF), greater than 10, indicating the high multicollinearity.

TABLE 9.4: Multicollinearity Assumption Checking of MLR in Dataset 1

Cluster	Variance Inflation Factor
0	8.54
1	22.52
2	59.82
3	44.10

9.3.5.5 *Independence*

Assumption 5: No autocorrelation.

$$\frac{\text{Performing Durbin-Watson Test}}{\text{Durbin-Watson}: 1.8351480272859468}$$

If the value obtained is between 0 to 2, we conclude that the residuals are positively correlated. If the value obtained is between 2 to 4, we conclude that the residuals are negatively correlated.

From the Durbin–Watson test, we got the value of 1.84 (between 1.5 and 2.5), which led to no autocorrelation between residuals.

9.3.6 Discussion in Assumptions Checking of Multiple Linear Regression

Upon conducting a comprehensive assessment of the assumptions, we have determined that the prerequisites for linearity, normality, homoscedasticity, and independence have all been satisfied. Regrettably, the assumption pertaining to multicollinearity has not been met. This deficiency arises due to elevated VIF values which signify a noteworthy correlation among the independent variables. Given this breach of assumption, the reliability of the MLR model integrated with *K*-means clustering is compromised. Consequently, it is our considered opinion that the model exhibiting the second-best performance, namely the RFR in conjunction with *K*-means clustering, stands as the most robust choice. This particular model emerges as the optimal candidate for predicting the fat depth of lamb carcasses at the C-site, considering the shortcomings encountered with the previously mentioned MLR approach.

9.4 CONCLUSION

The findings of this investigation indicate that the RFR model with *K*-means clustering emerges as the optimal model for predicting the fat depth of lamb carcasses at the C-site. While the MLR model with *K*-means clustering exhibited superior accuracy metrics such as MSE, RMSE, R -squared (R^2), adjusted R-squared (adjusted R^2), and MAE, it failed to meet the assumption of multicollinearity. Consequently, RFR with K-means clustering, which attained the second-best performance, was chosen as the most suitable model for this fat depth prediction task.

Therefore, the conclusion drawn is that RFR outperforms MLR in this context, mainly due to its fewer assumption requirements. Given the limited sample size of the dataset in this study, it is advisable to conduct further training and validation to enhance the ability of model to predict the fat depth of lamb carcasses at the C-site.

ACKNOWLEDGMENT

This study was undertaken through the Advanced Livestock Measurement Technologies Project (ALMTech) and funded by the Department of Agriculture, Rural Research and Development for Profit program and Meat and Livestock Australia.

REFERENCES

1. Polkinghorne, R., Philpott, J., Gee, A., Doljanin, A., & Innes, J. (2008). Development of a commercial system to apply the Meat Standards Australia grading model to optimise the return on eating quality in a beef supply chain. *Australian Journal of Experimental Agriculture*, *48*(11), 1451. https://doi.org/10.1071/ea05181
2. Bzdok, D., Altman, N., & Krzywinski, M. (2018). Statistics versus machine learning. *Nature Methods*, *15*(4), 233–234. https://doi.org/10.1038/nmeth.4642
3. Al-Jarrah, O. Y., Yoo, P. D., Muhaidat, S., Karagiannidis, G. K., & Taha, K. (2015). Efficient machine learning for big data: A review. *Big Data Research*, *2*(3), 87–93. https://doi.org/10.1016/j.bdr.2015.04.001
4. Hassibi, K. (2016). Machine Learning vs. Traditional Statistics: Different Philosophies, Different Approaches. DataScienceCentral.com. Retrieved January 20, 2022, from Data Science Central website: https://www.datasciencecentral.com/machine-learning-vs-traditional-statistics-different-philosophi-1/
5. Deoras, S. (2017). How Is Machine Learning Different from Statistics. Retrieved January 20, 2022, from Analytics India Magazine website: https://analyticsindiamag.com/machine-learning-different-statistics/
6. Breiman, L. (2001). Random forests. *Machine Learning*, *45*(1), 5–32. https://doi.org/10.1023/a:1010933404324
7. Fawagreh, K., Gaber, M. M., & Elyan, E. (2014). Random forests: From early developments to recent advancements. *Systems Science & Control Engineering*, *2*(1), 602–609. https://doi.org/10.1080/21642583.2014.956265

8. Li, Y., Zou, C., Berecibar, M., Nanini-Maury, E., Chan, J. C.-W., van den Bossche, P., & Omar, N. (2018). Random forest regression for online capacity estimation of lithium-ion batteries. *Applied Energy*, *232*, 197–210. https://doi.org/10.1016/j.apenergy.2018.09.182
9. Zhao, X., Yu, B., Liu, Y., Chen, Z., Li, Q., Wang, C., & Wu, J. (2019). Estimation of poverty using random forest regression with multi-source data: A case study in Bangladesh. *Remote Sensing*, *11*(4), 375. https://doi.org/10.3390/rs11040375
10. Zhao, J., Gu, S., & McDermaid, A. (2019). Predicting outcomes of chronic kidney disease from EMR data based on random forest regression. *Mathematical Biosciences*, *310*, 24–30. https://doi.org/10.1016/j.mbs.2019.02.001
11. Montgomery, D. C. (2013). *Introduction to Linear Regression Analysis*. (Fifth Edition). John Wiley & Sons Inc.
12. AraizaAguilar, J., RojasValencia, M., & AguilarVera, R. (2020). Forecast generation model of municipal solid waste using multiple linear regression. *Global Journal of Environmental Science and Management*, *6*(1), 1–14.
13. Trigo-González, M., Batlles, F. J., Alonso-Montesinos, J., Ferrada, P., del Sagrado, J., Martínez-Durbán, M., & Marzo, A. (2019). Hourly PV production estimation by means of an exportable multiple linear regression model. *Renewable Energy*, *135*, 303–312. https://doi.org/10.1016/j.renene.2018.12.014
14. Yang, S. J. H., Lu, O. H. T., Huang, A. Y. Q., Huang, J. C. H., Ogata, H., & Lin, A. J. Q. (2018). Predicting students' academic performance using multiple linear regression and principal component analysis. *Journal of Information Processing*, *26*(0), 170–176. https://doi.org/10.2197/ipsjjip.26.170
15. Yuchi, W., Gombojav, E., Boldbaatar, B., Galsuren, J., Enkhmaa, S., Beejin, B., & Allen, R. W. (2019). Evaluation of random forest regression and multiple linear regression for predicting indoor fine particulate matter concentrations in a highly polluted city. *Environmental Pollution*, *245*, 746–753. https://doi.org/10.1016/j.envpol.2018.11.034
16. Pahlavan-Rad, M. R., Dahmardeh, K., Hadizadeh, M., Keykha, G., Mohammadnia, N., Gangali, M., & Brungard, C. (2020). Prediction of soil water infiltration using multiple linear regression and random forest in a dry flood plain, eastern Iran. *CATENA*, *194*, 104715. https://doi.org/10.1016/j.catena.2020.104715
17. Čeh, M., Kilibarda, M., Lisec, A., & Bajat, B. (2018). Estimating the performance of random forest versus multiple regression for predicting prices of the apartments. *ISPRS International Journal of Geo-Information*, *7*(5), 168. https://doi.org/10.3390/ijgi7050168
18. Marimuthu, J., Loudon, K. M. W., & Gardner, G. E. (2021b). Ultrawide band microwave system as a non-invasive technology to predict beef carcase fat depth. *Meat Science*, *179*, 108455. https://doi.org/10.1016/j.meatsci.2021.108455
19. Marimuthu, J., Loudon, K. M. W., & Gardner, G. E. (2021a). Prediction of lamb carcase C-site fat depth and GR tissue depth using a non-invasive portable microwave system. *Meat Science*, *181*, 108398. https://doi.org/10.1016/j.meatsci.2020.108398

20. Wold, S., Esbensen, K., & Geladi, P. (1987). Principal component analysis. *Chemometrics and Intelligent Laboratory Systems*, *2*(13), 37–52.
21. James, G., Witten, D., Hastie, T., & Tibshirani, R. (2013). *An Introduction to Statistical Learning: with Applications in R*. Springer.
22. Syakur, M., Khotimah, B., Rochman, E., & Satoto, B. D. (2018). Integration k-means clustering method and elbow method for identification of the best customer profile cluster. *IOP Conference Series: Materials Science and Engineering*, *336*(1), 012017.

CHAPTER 10

Intelligent Identification System for MOOC Security

Ling Hang Yek, Chin Kim On,
Mohd Hanafi Ahmad Hijazi, Chin Pei Yee,
Chai Soo See, and Chi Jing

10.1 INTRODUCTION

Face recognition is a powerful technique capable of identifying or verifying individuals based on their facial features captured in photographs or videos. Face recognition systems can be implemented in a variety of ways. In general, it is a system that compares selected facial features from a picture to faces in a database. Face recognition is also known as a biometric artificial intelligence method because it confirms a person's identity based on facial textures, shapes, and contours without human intervention. Face recognition has become increasingly popular in recent years for a variety of purposes, including unlocking smartphones [1], finding missing persons [2], assisting blind people [3], identifying users on social networking sites [4], taking online examinations [5], managing attendance [6], and recognizing facial expressions [7]. Because of its contactless and noninvasive nature, face recognition has been frequently employed in security and surveillance systems. However, its performance is slightly poorer than other biometric recognition systems. Researchers have proposed different algorithms and methods to improve existing face recognition systems. However, fully implementing a reliable face recognition system remains challenging due to various environmental conditions.

DOI: 10.1201/9781003400387-10

The algorithms applied are the most important consideration, followed by the hardware used. Furthermore, most of the research findings have been presented without actual platform implementation. As a result, the outcomes are debatable.

Convolutional neural networks (CNNs) have gained popularity in visual image analysis due to their superior classification and recognition capabilities [6]. Unlike traditional neural networks, CNNs utilize convolution instead of general matrix multiplication in one or more of their layers, allowing them to excel in various tasks. Each neuron in each layer of a CNN is connected to all neurons in the next layer. The connection pattern of neurons in a CNN is inspired by biological processes, and these connected neurons can lead to data overfitting [8]. CNNs preprocess images differently than other image classification methods. The network removes unnecessary components with feature engineering to increase the performance of the algorithms. CNNs consist of three layers: an input layer, an output layer, and hidden layers. The hidden layers involve a series of convolution layers that convolve when multiplied. The input image is processed by the hidden layers, which perform convolution calculations, subsampling, and conclude with a fully connected layer for further feature extraction. Finally, the network generates an output. However, CNN models have some drawbacks. For example, they can have difficulty classifying images with different positions, and their performance can degrade when the number of images in the server is increased.

In addition to CNN, other researchers have shown that template matching [9] and geometric-based [10] methods can also be effective in face recognition. Template matching works by comparing two images and determining their similarity. It is a common technique in image processing for tasks such as feature extraction, edge detection, and object extraction. Template matching can be divided into two types: feature-based and template-based. Feature-based comparison compares the edges and corners of the object and template images. The goal of the comparison is to locate the location that most closely matches the template image. Template-based matching, on the other hand, uses the entire template to identify the best match. It focuses on unique features such as eyes, nose, mouth, chin, and the shape of the face. However, the recognition process can be challenging due to potential challenges such as scaling, rotation, illumination, and others.

The geometric-based approach is a facial recognition algorithm that compares and recognizes faces using fiducial points. Fiducial points can be automatically detected or manually extracted. Automatic detection is usually preferred and used, but the detected points are then manually corrected by the operator. The geometric features could be the corners, edges, and area of the image formed by the detected points. The features can be extracted via corner detection, curve fitting, edge detection, and global structure extraction. Faces, eyes, noses, mouths, and ears are formed from these features. These facial traits are compared to the extracted data in the dataset to recognize faces.

In this research, a standalone system with template matching, geometric-based, and CNN algorithms is applied to a web-based massive open online course (MOOC) platform. Most students widely use the MOOC platform for attending online courses. Registered students can log in to any MOOC platform to download teaching and learning materials, attempt quizzes and tests, submit assignments, share their thoughts in the forum, and so on. However, the safety and security of the MOOC system remains a challenge. Others may view the content of the MOOC after the owner has logged in or forgotten to log out. An intelligent system with face recognition is required to prevent others from taking the quizzes, tests, or assignments assigned by the lecturer. It also prevents others from viewing the test/exam question content during the test/exam session.

This research is divided into two parts. The first part is to design and develop an automated login system without modifying any existing MOOC platform. The second part involves the design and embedding of the geometric-based approach, template matching method, and CNN algorithm to perform automated face recognition. Finally, the performance of the algorithms is compared and analyzed.

10.2 RELATED RESEARCH WORKS

10.2.1 Template Matching

Template matching is a simple and fast method used in object detection studies by changing the template used without a time-consuming training process. However, it is not popular because traditional template matching is not reliable and robust. Recently, researchers have proposed various ways to improve the template matching algorithm to fit the robustness. Some of them have proven that template matching performs well in vision sensor systems for surveillance [11].

In ref [12], the authors proposed a face detection technique based on template matching and color segmentation. They used eye and face templates to find areas with high similarity. Then, they derived a mask using color segmentation. Next, they used texture filtering and a variety of binary operations to clean the derived mask. The false positives were then edited and removed using the derived mask. The outcome then went through a clustering process. In the clustering process, points that were less than a certain Euclidean distance from each other were clustered into one point. The above process was repeated with a variety of scales or resolutions. Finally, they used the results to reconstitute the mask. The proposed method achieved the best average detection rate of 89%.

In ref [13], the authors proposed an almost similar template matching method for comparing facial features. The accuracy rate was 86.11%. The method involves face region extraction, detection, and selection of the iris. For face region extraction, the researchers selected colored images of men with beards, mustaches, and facial expressions from the AR face database. They also selected colored images of women with long hair and facial expressions from the AR face database. Each head-and-shoulder photo selected had a plain background and a head rotation of fewer than 30 degrees. The images were then used to develop a skin color model. They converted the photos to grayscale images for iris recognition and selection. Then, they used grayscale closing to extract the intensity valleys. Illumination normalization and light spot removal were applied to improve the image quality.

In ref [14], the authors proposed skin color detection and template matching for face recognition. They started by differentiating the skin and non-skin areas in the input images using the skin color detector. The objective was to remove all non-facial areas based on template matching. The accuracy of template matching was increased after excluding all non-facial areas. The template matching method used an eigenfaces trained image to locate the mouth, nose, and eyes. Lastly, the recognition involved image segmentation, edge detection, and template matching. This method did not perform as expected and achieved the best recognition rate of 62.5%.

In ref [15], the authors divided the proposed approach into four parts: wavelet decomposition, edge image extraction, feature extraction, and template matching. To start, they decomposed the input image using the wavelet decomposition method. The edge images (template images) were then created using the decomposed subimages. Next, they used template images to estimate the shapes of faces. Twenty-eight features were then calculated for each face and saved in the feature database. Finally, they mapped the faces using the feature database. The proposed method achieved the best recognition rate of 84%, but the overall processing time was slightly higher than other template matching methods.

In ref [16], the authors proposed template matching and appearance-based methods for face recognition. The authors divided the proposed approach into several steps. They first identified human faces in video frames using the Haar classifier. Then, they extracted facial features such as eyes, nose, and mouth using the Otsu threshold method. Lastly, the faces were then recognized using the Hu moment feature set. Interestingly, the overall best recognition rate was 91%.

10.2.2 Geometric-Based Model

During the past decades, two-dimensional (2D) face recognition methods have achieved remarkable results in face recognition. However, these methods are still weak in handling recognition tasks in various illumination, pose, expression, occlusion, and disguise situations. In fact, geometric-based methods may outperform 2D face recognition due to their advantage in robustness.

Local feature-based methods are one of the most commonly used geometric-based methods. These methods involve a keypoint detector called scale-invariant feature transformation (SIFT). In their study in ref [17], researchers utilized SIFT to identify significant keypoints in a 3D depth image. These keypoints were then utilized to measure the variations in facial depth within their respective neighborhoods. To enhance accuracy, they incorporated local covariance descriptors and employed Riemann kernel sparse coding techniques. However, the algorithm's performance could be reduced due to facial expressions. The presence of occlusions would also reduce the accuracy rate. The dataset used was FRGC V2 and the proposed method achieved a 97.3% accuracy rate.

In a recent work in ref [18], the authors introduced a novel approach based on the wave kernel signature that incorporates geometry and local shape descriptors. This method was specifically designed to address the challenges posed by facial expressions in geometric-based face recognition methods. The proposed method has greatly improved other geometric-based methods in terms of time consumption and face expression issues. However, the proposed method relied heavily on the database. The performance was low with missing data. The presence of occlusions remains a challenge. The researchers improved the database for GavabDB, and the proposed method achieved a 99.18% accuracy rate.

Study in ref [19] proposed a local binary pattern (LBP) based method to extract 3D depth image features. Then, the support vector machine (SVM) algorithm was used to classify the image features. The feature extraction time of each depth map in Texas 3D was reduced to 0.1856 seconds, while other researchers needed 23.54 seconds. However, the algorithm's performance could be low with various pose and occlusion problems. The proposed algorithm achieved a 96.83% accuracy rate.

10.2.3 Convolution Neural Networks

CNN has significantly improved 2D facial recognition due to its robustness. There are a variety of deep neural networks, and CNN is the most commonly used one. CNN is a type of neural network that uses convolution instead of conventional matrix multiplication in at least one of its layers. The word "convolution" means that the neural network uses a mathematical operation called convolution.

In ref [8], the researchers divided the proposed algorithm into three main parts. First, the input images were resized into $16 \times 16 \times 1$, $16 \times 16 \times 3$, $32 \times 32 \times 1$, $32 \times 32 \times 3$, $64 \times 64 \times 1$, and $64 \times 64 \times 1$. Then, they designed a CNN structure containing five convolutional layers and three max pooling layers. Next, they used the Softmax classifier to classify all the features. The Georgia Tech face dataset was used, and the best recognition rate was 98.8%.

In ref [20], the researchers first sent the gallery images to undergo image preprocessing to improve the image quality. After that, they determined the similarity of the gallery images and reference-based feature vectors. Then, the gallery images were arranged based on their similarity. Lastly, the gallery images were sent for training. The ORL dataset was used, and the proposed CNN achieved a 98.3% accuracy rate.

Ref [21] proposed a two-stream CNN. The two-stream CNN specializes in the classification and localization of single face detection. The proposed method can output the human faces in the input images. The authors also used a cascade of Region of Interest Network and two-stream CNN for multi-object detection. The author used the ChokePoint dataset, and the testing accuracy rate was 91.0%.

Ref [22] proposed an unconstrained face verification method using CNN. The authors used the CASIA-WebFace and IJB-A datasets for face detection in the training step. The applications of the dataset can simplify the localization and alignment of each face. The CNN model was trained with the CASIA-WebFace dataset, and the Joint Bayesian metric was derived using the IJB-A dataset. In the testing phase, pairs of test image sets were input, and a similarity score was generated based on CNN features and learning metrics.

Ref [23] proposed a face recognition with pose and illumination method using CNN. They combined the convolution and resampling layer to produce a simplified version of CNN. The combination had adapted to face recognition for 100 classes. It also incorporated a partial connection between the first and second layers. The reason for doing this was to make sure that all different features were involved in the training process. This combination required only one stage to complete the convolution and subsampling, whereas CNN needs two stages.

10.3 THE PROPOSED METHOD

10.3.1 Machine Learning Method

Almost 100 volunteers participated in the experiments. The captured photos included head and shoulder parts. Later, template images were created based on appearance-based methods. The images were then converted to grayscale to remove extraneous information. In the face recognition process, an input image was captured and preprocessed before applying template matching. Finally, the accuracy of the matching was calculated and analyzed.

In geometric modeling, the captured photos were cropped to the same size and then converted to grayscale. The geometric features were then extracted from the grayscale images. The extracted features were used for training purposes. Finally, the training results were collected and analyzed.

Almost similar steps were applied with the CNN algorithm. The captured photos were cropped and converted to grayscale in the segmentation process. Only the facial features remained after the segmentation process to reduce the information involved in the CNN training process. Then, the trained model was used for the recognition process. Figure 10.1 shows the proposed model used in this study.

10.3.2 The Proposed Identification Model

A program was created using the SCREEN_SEARCH library. The program is designed to automatically identify the SmartV3UMS website by comparing the URL and screenshot with pre-saved information. Then, the program will turn on the camera and automatically identify the user using face recognition methods. The login process will start after the user is identified. The user must first configure the username and password when registering the program for the first time. The computer's MAC address will

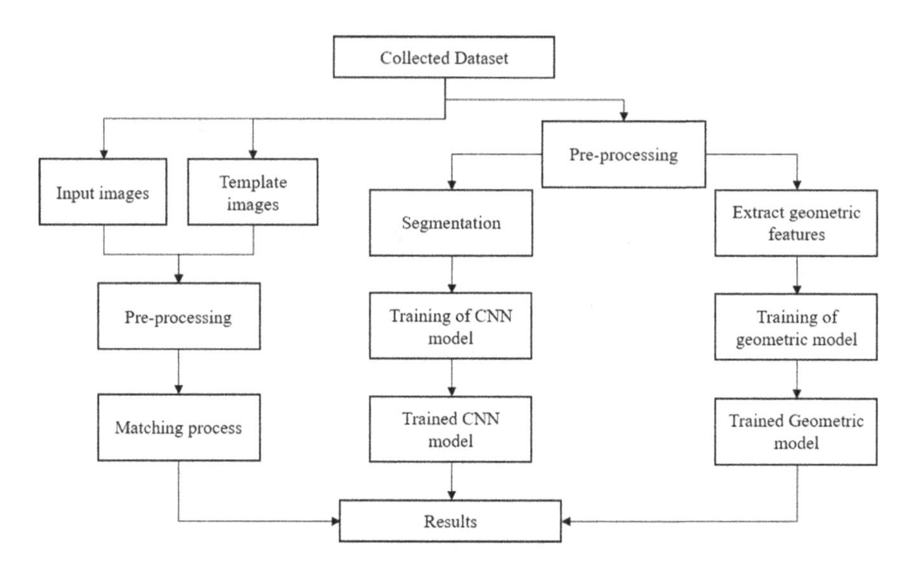

FIGURE 10.1 The proposed model for face recognition purpose.

be collected and mapped to the user to strengthen the login process. The user must confirm whether they want to continue logging in to SmartV3UMS within one minute. Otherwise, the login process will be canceled. After logging in to SmartV3UMS, the program will automatically minimize the browser within 10 seconds if another user is detected or the identified user (the account owner) is missing from the screen for more than 10 seconds. Figure 10.2 shows the overall process of the proposed model.

10.4 RESULTS AND DISCUSSIONS

A set of preliminary experiments were carried out to identify the performance of template matching, geometry-based, and CNN algorithms. The proposed CNN parameter settings in ref [23] were referred to, and the average testing results are shown in Table 10.1.

The results show that the template matching algorithm performed well, with an average recognition rate of 89.06%. Both the CNN and geometry-based algorithms achieved much better results in terms of accuracy than the template matching algorithm. Surprisingly, the CNN algorithm achieved an average accuracy rate of 93.33%, compared to the accuracy rate of 93.06% of the geometry-based algorithm. However, many researchers have shown that the CNN algorithm outperforms other facial recognition algorithms. The configuration of CNN's training parameters is probably required due to the different datasets used.

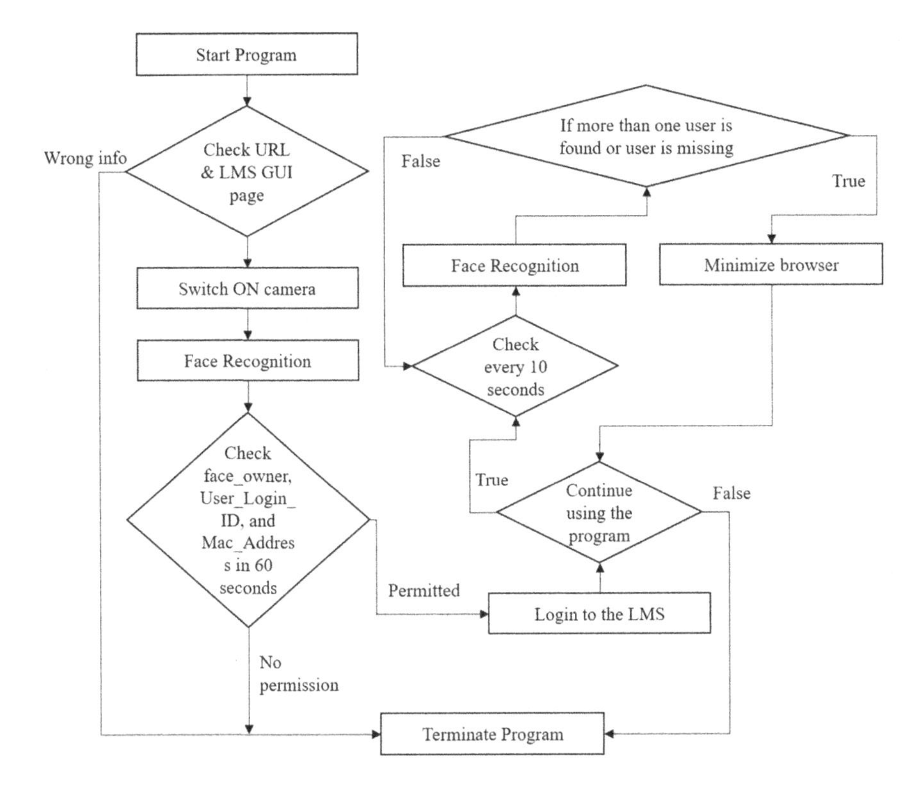

FIGURE 10.2 The proposed identification model.

TABLE 10.1 Comparison of Testing Results

	Template Matching	Geometric Based	CNN
	First 100 tests		
Accuracy result (%)	89.00	93.00	94.00
Recognition time (s)	1.88	1.86	2.62
	200 tests		
Accuracy result (%)	87.50	93.50	93.00
Recognition time (s)	2.09	1.98	2.88
	300 tests		
Accuracy result (%)	90.67	92.67	93.00
Recognition time (s)	1.79	1.97	3.32
	Overall results		
Average accuracy (%)	**89.06%**	**93.06%**	**93.33%**
Average duration	**1.92s**	**1.94s**	**2.94s**

TABLE 10.2 Comparison of CNN Training Results

Epoch		**10**			**20**			**30**	
Experiments	1	2	3	1	2	3	1	2	3
Batch sizes	16	16	16	16	16	16	16	16	16
Learning rate	0.0004	0.0004	0.0004	0.0004	0.0004	0.0004	0.0004	0.0004	0.0004
Accuracy	0.8301	0.8108	0.8304	0.9717	0.9804	0.9694	0.9822	0.9701	0.9748
Average accuracy		**82.38%**			**97.38%**			**97.57%**	

Table 10.1 also shows that the template matching algorithm performed slightly faster than the geometry-based and CNN algorithms. This is probably because the geometry-based algorithm required more information than the template matching algorithm, thus increasing the matching time. For CNN, the average recognition time increased for every 100 tests, and the recognition rate was slightly reduced. This is probably due to the increasing amount of data captured over time.

Next, several experiments were conducted with CNN to identify the best configuration for training and testing results. The experiments involved a range of 10–60 epochs. The batch sizes were varied: 8, 16, 32, 64, 128, and 192. Learning rates of 0.01, 0.001, 0.0001, 0.0002, 0.0003, 0.0004, and 0.0005 were used. The results showed that the combination of 20 and 30 epochs, batch size 16, and learning rate 0.0004 generated the highest experimental results. These results are simplified and tabulated in Table 10.2.

Table 10.2 shows that the CNN testing accuracy was increased from 93.73% (in Table 10.1) to 97.10% (in Table 10.2). This proves that the CNN accuracy performance can be improved with the right configuration.

Another 300 tests were conducted with the CNN configuration mentioned above to determine the performance of the settings. The overall testing results were improved, achieving an average testing rate of 94.61% with the new configuration. However, the average recognition time was about 3.62 seconds, slightly higher than the preliminary experiments. The results are shown in Table 10.3.

TABLE 10.3 CNN Testing Results

	Average Accuracy (%)	**Average Recognition Time (s)**
First 100 tests	95.00	3.83
200 tests	94.50	3.36
300 tests	94.33	3.66
Average results	**94.61**	**3.62**

10.5 CONCLUSION AND FUTURE WORKS

In summary, a simple and fast face recognition standalone application has been created in this project. However, it is not yet integrated into the SmartV3UMS MOOC system, as formal approval from the authorized department is required first. The geometric-based approach, template matching method, and CNN algorithm performances were compared and analyzed. Surprisingly, the CNN performed slightly better than the geometric-based and template matching methods. However, the algorithms' performance varied due to different hardware used. Therefore, the CNN performance in actual testing is still debatable.

The template matching method recognizes faces using extracted features. The template matching accuracy highly depends on the template images used. Face recognition will fail if the intensity difference between the faces and the surrounding region is too small. Failure may also happen if the face image is obscured, if there is an image with a thick beard, moustaches, or conditions with mouth open. Illumination and hardware used could be another factor of failure.

ACKNOWLEDGMENT

This work was partly supported by the Universiti Malaysia Sabah under SDN Grant SDN51-2019.

REFERENCES

1. Y. Shen, M. Yang, B. Wei, C. T. Chou, and W. Hu. Learn to Recognise: Exploring Priors of Sparse Face Recognition on Smartphones. IEEE Transactions on Mobile Computing, vol. 16, no. 6, pp. 1705–1717. 2016. https://doi.org/10.1109/TMC.2016.2593919
2. N. Gholape, A. Gour, and S. Mourya. Finding Missing Person Using ML, AI. Journal of Medical Engineering & Technology, vol. 3, pp. 1517–1520, 2021.
3. F. Al-Muqbali, N. Al-Tourshi, K. Al-Kiyumi, and F. Hajmohideen. Smart Technologies for Visually Impaired: Assisting and conquering infirmity of blind people using AI Technologies. In IEEE 2020 12th Annual Undergraduate Research Conference on Applied Computing (URC), pp. 1–4, 2020. https://doi.org/10.1109/URC49805.2020.9099184
4. V. Singh, R. Shanmugam, and S. Awasthi. Preventing Fake Accounts on Social Media Using Face Recognition Based on Convolutional Neural Network. In Sustainable Communication Networks and Application, pp. 227–241, Springer, Singapore, 2021. https://doi.org/10.1007/978-981-15-8677-4_19
5. J. Valera, J. Valera, and Y. Gelogo. A Review on Facial Recognition for Online Learning Authentication. In IEEE 2015 8th International Conference

on Bio-Science and Bio-Technology (BSBT), pp. 16–19, 2015. https://doi.org/10.1109/BSBT.2015.15
6. S. K. Tiwari, V. Fande, G. N. Patil, K. Kalbande, and D. M. Khanapurkar. A Novel Approach for Attendance Monitoring System with Face Mask Detection. In AIP Conference Proceedings, vol. 2424, no. 1, 2022. https://doi.org/10.1063/5.0077093
7. E. G. Moung, C. C. Wooi, M. M. Sufian, C. K. On, and J. A. Dargham. Ensemble-Based Face Expression Recognition Approach for Image Sentiment Analysis. International Journal of Electrical & Computer Engineering, vol. 12, no. 3, 2022.
8. M. Coşkun, A. Uçar, Ö Yildirim, and Y. Demir. Face Recognition Based on Convolutional Neural Network. In IEEE 2017 International Conference on Modern Electrical and Energy Systems, pp. 376–379, 2017. https://doi.org/10.1109/MEES.2017.8248937
9. R. Brunelli. Template Matching Techniques in Computer Vision: Theory and Practice. John Wiley & Sons. 2009.
10. N. Osia, and T. Bourlai. A Spectral Independent Approach for Physiological and Geometric Based Face Recognition in the Visible, Middle-Wave and Long-Wave Infrared Bands. Image and Vision Computing, vol. 32, no. 11, pp. 847–859, 2014.
11. Y. Han. Reliable Template Matching for Image Detection in Vision Sensor Systems. Sensors. 2021; 21(24):8176. https://doi.org/10.3390/s21248176
12. S. T. Y. Ping, C. H. Weng, and B. Lau. Face Detection Through Template Matching and Color Segmentation. Nevim: Nevim, vol. 89, 2003.
13. T. Y. Chai, M. Rizon, and S. S. Woo, Facial Features for Template Based Face Recognition. American Journal of Applied Science, 2009; vol. 11, no. 7, pp: 1897–1901. https://thescipub.com/abstract/ajassp.2009.1897.1901
14. T. Smita, S. Varsha, and S. Sanjeev. Face Detection Using Combined Skin Color Detector and Template Matching Method. International Journal of Computer Applications, vol. 26, no. 7, pp. 5–8, 2011.
15. K. Amandeep. Face Recognition Using Template Matching. International Journal of Computer Applications, vol. 115, no. 8, pp. 10–13, 2015.
16. L. Dongmei, and C. Wushan. Partial Matching Face Recognition Method for Rehabilitation Nursing Robots Beds. College of Mechanical Engineering, Shanghai University of Engineering Science, Shanghai China, 2016. https://doi.org/10.48550/arXiv.1508.00239
17. X. Deng, F. Da, and H. Shao. Efficient 3d Face Recognition Using Local Covariance Descriptor and Riemannian Kernel Sparse Coding. Computers & Electrical Engineering, vol. 62, pp. 81–91, 2017.
18. A. Abbad, K. Abbad, and H. Tairi. 3D Face Recognition: Multi-Scale Strategy Based on Geometric and Local Descriptors. Computers & Electrical Engineering, vol. 70, pp. 525–537, 2018.
19. L. Shi, X. Wang, and Y. Shen. Research on 3D Face Recognition Method Based on LBP and SVM. Optik, vol. 220, p. 165157, 2020.
20. P. Kamencay, M. Benco, T. Mizdos, and R. Radil. A New Method for Face Recognition Using Convolutional Neural Network. Advances in Electrical

and Electronic Engineering, vol. 15, no. 4, pp. 663–672, 2017. https://doi.org/10.15598/aeee.v15i4.2389

21. D. Chen, S. Zhang, W. Ouyang, J. Yang, and Y. Tai. Person Search Via a Mask-Guided Two-Stream CNN Model. In Proceedings of the European Conference on Computer Vision (ECCV), pp. 734–750, 2018.
22. J. C. Chen, V. M. Patel, and R. Chellappa. Unconstrained Face Verification Using Deep Cnn Features. In 2016 IEEE Winter Conference on Applications of Computer Vision, pp. 1–9, 2016. https://doi.org/10.1109/WACV.2016.7477557
23. A. R. Syafeeza, M. Khalil-Hani, S. S. Liew, and R. Bakhteri. Convolutional Neural Network for Face Recognition With Pose and Illumination Variation. Faculty of Electrical Engineering, Malaysia, vol. 6, no. 1, pp. 44–57, 2014.

CHAPTER 11

Low Illumination Surveillance for Object Detection and Recognition Using Deep Learning Methods

Gan Jun Ming, Chin Kim On, Samsul Ariffin Abdul Karim, Ervin Gubin Moung, and Chi Jing

11.1 INTRODUCTION

Artificial intelligence-powered computer vision is a hot research topic at present because of its potential applications. Image recognition, object detection, image super-resolution, and image generation are some of the major topics in computer vision. All of these applications, from the most basic, such as automotive, to the most complex, such as healthcare, rely on the computer to process, understand, and analyze the situation in order to make a wise decision. In simple terms, computer vision is a technology that teaches computers to replicate human eyes and brain.

Image processing is different from computer vision. Image processing is a subset of computer vision. It is a technique for enhancing an image or extracting relevant information from an image by performing operations such as smoothing, sharpening, contrast adjustment, and stretching. Due to the need for high-quality images, it is necessary to improve low-light images with a limited dynamic range that are produced by computer

DOI: 10.1201/9781003400387-11

vision due to inadequate lighting. There is a growing body of research that shows that image quality can be improved using machine learning algorithms [1]. Deep learning has emerged as a cutting-edge new technology across industries due to its remarkable achievements [2]. Convolutional Neural Network (CNN) [3], Region-based Convolutional Neural Network (R-CNN) [4], and You Only Look Once (YOLO) [5] are well-known deep learning methods used in computer vision. Deep learning is a type of machine learning that teaches computers to learn from experiences, like humans do. It does this by processing data through multiple layers of a neural network. Each layer in the neural network passes a representation of the data to the next layer.

Deep learning has been used effectively in object detection and recognition. Object detection and recognition research is important in a number of industries, including security, defense, and healthcare. Object detection and recognition technologies can be used to verify or identify an object in a digital image or video. A computer algorithm analyzes and matches the distinctive properties of an object, transforms them into a mathematical representation, stores the data, and then compares it with the data already stored in the database for identification purposes.

Object detection and recognition in low-light environments is a challenging task due to variations in viewpoint, deformation, illumination, occlusion, background clutter, and intra-class variability. This study explores the application of deep learning, specifically the YOLOv3 model, for object detection and recognition in low-light settings. To improve the performance of YOLOv3, a systematic approach to fine-tuning was implemented. Finally, a comparison was conducted to determine whether image enhancement techniques are necessary prior to training YOLOv3.

In rest of the chapter, Section 2 discusses the research works related to the proposed algorithms. Section 3 explains the methodology used in the study. Section 4 discusses the research findings and Section 5 concludes the research works.

11.2 RELATED RESEARCH WORKS

The artificial neural networks (ANNs) are made up of many interconnected nodes called neurons and these mimic human brain. The developed ANNs learn from inputs/datasets and optimize the outputs. The input is loaded into the input layer as a multidimensional vector. The hidden layers then take the previous layer's decisions and weigh how a stochastic change within itself affects the final output. This process is known as the learning process.

Various types of ANNs have been proposed since the past few decades, and the application of ANNs has achieved remarkable results in robotics, biometric security, stock market prediction, gaming, and other fields. In 2014, deep learning was introduced. The model is having deeper topology compared to the ANNs. No doubt, it has become a popular model nowadays in object detection and recognition due to its robustness and capability. Deep learning usually consists of multiple hidden layers stacked with each other. The popular deep learning methods are CNN, R-CNN, and YOLO.

11.2.1 Deep Learning Methods

The authors introduced a novel object detection method with VOC 2007 dataset in ref [6]. The proposed method used R-CNN selective search to highlight and select regions. The model does not incorporate multi-scale input, instead it focuses on a single scale for processing. Stochastic gradient descent (SGD) and backpropagation are used to train the model. The applied loss function consists of hinge loss for classification and bounding box regression (BBR) for accurate localization. The model also incorporates a Softmax layer for probability estimation. In terms of performance, the testing mean average precision (mAP) achieved a score of 62.4. However, the processing speed of the model is relatively slow, with 0.03 frames per second (FPS). In subsequent work in ref [7], the authors proposed Fast R-CNN algorithm, resulting in an enhanced performance. They used SDG without backpropagation, which led to an impressive achievement of 65.7 mAP. The processing speed was also slightly improved, reaching a frame rate of 0.06 FPS. Notably, the loss function was modified to incorporate class log loss in addition to BBR, as opposed to the previous hinge loss + BBR approach.

In 2016, the Faster R-CNN algorithm was proposed by the authors of ref [8]. This algorithm incorporates a region proposal network (RPN) to enhance the performance of object detection. It incorporates multi-scale input which allows it to better adapt to different image scales. SGD is used as the learning method which enables efficient optimization of the model. The loss function employed is class log loss + BBR which enhances the accuracy of object classification and localization. The inclusion of Softmax layer further enhances model's capabilities. During testing, the Faster R-CNN achieved an mAP of 69.9 on the VOC 2007 dataset and 51.9 on the COCO dataset. With a frame rate of 0.4 FPS, the Faster R-CNN demonstrates efficient processing speed while maintaining high accuracy in object detection.

The YOLO algorithm was proposed by Joseph in 2016 using VOC 2007 + 2012 dataset [9]. The YOLO algorithm is a popular approach for real-time object detection. Unlike other algorithms, YOLO does not use a region proposal mechanism. Instead, it takes a single image as input and divides it into a grid, predicting bounding boxes and class probabilities directly. YOLO does not consider multiple scales during the detection process. SGD is used as the learning method to optimize the model parameters. The loss function employed is a combination of class sum-squared error loss, BBR, object confidence, and background confidence, which ensures accurate detection and classification. The inclusion of a Softmax layer further enhances the model's capabilities. In testing, YOLO achieved an mAP of 63.6 with 45 FPS, and YOLO exhibits a high processing speed, rendering it well-suited for real-time applications.

In the same year, Joseph introduced YOLOv2 [10], an improved version of the YOLO algorithm. YOLOv2 demonstrated significant advancements in object detection performance. It achieved an mAP of 78.6 on the VOC 2007 + 2012 dataset and 48.1 on the COCO dataset. In addition, YOLOv2 showcased a competitive processing speed, operating at 40 FPS. These advancements in both accuracy and speed made YOLOv2 a highly effective solution for real-time object detection applications.

In 2018, Joseph introduced YOLOv3 [11]. YOLOv3 brought further improvements in object detection and recognition. The algorithm utilized a comprehensive loss function that combined class sum-squared error loss, BBR, object confidence, and background confidence. With the inclusion of a Softmax Layer, YOLOv3 enhanced the precision and accuracy of object classification. In terms of performance, YOLOv3 achieved a testing mAP of 57.9 on the COCO dataset and an impressive 88.2 on the PASCAL VOC 2012 dataset. In addition, YOLOv3 demonstrated exceptional processing speed, operating at 78 FPS. These advancements made YOLOv3 a highly effective and efficient solution for real-time object detection tasks.

11.3 METHODOLOGY

11.3.1 Image Acquisition

In this project, we utilized the ExDark image dataset which comprises a total of 7,363 images across 12 classes. Table 11.1 shows the details of the dataset.

For the experiment, the dataset was divided into three sets: training data (250 images per class), testing data (2,500 images), and validation data (150 images per class).

TABLE 11.1 ExDark Image Dataset (List of Data from ExDark)

Objects	Total Data
Bicycle	652 images
Boat	679 images
Bottle	547 images
Bus	547 images
Car	638 images
Cat	735 images
Chair	648 images
Cup	519 images
Dog	801 images
Motorbike	503 images
People	609 images
Table	505 images

11.3.2 Image Preprocessing

Image preprocessing consists of two parts: image enhancement and feature extraction. The image enhancement phase involves using the Python-PILLOW library, which is an image processing library that supports image manipulation, opening, and saving. It enables the enhancement of images by adjusting brightness, contrast, and color noise to improve the learning capabilities of the deep learning neural network. Various functions, such as blur, contour, detail, edge enhance, emboss, find edge, smooth, and sharpen, are used for image enhancement. To ensure smooth training of the deep neural network, all the collected images need to adhere to the instructions and requirements of Faster R-CNN and YOLO based on the TensorFlow API. Dataset analysis is performed using the LabelImg tool, a graphical image annotation tool that allows labeling object bounding boxes in images. LabelImg supports PASCAL VOC XML or YOLO text file formats, enabling data compatibility with Faster R-CNN and YOLO neural networks.

During the dataset review, images with low illumination were adjusted for brightness and contrast using the PILLOW library in Python. This process improved the clarity of the objects in the images and facilitated easier labeling. Images that showed minimal difference after enhancement were removed, while additional images were added to the dataset through smartphone captures or downloads from the internet to maintain balance. Ultimately, the dataset consisted of 5,686 images across 20 classes.

After labeling the images, the ground-truth information was defined, consisting of a total of 10,830 objects belonging to 19 classes. These classes include persons, cars, bottles, chairs, bicycles, cups, cats, dogs, buses, motorbikes, tables, laptops, monitors, signboards, traffic lights, books, bags, smartphones, and rubbish bins.

11.3.3 Data Filtering

During the labeling process, some images may have outliers in the bounding box information, such as negative values for the x_min coordinate. These outliers are considered abnormal and not allowed in the neural network. To address this issue, the dataset is checked for such anomalies, and any affected data is removed and the corresponding images are relocated to a separate folder.

11.3.4 Object Detection and Recognition

The captured image or video is converted into frames and fed into the neural network. A Python program is created with a pre-trained neural network model. If the confidence score is less than 0.5, the detection process is terminated. However, if the confidence score exceeds 0.5, a bounding box is drawn and the object information is displayed on the frame. Figure 11.1 illustrates the process of detection and recognition.

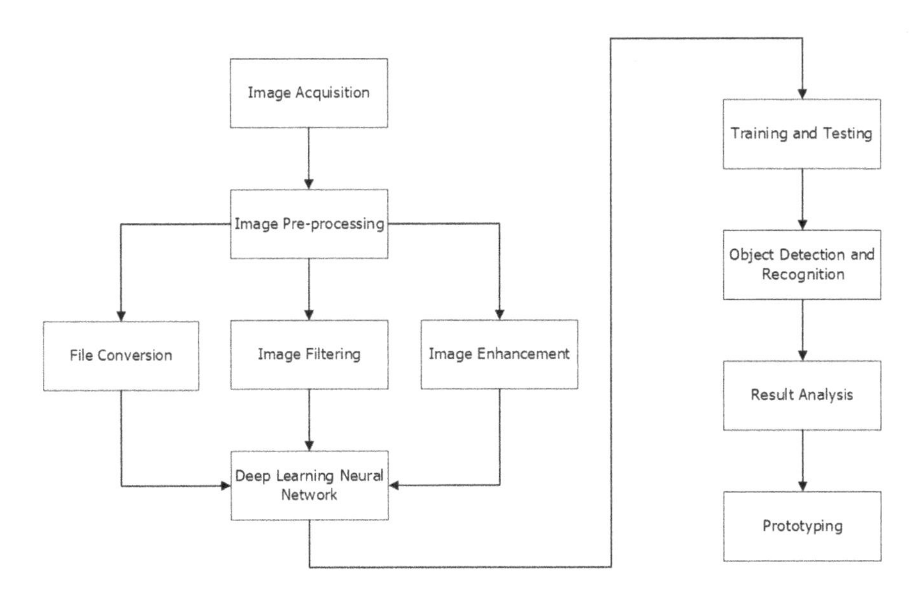

FIGURE 11.1 The proposed methodology.

11.3.5 Performance Metric

Each model's performance was evaluated using a validation and testing dataset. The mAP was used to measure and analyze the overall result. The ground truth consists of the image, the classes of the objects in it, and the true bounding box of each object in that image. Intersection over Union (IoU) metrics were used to determine the correctness of a given bounding box. It is calculated as the ratio of the intersection between the predicted boxes and the ground truth to the union of the two. The F-score was calculated based on the Precision and Recall formulas.

11.4 RESULTS AND DISCUSSIONS

11.4.1 Comparison of Faster R-CNN and YOLOv3

In this experiment, the main difference was the dataset shuffling, which was done to compare the performance of Faster R-CNN and YOLOv3. The training process used a GTX 1050TI GPU with limited memory, leading to a batch size of 1. The dataset was split into 75% for training and 25% for testing dataset. Prediction of anchor boxes was carried out using the K-means clustering algorithm.

The experiment yielded predictions for both the aspect ratios of images and the anchor boxes. The aspect ratios of the images were as follows: 0.35, 0.38, 0.45, 0.48, 0.89, 0.93, 0.94, 1.08, and 1.19. These values represent the width-to-height ratios of the images in the dataset. On the other hand, the anchor box predictions were obtained as follows: 69.55, 181.35, 235.34077618, 198.44812403, 22.75, 51.09638243, 149.90625, 139.04814305, 40.3, 113.73399015, 56.55, 60.30769231, 284.740625, 319.3650733, 126.815, 266.48843416, 91.52, and 98.45582724. Object detection algorithms utilize these anchor box values to define the bounding boxes around objects of interest in images.

The preliminary results indicate that Faster R-CNN achieved an mAP of 35.71%, while YOLOv3 achieved 39.18%. After shuffling the dataset, Faster R-CNN achieved an mAP of 38.20%, and YOLOv3 achieved 42.19%. These findings demonstrate that YOLOv3 outperformed Faster R-CNN as the current state-of-the-art deep learning algorithm for object detection and recognition. The superior performance of YOLOv3 can be attributed to several factors, including the number of epochs, mini batch size, stability, and optimal learning rate.

TABLE 11.2 Fine-Tuning Epochs (Own Data. No Copyright Issue)

Epochs	20	30	40	50
YOLOv3	47.01 mAP	50.77 mAP	53.65 mAP	55.25 mAP

11.4.2 Fine-Tuning Epochs

We used the parameter setting proposed in ref [12]. We varied the number of epochs in relation to the mini batch size, testing four different epochs: 20, 30, 40, and 50. Table 11.2 shows the training results and demonstrates the improvements achieved.

The results presented in Table 11.2 show an increasing trend in mAP as the number of epochs is varied in different experiments. Notably, the best result of 55.25% mAP was achieved with 50 epochs. However, it is important to note that while the overall mAP increased with the increase in epoch number, there was a decrease in the detection performance for some objects. This suggests that the stability of the YOLOv3 architecture may not be consistent. Therefore, we investigated the stability of the YOLOv3 architecture in the next experiment.

11.4.3 YOLOv3 Stability Improvement

The inconsistency observed in the experiments can be attributed to the nature of YOLOv3's supervised learning approach. YOLOv3 uses SGD as its learning method which automatically learns a mapping function from the input data. During the learning process, small algorithmic decisions are made which can vary randomly. As a result, there may be slight variations in the precision-recall rate for each class, leading to inconsistent mAP values. To address this, the experiment was conducted using the parameter setting as shown in Table 11.3.

It can be observed that there is a marginal change in the mAP during the training process when the duration of both the first and second stage training is extended. The experiment used 100 epochs, with 50 epochs for the first stage training and 50 epochs for the second stage training. This resulted in a higher mAP of 55.18% for the YOLOv3. However, it is worth noting that the results across different epochs exhibit a slight fluctuation of approximately 1% to 2% in the mAP. This phenomenon prompts further investigation to determine the underlying factors contributing to this observation within the context of the experiment.

TABLE 11.3 Parameter Setting for Stability Investigation (Own Data. No Copyright Issue)

Train-test split	**Train set (80%), test set (20%)**
Batch size	1
Decay	0.9995
Warm-up epoch	10 epochs
Total training epochs	20, 40, 60, 80, 100 epochs
Learning rate	0.0001 to 0.000001
Training first stage epoch	10, 20, 30, 40, 50 epochs
Training input size	Random
IoU threshold	0.45
Confidence score value	0.3
Pretrained weight	YOLOv3-608 with COCO travel

The next experiment focused on the learning rate during the SDG. The default learning rate of 0.01 were tested against a range of values from 0.0001 to 0.000001 to determine the optimal learning rate for the low-illumination dataset. This investigation aims to improve the stability and consistency of the YOLOv3 model's performance in object detection and recognition tasks.

The result is tabulated in Table 11.4 which shows the results of the impact of different learning rate ranges on the mAP values. The highest mAP% of 55.59 was achieved for the learning rate range of 0.00001 to 0.0000001, followed by 55.18 for the range of 0.0001 to 0.000001. Conversely, the lowest mAP% of 25.67 was observed for the range of 0.000001 to 0.00000001. However, these results do not provide strong evidence to demonstrate a significant effect of learning rate in YOLOv3. To further investigate the effect of learning rate, we conducted t-tests with equal variances to determine if there is a significant difference between the learning rate ranges of 0.0001–0.000001 and 0.00001–0.0000001. The t-test results indicate that the learning rate range of 0.00001–0.0000001 is not significantly different from the learning rate range of 0.0001–0.000001. Therefore, we can accept the null hypothesis that there is no significant difference between the two learning rate ranges.

TABLE 11.4 Learning Rate Comparison (Own Data. No Copyright Issue)

Learning Rates	**0.01–0.0001**	**0.001–0.00001**	**0.0001–0.000001**	**0.00001–0.0000001**	**0.000001–0.00000001**
YOLOv3	49.73	54.79	55.18	55.59	25.67

11.4.4 Summary of CNN Comparison

YOLOv3 achieves optimal performance in object detection and recognition tasks when trained with specific settings. These settings include:

- 50 epochs for the first stage training,
- 50 epochs for the second stage training,
- A learning rate ranging from 0.00001 to 0.0000001, and
- A confidence score value of 0.2.

However, slight adjustments may be required based on the specific training environment and dataset being used. For example, the experiment on the confidence score value revealed that a value of 0.2 achieves higher results with an IoU of 0.45.

11.4.5 Testing with Different Image Enhancement Algorithms

The objective of this experiment is to assess the effectiveness of training with a low-illumination dataset in detecting and recognizing objects in both low-illumination and normal environments. To achieve this, we compare the model's inference on normal low-illumination images to those enhanced using the Retinex theory, Multi-Scale Retinex with Color Restoration (MSRCR), the Low-light Image Enhancement via Illumination Map Estimation (LIME) algorithm, and the Multi-scale Boosted Local Laplacian-based Enhancement Network (MBLLEN) algorithm. The goal is to determine the impact of these image enhancements on the model's performance in low-illumination conditions. Table 11.5 presents the experimental results for ten objects in each class during the testing phase.

Table 11.5 shows that image enhancement techniques such as Retinex theory, LIME, and MBLLEN can improve image quality under low-illumination conditions. However, the MSRCR technique did not yield significant improvements. Interestingly, when the models were specifically trained on low-illumination images, they demonstrated accurate detection performance even without the use of image enhancement techniques. This suggests that proper training and fine-tuning of the neural networks are crucial factors for achieving good performance. It is important to note that poorly optimized settings can adversely affect the performance of trained neural networks. Therefore, while image enhancement techniques can be

TABLE 11.5 Testing Results with Different Image Enhancement Algorithms (Own Data. No Copyright Issue)

Objects	Recognition Rate (%)				
	YOLOv3	YOLOv3 + Retinex	YOLOv3 + MSRCR	YOLOv3 + LIME	YOLOv3 + MBLLEN
Bus	90	90	70	90	90
Bicycle	100	100	90	100	100
Cat	90	90	70	90	90
Bottle	100	100	90	100	100
Car	100	100	80	100	100
Monitor	90	100	90	100	100
Chair	100	100	100	90	100
Laptop	100	90	90	100	100
Cup	100	100	80	100	100
Table	100	100	100	100	90
Person	90	100	100	100	100
Signboard	100	100	100	100	90
Dog	90	100	70	100	100
Traffic light	100	100	100	90	100
Bag	100	100	70	100	90
Handphone	100	100	100	100	100
Motorbike	100	90	90	100	100
Rubbish bin	90	90	60	80	90
Book	100	100	100	100	90

beneficial, they may not be necessary if the neural networks are trained effectively with appropriate settings.

11.5 CONCLUSION AND FUTURE WORKS

The main contribution of this chapter is the investigation of applying a low-illumination dataset to a deep learning neural network, with a specific focus on the stability of YOLOv3. The ExDark dataset, which is based on a low-illumination environment, was used for the learning process. Parameter tuning for enhancement was designed to improve the learning process. Later, the performance of YOLOv3 was tested with different image enhancement algorithms, such as Retinex theory, MSRCR, LIME, Retinexnet, and the MBLLEN algorithm, for comparison.

There are some limitations to this project. First, there is a limitation based on YOLOv3. When testing on the low-illumination dataset, if multiple objects are close to each other in an image and have inconsistent sizes, there may be issues with missed detection and recognition. YOLOv3 struggles to detect extremely small and tiny objects in images or videos. In other words, the model trained in this project can detect objects of normal size, but it may fail to detect objects that are too small. In addition, there is a limitation regarding the prototyping of the surveillance system. The transfer of video streaming from the Raspberry Pi to the HTTP requesting web causes high latency, resulting in a delay of approximately 10 seconds in the video streaming.

In future research, there is a need to enhance the low-illumination dataset to match the scale of the PASCAL VOC dataset in order to achieve better results. Furthermore, it would be beneficial to explore the latest algorithms, such as the enhanced version of the YOLO family, YOLOv6.

ACKNOWLEDGMENT

This work was partly supported by the Universiti Malaysia Sabah under SDN Grant SDN51-2019.

REFERENCES

1. Hayat, S., Kun, S., Tengtao, Z., Yu, Y., Tu, T., & Du, Y. (2018). A deep learning framework using convolutional neural network for multi-class object recognition. 2018 IEEE 3rd International Conference on Image, Vision and Computing (ICIVC) (pp. 194–198). IEEE.
2. Liu, L., Ouyang, W., Wang, X., Fieguth, P., Chen, J., Liu, X., & Pietikäinen, M. (2020). Deep learning for generic object detection: A survey. International Journal of Computer Vision, 128, 261–318.
3. Chavda, H. K., & Dhamecha, M. (2017). Moving object tracking using PTZ camera in video surveillance system. 2017 International Conference on Energy, Communication, Data Analytics and Soft Computing (ICECDS) (pp. 263–266). IEEE.
4. Kanimozhi, S., Gayathri, G., & Mala, T. (2019). Multiple real-time object identification using single shot multi-box detection. 2019 International Conference on Computational Intelligence in Data Science (ICCIDS) (pp. 1–5). IEEE.
5. Kumar, S., Balyan, A., & Chawla, M. (2017). Object detection and recognition in images. Computer Science & Engineering Department, Maharaja Surajmal Institute of Technology, New Delhi, India. International Journal of Engineering Development and Research 2017. pp. 1029–1034.
6. Girshick, R., Donahue, J., Darrell, T., & Malik, J. (2014). Rich feature hierarchies for accurate object detection and semantic segmentation Tech report (v5). UC Berkeley. ArXiv e-prints.
7. Girshick, R. (2015). Fast R-CNN. In Proceedings of the IEEE International Conference on Computer Vision (pp. 1440–1448).
8. Ren, S., He, K., Girshick, R., & Sun, J. (2015). Faster R-CNN: Towards real-time object detection with region proposal networks. In Proceedings

of the 28th International Conference on Advances in Neural Information Processing Systems (pp. 91–99).

9. Redmon, J., Divvala, S., Girshick, R., & Farhadi, A. (2016). You only look once: Unified, real-time object detection. In Proceedings of the IEEE Conference on Computer Vision and Pattern Recognition (pp. 779–788).
10. Redmon, J., & Farhadi, A. (2017). YOLO9000: Better, faster, stronger. In Proceedings of the IEEE Conference on Computer Vision and Pattern Recognition (pp. 7263–7271).
11. Redmon, J., & Farhadi, A. (2018). Yolov3: An incremental improvement. https://doi.org/10.48550/arXiv.1804.02767.
12. Qu, H., Yuan, T., Sheng, Z., & Zhang, Y. (2018). A pedestrian detection method based on yolov3 model and image enhanced by retinex. In 2018 11th International Congress on Image and Signal Processing, Biomedical Engineering and Informatics (CISP-BMEI) (pp. 1–5). IEEE.

CHAPTER 12

Intelligent System Design for the Solutions of Nonlinear Diffusion in the Two-Dimensional Porous Medium

Jackel Vui Lung Chew, Elayaraja Aruchunan, Andang Sunarto, and Jumat Sulaiman

12.1 INTRODUCTION

An intelligent system is an advanced computer system that can collect data from its surrounding, analyze them, and produce the desired result. It can also collaborate with agents like users or other computer systems or learn from adaptation using initial data. Its application and integration with other fields have benefited academics and industries. In recent years, scientific machine learning has emerged as a new and fast-developing field combining scientific computing and machine learning. One of the aspects of scientific machine learning that the increasing number of researchers focus on is enhancing algorithms to solve partial differential equations (PDEs) with machine learning techniques [1–4].

Although traditional methods such as finite difference method and finite element method are frequently used by mathematicians to solve PDEs, the data collected from numerical and physical science experiments serve as the basis for developing an intelligent system. The previously

DOI: 10.1201/9781003400387-12

mentioned traditional methods are undeniably straightforward to implement, and the mathematical theories behind these methods, such as stability and convergence, are very well proven. However, these traditional methods are intensive in computation and need expensive supercomputers to run the computation. It is time-consuming to develop better improvements on these methods. In addition, the accuracy of the solution depends heavily on the fineness of the solution grid. The solution procedure needs to use a finer grid for the numerical computation in order to obtain more accurate solutions. Thus the process takes more intense computations and wastes valuable time.

Following the emergence of scientific machine learning and the rising issues with traditional methods, this chapter aims to develop an efficient algorithm using the explicit decoupled group successive over-relaxation iterative method. The proposed algorithm becomes the main component of designing an intelligent system for solving nonlinear diffusion problems in the two-dimensional porous medium. This chapter is organized as follows: Section 2 illustrates the targeted nonlinear problem in brief; Section 3 presents the numerical method to solve the targeted nonlinear problem; and Section 4 shows the design of an intelligent system to obtain the solutions to nonlinear problems in the two-dimensional porous medium. Section 5 shows the numerical results and discusses the computational efficiency and a concluding remark is stated in Section 6.

12.2 NONLINEAR DIFFUSION IN A TWO-DIMENSIONAL POROUS MEDIUM

This chapter focuses on the nonlinear diffusion equation in a two-dimensional porous medium that is expressed in the form of

$$u_t = \alpha\left(\left(D(u)u_x\right)_x + (D(u)u_y)_y\right), \tag{12.1}$$

where α and $D(u)$ represent the constant and diffusion function, respectively. The value of α can be determined by considering either physical or mathematical properties to the targeted nonlinear problem. The calculation to get the actual value of α depends on the physical system, the nature of phenomenon, and the constraint imposed on the main equation. For the sake of simplicity, this study emphasizes on arbitrary numbers to substitute into α. Moreover, Eq. (12.1) exists in various applications in natural sciences and the equation has demonstrated an excellent fitting to many natural phenomena. For instance, David and Ruan [5] studied a mathematical model of tumor growth that describes the evolution of the cell population density using Eq. (12.1) with source terms. They also investigated the stability of upwind finite difference approximation to this model variant. Then, Muller and Sonner [6] expanded the theory of the well-posedness of a class of Eq. (12.1), which was motivated by the importance of the biofilm growth model. In addition, Mary et al. [7] investigated the nonlinear problem in unsteady isothermal gas flow through a semi-infinite medium formulated using Eq. (12.1). They proposed an analytical approach to obtain the expression of gas flow. In another article, Pop et al. [8] introduced a nonstandard line method for solving Eq. (12.1) in the unsteady gas flow in a nano-porous medium setting. Besides that, Palencia [9] proposed an analytical approach to model flame propagation using Eq. (12.1). They described the characteristics of a flame using the porous medium equation theory [10]. The solution of Eq. (12.1) contributes significantly to phenomena that involve nonlinear diffusion in a porous medium setting. Hence, using the knowledge of the advanced computer, an intelligent system that can compute the solution of Eq. (12.1) is crucial so that complex nonlinear diffusion in a porous medium can be solved accurately and fast.

12.3 NUMERICAL METHOD

As the main component of an intelligent system for solving nonlinear diffusion problems in the two-dimensional porous medium, this section discusses the numerical method used to develop the important computational algorithm. This study proposes a numerical method that combines the explicit decoupled group (EDG) strategy and successive over-relaxation (SOR) iterative method. The EDG strategy is one of the most effective methods for reducing the computing difficulty of solving a large-scale system of equations resulting from the discretization of PDEs [11–16]. Thus, this method uses the EDG computation strategy and then extends its performance by adding the SOR linear solver.

To begin the construction of the so-called EDGSOR method for solving Eq. (12.1), let the solution of Eq. (12.1) be denoted by $u(x,y,t)$. Then, we aim to approximate the solution under the following problem specification.

$$u\left(x,y,t_0\right)=F_0, 0\leq x\leq X, 0\leq y\leq Y, \tag{12.2}$$

$$u(0,y,t)=F_a, u(X,y,t)=F_b, 0\leq y\leq Y, 0\leq t\leq t_f, \tag{12.3}$$

and

$$u(x,0,t)=F_c, u(x,Y,t)=F_d, 0 \le x \le X, 0 \le t \le t_f. \tag{12.4}$$

Eq. (12.2) represents the initial condition imposed to the problem which also acts as the initial data values for the diffusion process. Meanwhile, Eqs. (12.3) and (12.4) set the boundaries for the diffusion process. Dirichlet boundary conditions are used in this study so that the values of $u(x,y,t)$ are fixed on the boundary. The purpose of using Dirichlet boundary condition is to ensure that the behavior of the solutions can be limited on the boundary.

To derive the required finite difference-based approximation to the solution of Eq. (12.1), let the solution of Eq. (12.1) be distributed throughout an available domain and denoted by $u(x_p, y_q, t_n)$ where $p=1,2,\ldots,M-1$, $q=1,2,\ldots,M-1$, and $n=1,2,\ldots,N-1$. Then, let the approximation to the solution be represented by $U^n_{p,q}$. This method uses a uniform two-dimensional mesh with the spatial step size $h=(b-a)/M=(d-c)/M$ and temporal step size $k=(t_f-t_0)/N$. Adopting the finite difference scheme proposed by Abdullah [11], a nonlinear finite difference-based approximation to Eq. (12.1) can be derived into:

$$\begin{aligned} G^n_{p,q} &= U^n_{p,q} - A_1(U^n_{p,q})^m U^n_{p+1,q+1} + 2A_1(U^n_{p,q})^{m+1} \\ &- A_1(U^n_{p,q})^m U^n_{p-1,q-1} - A_2 m(U^n_{p,q})^{m-1}(U^n_{p+1,q+1})^2 \\ &+ 2A_2 m(U^n_{p,q})^{m-1} U^n_{p+1,q+1} U^n_{p-1,q-1} - A_2 m(U^n_{p,q})^{m-1}(U^n_{p-1,q-1})^2 \\ &- A_3(U^n_{p,q})^m U^n_{p-1,q+1} + 2A_3(U^n_{p,q})^{m+1} - A_3(U^n_{p,q})^m U^n_{p+1,q-1} \\ &- A_4 m(U^n_{p,q})^{m-1}(U^n_{p-1,q+1})^2 + 2A_4 m(U^n_{p,q})^{m-1} U^n_{p-1,q+1} U^n_{p+1,q-1} \\ &- A_4 m(U^n_{p,q})^{m-1}(U^n_{p+1,q-1})^2 - U^{n-1}_{p,q}, \end{aligned} \tag{12.5}$$

where $A_1 = A_3 = \alpha k/2h^2$ and $A_2 = A_4 = \alpha k/8h^2$.

Then, taking all grid points from the available domain, a large-scale system of nonlinear equations can be formed,

$$\underset{\sim}{G}^n = 0, \tag{12.6}$$

where $\underset{\sim}{G}^n = \left(G^n_{1,1},\ldots,G^n_{1,M-1},\ldots,G^n_{M-1,1},\ldots,G^n_{M-1,M-1}\right)$. Eq. (12.6) is the cause of extensive computation in solving Eq. (12.1). The selection of the method to solve Eq. (12.6) plays an important in achieving accurate solutions.

This study attempts to obtain the solution of Eq. (12.6) using the EDGSOR method with a linearization approach. Eq. (12.6) can be transformed into a system of linear equations using Newton's linearization procedure, which yields:

$$J_G^n \underset{\sim}{W}^n = -\underset{\sim}{G}^n, \tag{12.7}$$

where J_G^n is a $(M-1)\times(M-1)$ coefficient matrix obtained by computing the Jacobian matrix based on Eq. (12.6), $\underset{\sim}{W}^n$ is the corrector vector that needs to be solved through iteration, and $-\underset{\sim}{G}^n$ serves as the right-hand values of the whole system of equations. In addition, the corrector vector in Eq. (12.7) is obtained using the following equation:

$$\left(\underset{\sim}{W}^n\right)^{(i)} = \left(\underset{\sim}{U}^n\right)^{(i)} - \left(\underset{\sim}{U}^n\right)^{(i-1)}, i = 1,2,\ldots, \tag{12.8}$$

where $\left(\underset{\sim}{W}^n\right)^{(i)} = \left(W_{1,1}^n,\ldots, W_{1,M-1}^n,\ldots,W_{M-1,1}^n,\ldots,W_{M-1,M-1}^n\right)^{(i)}$ and $\left(\underset{\sim}{U}^n\right)^{(i)} = \left(U_{1,1}^n,\ldots,U_{1,M-1}^n,\ldots,U_{M-1,1}^n,\ldots,U_{M-1,M-1}^n\right)^{(i)}$.

Next, let us consider two cases to form decoupled groups of equations.
Case 1: Take the equations

$$G_{p,q}^n = a_{p,q}W_{p+1,q-1}^n + b_{p,q}W_{p-1,q-1}^n + c_{p,q}W_{p,q}^n + d_{p,q}W_{p+1,q+1}^n + e_{p,q}W_{p-1,q+1}^n, \tag{12.9}$$

and

$$\begin{aligned} G_{p+1,q+1}^n &= a_{p+1,q+1}W_{p+2,q}^n + b_{p+1,q+1}W_{p,q}^n + c_{p+1,q+1}W_{p+1,q+1}^n \\ &\quad + d_{p+1,q+1}W_{p+2,q+2}^n + e_{p+1,q+1}W_{p,q+2}^n. \end{aligned} \tag{12.10}$$

An equivalent 2×2 matrix equation can be formed as:

$$A_1^n \underset{\sim}{W}_1^n = \underset{\sim}{S}_1^n, \tag{12.11}$$

where

$$A_1^n = \begin{bmatrix} c_{p,q} & d_{p,q} \\ b_{p+1,q+1} & c_{p+1,q+1} \end{bmatrix}^n, \tag{12.12}$$

$$\underset{\sim}{W}_1^n = \begin{bmatrix} W_{p,q} \\ W_{p+1,q+1} \end{bmatrix}^n, \tag{12.13}$$

$$\underset{\sim}{S}_1^n = \begin{bmatrix} S_{p,q} \\ S_{p+1,q+1} \end{bmatrix}^n, \tag{12.14}$$

$$S_{p,q} = G_{p,q} - a_{p,q}W_{p+1,q-1} - b_{p,q}W_{p-1,q-1} - e_{p,q}W_{p-1,q+1}, \tag{12.15}$$

and

$$S_{p+1,q+1} = G_{p+1,q+1} - a_{p+1,q+1}W_{p+2,q} - d_{p+1,q+1}W_{p+2,q+2} - e_{p+1,q+1}W_{p,q+2}. \tag{12.16}$$

Case 2: Take another two equations

$$\begin{aligned} G_{p+1,q}^n = a_{p+1,q}W_{p+2,q-1}^n + b_{p+1,q}W_{p,q-1}^n + c_{p+1,q}W_{p+1,q}^n + d_{p+1,q}W_{p+2,q+1}^n \\ + e_{p+1,q}W_{p,q+1}^n, \end{aligned} \tag{12.17}$$

and

$$\begin{aligned} G_{p,q+1}^n = a_{p,q+1}W_{p+1,q}^n + b_{p,q+1}W_{p-1,q}^n + c_{p,q+1}W_{p,q+1}^n + d_{p,q+1}W_{p+1,q+2}^n \\ + e_{p,q+1}W_{p-1,q+2}^n. \end{aligned} \tag{12.18}$$

An equivalent 2×2 matrix equation can be formed into

$$A_2^n \underset{\sim}{W}_2^n = \underset{\sim}{S}_2^n, \tag{12.19}$$

where

$$A_2^n = \begin{bmatrix} c_{p+1,q} & e_{p+1,q} \\ a_{p,q+1} & c_{p,q+1} \end{bmatrix}^n, \tag{12.20}$$

$$\underset{\sim}{W}_2^n = \begin{bmatrix} W_{p+1,q} \\ W_{p,q+1} \end{bmatrix}^n, \tag{12.21}$$

$$\underset{\sim}{S}_2^n = \begin{bmatrix} S_{p+1,q} \\ S_{p,q+1} \end{bmatrix}^n, \tag{12.22}$$

$$S_{p+1,q} = G_{p+1,q} - a_{p+1,q}W_{p+2,q-1} - b_{p+1,q}W_{p,q-1} - d_{p+1,q}W_{p+2,q+1}, \quad (12.23)$$

and

$$S_{p,q+1} = G_{p,q+1} - b_{p,q+1}W_{p-1,q} - d_{p,q+1}W_{p+1,q+2} - e_{p,q+1}W_{p-1,q+2}. \quad (12.24)$$

Hence, by finding the inverse of both Eqs. (12.11) and (12.19) and adding SOR linear solver, the proposed EDGSOR method to solve Eq. (12.1) can be expressed as:

$$\left(\underset{\sim}{W_1^n}\right)^{(i)} = (1-\omega)\left(\underset{\sim}{W_1^n}\right)^{(i-1)} + \omega\left(A_1^n\right)^{-1}\underset{\sim}{S_1^n}, \quad (12.25)$$

and

$$\left(\underset{\sim}{W_2^n}\right)^{(i)} = (1-\omega)\left(\underset{\sim}{W_2^n}\right)^{(i-1)} + \omega\left(A_2^n\right)^{-1}\underset{\sim}{S_2^n}, \quad (12.26)$$

where $1.0 < \omega < 2.0$ is a preset parameter to optimize the convergence of the solutions.

12.4 DESIGN OF INTELLIGENT SYSTEM

This section discusses the intelligent system design for solving Eq. (12.1) under the problem specifications shown in Eqs. (12.2), (12.3), and (12.4). The design begins with the fundamental architecture based on the requirements and formulated equations. This study uses C++ to convert the architecture into an executable program before the proposed intelligent system is designed. Figure 12.1 shows the architecture of the EDGSOR method

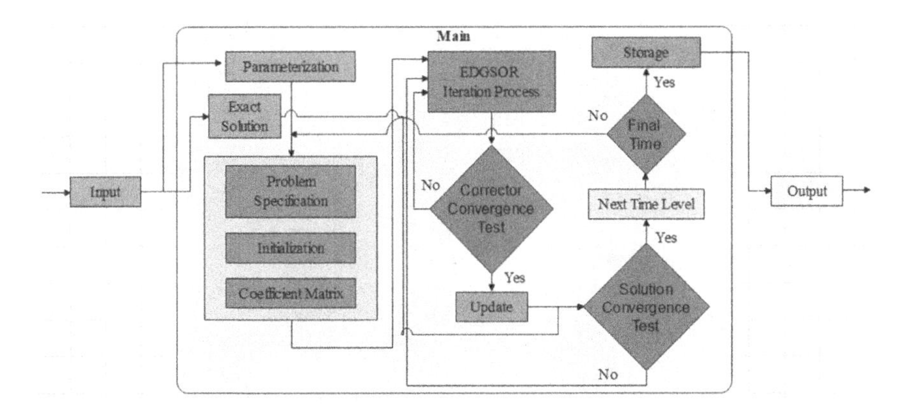

FIGURE 12.1 Fundamental architecture of the EDGSOR method.

TABLE 12.1 EDGSOR Algorithm

Set the values of F_0, F_a, F_b, F_c, and F_d
Set initial values $(\underset{\sim}{U}^n)^{(i=1)} = 1.0$, and $\left(\underset{\sim}{W}_1^n\right)^{(i=1)} = \left(\underset{\sim}{W}_2^n\right)^{(i=1)} = 0$
While $n < N$ Do
    While $\left|\left(\underset{\sim}{U}^n\right)^{(i)} - \left(\underset{\sim}{U}^n\right)^{(i-1)}\right| > 10^{-10}$ Do
        While $\left|\left(\underset{\sim}{W}^n\right)^{(i)} - \left(\underset{\sim}{W}^n\right)^{(i-1)}\right| > 10^{-10}$ Do
            $\left(\underset{\sim}{W}_1^n\right)^{(i)} = (1-\omega)\left(\underset{\sim}{W}_1^n\right)^{(i-1)} + \omega\left(A_1^n\right)^{-1} \underset{\sim}{S}_1^n$
            $\left(\underset{\sim}{W}_2^n\right)^{(i)} = (1-\omega)\left(\underset{\sim}{W}_2^n\right)^{(i-1)} + \omega\left(A_2^n\right)^{-1} \underset{\sim}{S}_2^n$
            Run inner i++
        End
        Compute $\left(\underset{\sim}{W}^n\right)^{(i)} + \left(\underset{\sim}{U}^n\right)^{(i-1)} = \left(\underset{\sim}{U}^n\right)^{(i)}$
        Total $i = i +$ total inner i
    End
End
Display output.

to solve Eq. (12.1). Then, this study refers to Figure 12.1 to develop a C++ executable program using Dev-C++ software. The algorithm used in the developed program can be referred to in Table 12.1. Based on Table 12.1, the EDGSOR algorithm consists of two iteration cycles namely the inner and outer cycles. The inner cycle runs the EDGSOR method to obtain the minimized values of $\underset{\sim}{W}_1^n$ and $\underset{\sim}{W}_2^n$, while the outer cycle computes the numerical solutions after convergence criterion is satisfied.

Next, the design of the proposed intelligent system to solve Eq. (12.1) using the developed EDGSOR method is shown in Figure 12.2. Based on Figure 12.2, the proposed system starts with the data source which

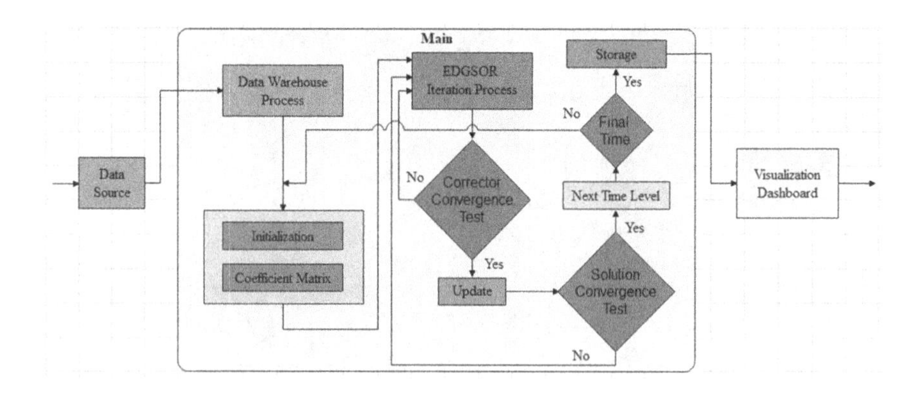

FIGURE 12.2 Proposed intelligent system using the EDGSOR method.

provides the required input. The data source can be obtained from a physical science experiment involving nonlinear diffusion processes. Then, the obtained data is transferred to the data warehouse through various processes according to the need, such as data extraction from one or more sources, cleaning to remove any noise, and reducing redundancy and loading. Next, "clean" data from the warehouse undergo the initialization of dummy solution and right-hand side values and construct a large-scaled coefficient matrix. EDGSOR module will be run to compute the solution before the solution is visualized on a dashboard. Due to unavailability and limited access to the required data source, this study will only investigate the computational and program efficiency using the execution of the developed C++ program of the proposed method.

12.5 NUMERICAL RESULTS AND FINDINGS

This section presents the numerical results of solving three test nonlinear diffusion problems that are described in Problems 12.1, 12.2, and 12.3 using the proposed method. The performance of the EDGSOR method is compared against Chew et al.'s method [15], Lung et al.'s method [17], and the benchmark, Newton–Gauss–Seidel (NGS) method. Comparing these methods focuses on both the computational efficiency and program efficiency. This study uses one of the common metrics to compare the program efficiency (PE) with the formula given as:

$$PE = \frac{\text{size of the developed program in Kilobyte (KB)}}{\text{line if codes}} \times 100\%. \quad (12.27)$$

The calculated PE for the EDGSOR, Chew et al.'s method [15], Lung et al.'s method [17], and NGS methods are tabulated in Table 12.2.

Next, the computational efficiency of the method to solve the problems is analyzed using the metrics such as the maximum number of iterations until program completion (i_{max}) and total program running time in seconds (s). Besides that, the most significant magnitude of absolute error (ε_{max}) is

TABLE 12.2 Comparison in Program Efficiency

Method	Line of Codes	Program Size (KB)	PE
NGS	487	365	74.95%
Chew et al. [15]	1127	405	35.94%
Lung et al. [17]	1142	406	35.55%
EDGSOR	1142	407	35.64%

recorded to determine the accuracy of approximate solutions compared to exact solutions. In addition, each problem is solved using sizes of the two-dimensional grid, such as $M \times M = 16\times16$, 32×32, 64×64, 128×128, and 256×256. The purpose of doing this is to make sure the method works at different domain complexities. Each method's computational efficiency is tabulated in Tables 12.3, 12.4, and 12.5, corresponding to Problems 12.1, 12.2, and 12.3, respectively.

PROBLEM 12.1 [18]

Let $\alpha = 1$ and $D(u) = 0.2u$ in Eq. (12.1) with the following initial and boundary conditions:

$$F_0 = x + y, 0 \le x \le 1, 0 \le y \le 1, \tag{12.28}$$

$$F_a = y + 0.4t, F_b = 1 + y + 0.4t, 0 \le y \le 1, 0 \le t \le 1, \tag{12.29}$$

and

$$F_c = x + 0.4t, F_d = 1 + x + 0.4t, 0 \le x \le 1, 0 \le t \le 1. \tag{12.30}$$

We use the following exact solution to compare the accuracy of the solutions,

$$u(x,y,t) = x + y + 0.4t. \tag{12.31}$$

PROBLEM 12.2 [18]

Next, let considers $\alpha = 1$, $D(u) = 0.2u^2$, and the following conditions:

$$F_0 = \sqrt{5x + 5y}, 0 \le x \le 1, 0 \le y \le 1, \tag{12.32}$$

$$F_a = \sqrt{5y + 5t}, F_b = \sqrt{5 + 5y + 5t}, 0 \le y \le 1, 0 \le t \le 1, \tag{12.33}$$

and

$$F_c = \sqrt{5x + 5t}, F_d = \sqrt{5 + 5x + 5t}, 0 \le x \le 1, 0 \le t \le 1. \tag{12.34}$$

The exact solution can be expressed in the form of

$$u(x,y,t) = \sqrt{5x + 5y + 5t}. \tag{12.35}$$

PROBLEM 12.3 [19]

Now, let $\alpha = 1$, $D(u) = u^5$, and the following conditions:

$$\begin{aligned} F_0 &= \sqrt[4]{0.8x + 0.8y}, 0 \le x \le 1, 0 \le y \le 1, \\ F_a &= \sqrt[4]{0.8y + 1.6t}, F_b = \sqrt[4]{0.8 + 0.8y + 1.6t}, \end{aligned} \tag{12.36}$$

$$0 \le y \le 1, 0 \le t \le 1, \tag{12.37}$$

and

$$\begin{aligned} F_c &= \sqrt[4]{0.8x + 1.6t}, F_d = \sqrt[4]{0.8 + 0.8x + 1.6t}, \\ &0 \le x \le 1, 0 \le t \le 1. \end{aligned} \tag{12.38}$$

The exact solution is

$$u(x, y, t) = \sqrt[4]{0.8x + 0.8y + 1.6t}. \tag{12.39}$$

Based on Table 12.2, it can be observed that a more substantial number of lines of code gives a lower percentage of PE. Then, the value of PE of the EDGSOR method is smaller by half compared to the NGS method and comparable to Chew et al.'s [15] and Lung et al.'s method [17]. Although the EDGSOR method is less efficient than the NGS method in terms of PE, the computational efficiency of the EDGSOR method is much superior to the NGS method because of the competitive strength of the EDG strategy with the SOR iterative process.

Next, based on the numerical outputs tabulated in Tables 12.3, 12.4, and 12.5, the study found that the EDGSOR method requires fewer iterations upon the program completion compared to Chew et al.'s [15], Lung et al.'s [17], and NGS method. Consequently, the EDGSOR program running time becomes significantly shorter than these tested methods. When the accuracy of the solution is observed after the convergence is reached, the maximum absolute errors of the EDGSOR method in solving Problem 12.1 are smaller than NGS and Chew et al.'s method [15] but almost equivalent to Lung et al.'s method [17]. This result means the efficacy of the approximation by EDGSOR and Lung et al.'s method [17] is the same for Problem 12.1. However, for Problems 12.2 and 12.3, the accuracy of the solution obtained using the EDGSOR method is the best among all tested methods. These findings illustrate the competitive efficiency level attained

TABLE 12.3 Comparison in Terms of Computational Efficiency in Solving Problem 12.1

	Iterative Method	16×16	32×32	64×64	128×128	256×256
i_{max}	NGS	136	436	1525	5462	19404
	Chew et al. [15]	60	184	633	2220	7915
	Lung et al. [17]	70	150	306	607	1204
	EDGSOR	46	97	200	400	786
s	NGS	0.76	2.93	19.59	360.89	4586.69
	Chew et al. [15]	0.30	2.36	7.86	178.05	1621.64
	Lung et al. [17]	0.34	2.16	6.45	68.01	419.75
	EDGSOR	0.24	1.20	5.51	41.22	263.76
ε_{max}	NGS	8.86 (10^{-11})	3.25 (10^{-10})	1.90 (10^{-9})	9.02 (10^{-9})	3.86 (10^{-8})
	Chew et al. [15]	1.96 (10^{-12})	1.22 (10^{-10})	5.89 (10^{-10})	2.62 (10^{-9})	1.27 (10^{-8})
	Lung et al. [17]	4.47 (10^{-12})	3.49 (10^{-12})	1.25 (10^{-11})	2.75 (10^{-11})	3.93 (10^{-11})
	EDGSOR	1.06 (10^{-12})	3.66 (10^{-12})	1.42 (10^{-11})	3.13 (10^{-11})	3.93 (10^{-11})

TABLE 12.4 Comparison in Terms of Computational Efficiency in Solving Problem 12.2

	Iterative Method	16×16	32×32	64×64	128×128	256×256
i_{max}	NGS	130	400	1380	4901	17458
	Chew et al. [15]	61	166	560	1977	7073
	Lung et al. [17]	74	155	313	621	1219
	EDGSOR	49	97	194	386	761
s	NGS	0.97	2.70	18.26	248.82	4243.79
	Chew et al. [15]	0.42	1.77	7.61	78.99	1382.19
	Lung et al. [17]	0.48	1.73	6.12	59.97	413.57
	EDGSOR	0.30	1.19	4.82	35.83	218.86
ε_{max}	NGS	7.57 (10^{-11})	2.31 (10^{-9})	1.31 (10^{-8})	4.95 (10^{-8})	1.75 (10^{-7})
	Chew et al. [15]	2.36 (10^{-12})	3.81 (10^{-10})	3.38 (10^{-9})	2.00 (10^{-8})	7.27 (10^{-8})
	Lung et al. [17]	4.08 (10^{-12})	2.44 (10^{-11})	1.16 (10^{-10})	1.90 (10^{-10})	3.18 (10^{-10})
	EDGSOR	3.23 (10^{-13})	5.83 (10^{-13})	2.16 (10^{-11})	3.02 (10^{-11})	3.45 (10^{-11})

TABLE 12.5 Comparison in Terms of Computational Efficiency in Solving Problem 12.3

	Iterative Method	**16×16**	**32×32**	**64×64**	**128×128**	**256×256**
i_{max}	NGS	739	2630	9478	34098	121649
	Chew et al. [15]	312	1073	3828	13786	49480
	Lung et al. [17]	214	430	835	1626	3178
	EDGSOR	164	323	628	1227	2405
s	NGS	0.98	9.51	113.63	1653.85	29234.80
	Chew et al. [15]	0.79	5.95	49.21	425.74	9269.68
	Lung et al. [17]	0.61	2.36	14.06	125.73	1026.12
	EDGSOR	0.54	1.90	9.64	86.23	676.13
ε_{max}	NGS	1.10 (10^{-9})	6.69 (10^{-9})	3.56 (10^{-8})	1.72 (10^{-7})	7.78 (10^{-7})
	Chew et al. [15]	1.55 (10^{-10})	1.42 (10^{-9})	7.01 (10^{-9})	4.64 (10^{-8})	2.61 (10^{-7})
	Lung et al. [17]	2.18 (10^{-11})	3.37 (10^{-11})	4.41 (10^{-11})	4.55 (10^{-11})	5.12 (10^{-11})
	EDGSOR	4.56 (10^{-12})	1.75 (10^{-11})	3.11 (10^{-11})	4.37 (10^{-11})	3.48 (10^{-11})

by the proposed EDGSOR method against Chew et al.'s method [15], Lung et al.'s method [17], and NGS method.

12.6 CONCLUSION

This chapter presented the systematic formulation of the finite difference approximation and the EDGSOR iteration. Then, the computational algorithm for solving nonlinear diffusion problems in the two-dimensional porous medium setting is developed according to the formulations. Based on the evaluation of the method efficiency based on the implementation in a computational software, this chapter found that the EDGSOR program efficiency is lower than the benchmark NGS method. However, it does not influence the EDGSOR's superiority in terms of computational efficiency against all tested methods such as Chew et al.'s method [15], Lung et al.'s method [17], and NGS method. This chapter also proposed the design of an intelligent system that uses the EDGSOR algorithm to solve nonlinear problems. Future work will improve the design of the intelligent system and develop the system based on the real-world problems.

REFERENCES

1. Regazzoni, F., Dedè, L., Quarteroni, A.: Machine learning for fast and reliable solution of time-dependent differential equations. J. Comput. Phys., 397, 108852, (2019). https://doi.org/10.1016/j.jcp.2019.07.050.

2. Samaniego, E., Anitescu, C., Goswami, S., Nguyen-Thanh, V.M., Guo, H., Hamdia, K., Zhuang, X., Rabczuk, T.: An energy approach to the solution of partial differential equations in computational mechanics via machine learning: Concepts, implementation, and applications. Comput. Methods Appl. Mech. Eng., 362, 112790, (2020). https://doi.org/10.1016/j.cma.2019.112790.
3. Cheung, K.C., See, S.: Recent advance in machine learning for partial differential equation. CCF Trans. High Perform. Comput., 3, 298–310, (2021). https://doi.org/10.1007/s42514-021-00076-7.
4. Raissi, M., Perdikaris, P., Karniadakis, G.E.: Physics-informed neural networks: A deep learning framework for solving forward and inverse problems involving nonlinear partial differential equations. J. Comput. Phys., 378, 686–707, (2019). https://doi.org/10.1016/j.jcp.2018.10.045.
5. David, N., Ruan, X.: An asymptotic preserving scheme for a tumor growth model of porous medium type. ESAIM: Math. Model. Num. Anal., 56(1), 121–150, (2022). https://doi.org/10.1051/m2an/2021080.
6. Muller, V.H., Sonner, S.: Well-posedness of singular-degenerate porous medium type equations and application to biofilm models. J. Math. Anal. Appl., 509(1), 125894, (2022). https://doi.org/10.1016/j.jmaa.2021.125894.
7. Mary, M.L.C., Saravanakumar, R., Lakshmanaraj, D., Rajendran, L., Lyons, M.E.G.: Mathematical modelling of unsteady flow of gas in a semi-infinite porous medium. Int. J. Electrochem. Sci., 17, 220619, (2022). https://doi.org/10.20964/2022.06.05.
8. Pop, D.N., Vrinceanu, N., Al-Omari, S., Ouerfelli, N., Baleanu, D., Nisar, K.S.: On the solution of a parabolic PDE involving a gas flow through a semi-infinite porous medium. Results Phys., 22, 103884, (2021). https://doi.org/10.1016/j.rinp.2021.103884.
9. Palencia, J.L.D.: Analytical assessments to model a flame propagation with a porous medium equation. Comput. Appl. Math., 41, 164, (2022). https://doi.org/10.1007/s40314-022-01878-3.
10. Vazquez, J.L.: The Porous Medium Equation. Oxford University Press, New York (2007).
11. Abdullah, A.R.: The four point explicit decoupled group (EDG) method: A fast Poisson solver. Int. J. Comput. Math., 38, 61–70, (1991). https://doi.org/10.1080/00207169108803958.
12. Abdi, N., Aminikhah, H., Sheikhani, A.H.R., Alavi, J., Taghipour, M.: An efficient explicit decoupled group method for solving two-dimensional fractional Burgers' equation and its convergence analysis. Adv. Math. Phys., 2021, 6669287, (2021). https://doi.org/10.1155/2021/6669287.
13. Abdi, N., Aminikhah, H., Sheikhani, A.H.R.: On rotated grid point iterative method for solving 2D fractional reaction–subdiffusion equation with Caputo-Fabrizio operator. J. Differ. Equ. Appl., 27(8), 1134–1160, (2021). https://doi.org/10.1080/10236198.2021.1965592.
14. Abdi, N., Aminikhah, H., Sheikhani, A.H.R.: High-order rotated grid point iterative method for solving 2D time fractional telegraph equation and its convergence analysis. Comput. Appl. Math., 40, 54, (2021). https://doi.org/10.1007/s40314-021-01451-4.

15. Chew, J.V.L., Sulaiman, J., Sunarto, A., Muhiddin, F.A.: Newton explicit decoupled group solution for two-dimensional nonlinear porous medium equation problems. In: Alfred, R., Lim, Y. (eds) Proceedings of the 8th International Conference on Computational Science and Technology. Lecture Notes in Electrical Engineering, vol 835. Springer, Singapore (2022). https://doi.org/10.1007/978-981-16-8515-6_25.
16. Dahalan, A.A., Saudi, A.: Pathfinding algorithm based on rotated block AOR technique in structured environment. AIMS Math., 7(7), 11529–11550, (2022). https://doi.org/10.3934/math.2022643.
17. Lung, J.C.V., Sulaiman, J., Sunarto, A.: The application of successive overrelaxation method for the solution of linearized half-sweep finite difference approximation to two-dimensional porous medium equation. IOP Conference Series: Materials Science and Engineering, 1088(1), 012002, (2021).
18. Polyanin, A.D., Zaitsev, V.F.: Handbook of Nonlinear Partial Differential Equations. Chapman & Hall, Boca Raton (2004).
19. Sommeijer, B.P., Houwen, P.: Algorithm 621: Software with low storage requirements for two-dimensional, nonlinear, parabolic differential equations. ACM Trans. Math. Softw., 10, 378–396, (1984). https://doi.org/10.1145/2701.356103.

CHAPTER 13

Improved False Position Method Based on Slope (IFPMS)

Timothy Christyan, Tasmi, Darra Funna, Muhammad Raidhy Mustafid, and Muhammad Zaki Almuzakki

13.1 INTRODUCTION

One of the interesting problems in mathematics is how to determine the root of a nonlinear equation. Finding roots of nonlinear equations is also utilized in various engineering problems such as predicting dependent variables, e.g., voltage, temperature, z-factor in petroleum engineering, etc.; for given independent variables like time and position. There exist two methods for root finding: the analytical method (i.e., using symbolic analysis to find the exact root) and the numerical method (i.e., finding the approximation of the root). Of the two methods, the numerical method is much more preferred than the analytical one, because the numerical method makes the root finding process for complex functions much easier and more computable. But, the current established numerical method for root finding has a problem where it can only find a single exact root; meanwhile the complexity of root finding increases significantly when faced with nonlinear function, e.g., function may have more than a single root.

Much research has been done to create a numerical method for root-finding algorithm that can overcome the shortcomings of currently established numerical methods, e.g., bisection method, false position method, etc. or using Monte Carlo method [1] or fuzzy logic method.

 DOI: 10.1201/9781003400387-13

The research by Reza et al. [2] propose a new algorithm called Improved Bisection Method based on Slope (IBMS). As the name suggests, it is an Improvement to the already established bisection method. This algorithm works well for finding multiple roots and cover many shortcomings of established numerical methods, but the fact that it focuses solely on the bisection method makes this algorithm inflexible and may result in long computation time for complex functions. In this chapter, we propose algorithm that work in a similar way to the mentioned IBMS algorithm [2], but using different closed numerical methods for root finding, that is false position method, i.e., Improved False Position Method based on Slope (IFPMS) algorithm.

The chapter is organized as follows: In Section 1, we explain briefly the purpose of this study and its related article. Section 2 explains all the algorithms and methods that will be used in our research. Section 3 focuses on the creation of our proposed IFPMS algorithm and compares the finished algorithm with the IBMS algorithm [2]. Lastly in Section 4 we will give the conclusion of the chapter.

The main objectives of the chapter are:

a. Propose IFPMS algorithm that works in similar way with the IBMS algorithm but use false position method;

b. Test the program of IFPMS and IBMS algorithm using some nonlinear function; and

c. Compare the test result of IFPMS and IBMS algorithms.

13.2 ALGORITHMS AND METHODS USED FOR RESEARCH

13.2.1 Bisection Method

Bisection method is a closed numerical method for root finding. Like all other closed methods, this method searches for single root inside a certain interval (between specified upper and lower limit). This method works with the assumption that if $f(x)$ is real function and continue inside interval $x_l \leq x \leq x_u$ where x_l is the lower limit and x_u is the upper limit, and $f(x_l)$, $f(x_u)$ have different sign, i.e., give negative result when multiplied, then the root of the function might be the middle value of the interval [3].

$$x_r = \frac{x_u + x_l}{2}, \tag{13.1}$$

with x_r as the current predicted root. One of the limits then will be switched with x_r, with condition:

1. If $f(x_l)f(x_r)>0$, then $x_l=x_r$
2. If $f(x_l)f(x_r)<0$, then $x_u=x_r$
3. If $f(x_l)f(x_r)=0$, then root $=x_r$

13.2.2 False Position Method

Also known as linear interpolation, false position method is another close method for root finding like bisection method. Difference between this method and bisection method is that instead of halving the interval $x_l \le x \le x_u$, the interval $x_l \le x \le x_u$ in false position method is reduced by using a chord that connects the points $f(x_l)$ and $f(x_u)$. The formula can be seen in Equation 13.2 [3].

$$x_r = x_u - \frac{f(x_u)(x_l - x_u)}{f(x_l) - f(x_u)}, \tag{13.2}$$

with x_r as the current predicted root. One of the limits then will be switched with x_r, with condition similar to that used in bisection method:

1. If $f(x_l)f(x_r)>0$, then $x_l=x_r$
2. If $f(x_l)f(x_r)<0$, then $x_u=x_r$
3. If $f(x_l)f(x_r)=0$, then root $=x_r$

13.2.3 Monte Carlo Method

Monte Carlo method or Monte Carlo experiment is a method for algorithms that use random sampling to get results and is often used for simulation and optimization problems [4]. Unlike the other two methods mentioned above, Monte Carlo method is not directly used for root finding problems, but instead can be used to help the development of other root-finding algorithms. This Monte Carlo method is also used in the IBMS algorithm.

13.2.4 IBMS Algorithm

IBMS is an algorithm for finding multiple roots of nonlinear function that was proposed by Reza et al. This algorithm works by combining the bisection method with Monte Carlo method. The Monte Carlo method is used to divide the interval for the bisection method into several smaller parts, then with the bisection method we will get root for each part if it exists [2].

The algorithm of IBMS is as per the following step:

1. Define x all random sampling between interval $x_l \leq x \leq x_u$ (Monte Carlo method).
2. To find roots that cross the x-axis, find x_i so $f(x_i)\,f(x_{i+1})$ lower than 0, for all x_i in x.
3. Bisection method is used to find root in interval $x_i \leq x_r \leq x_{i+1}$.
4. For the root that is tangent with x-axis, find x_i so

$$\left(f(x_i+\varepsilon)-f(x_i)\right)\left(f(x_{i+1}+\varepsilon)-f(x_{i+1})\right)<0$$

5. Do bisection method in interval $x_i \leq x_r \leq x_{i+1}$.
6. Make $x_i = x_{i+1}$ then repeat step 2 to step 5 for all x_i in x.

13.3 CONSTRUCTING IFPMS ALGORITHM

IFPMS algorithm uses the same method as IBMS algorithm, with the exception being that it uses false position method instead of bisection method. The algorithm works by combining the false position method with the Monte Carlo method. The Monte Carlo method is used to divide the interval for the bisection method into several smaller parts, then with the false position method we will get root for each part if it exists.

The algorithm of IFPMS mostly follows the same steps as IBMS algorithm:

1. Define x all random sampling between interval $x_l \leq x \leq x_u$ (Monte Carlo method).
2. To find roots that cross the x-axis, find x_i so $f(x_i)\,f(x_{i+1})$ lower than 0, for all x_i in x.
3. Do false position method with $x_i \leq x_r \leq x_{i+1}$ as an interval.
4. For the root that is tangent with the x-axis, find x_i so

$$\left(f(x_i+\varepsilon)-f(x_i)\right)\left(f(x_{i+1}+\varepsilon)-f(x_{i+1})\right)<0$$

TABLE 13.1 Pseudo Code for IFPMS Algorithm

```
func = #input function for root finding
lower_limit, upper_limit = #input interval
iteration = #input number of iteration
ε = #constant machine epsilon
root = #empty array for saving root value
partition_list = make_random_sampling(lower_limit, upper_limit,
    number_of_sample)
for each partition in partition_list:
    a = func(partition_now)
    b = func(partition_next)
    A = func(partition_now + ε * 10 ** 5) - a
    B = func(partition_next + ε * 10 ** 5) - b
    if (a * b) < 0 then # check for root that cross x axis
        root.add(IFPMS(func, partition_now, partition_next,
            iteration, tangent = False))
    else if (A * B) < 0 then #check for root that only tangent to x axis
        root.add(IFPMS(func, partition_now, partition_next,
            iteration, tangent = True))
show(root)
```

5. Do false position method with $x_i \le x_r \le x_{i+1}$ as an interval.
6. Make $x_i = x_{i+1}$ then repeat step 2 to step 5 for all x_i in x.

Pseudo code for IFPMS algorithm can be seen in Table 13.1.

13.4 COMPARING IFPMS WITH IBMS ALGORITHM

Using Hall and Yarborough correlation [5], and some use case functions, we try seeing the performance for both IBMS and IFPMS algorithm. For the sake of convenience, only the first ten iterations are performed and we limited the number of random sampling to a thousand. The exact root that we got using the Symbolab tool and graph of the testing function will also be given for comparison. The performance is shown for some cases below.

1. **Case 1**: Z-factor for Hall and Yarborough correlation.
   This correlation calculates with constant parameter pressure (P) = 200, temperature (T) = 150, gas gravity specification (γ_g) = 0.7. We use these parameters to get other parameters that we use to make main formula of z-factor which is

$$Z = \frac{P_{pr}}{y} \tag{13.3}$$

where y is obtained from equation below

$$-A_1 P_{pr} + \frac{y + y^2 + y^3 - y^4}{(1-y)^3} - A_2 y^2 + A_3 y^{A_4} = 0 \tag{13.4}$$

with

$$T_{pc} = 169.2 + 349.5\gamma_g - 74\gamma_g^{\;2} \tag{13.5}$$

$$P_{pc} = 756.8 + 131.07\gamma_g - 3.6\gamma_g^{\;2} \tag{13.6}$$

$$T_{pr} = \frac{T}{T_{pc}} \tag{13.7}$$

$$P_{pr} = \frac{P}{P_{pc}} \tag{13.8}$$

$$t = \frac{1}{T_{pr}} \tag{13.9}$$

$$A_1 = 0.06125\ t\ e^{1.2(t-1)^2} \tag{13.10}$$

$$A_2 = 14.76t - 9.76t^2 + 4.58t^3 \tag{13.11}$$

$$A_3 = 90.7t - 242.2t^2 + 42.4t^3 \tag{13.12}$$

$$A_4 = 2.18 + 2.82t \tag{13.13}$$

Curve of Equation (13.4) is shown in Figure 13.1.

From Figure 13.1, function (13.4) has two roots. Exact roots of function (13.4) are 0.626531 and 1. Interval for search roots with IBMS and IFPMS is: $-2 \le root \le 2$, with sampling 1000 and number iteration 10. Table 13.2 shows result comparison between IFPMS and IBMS for roots of function (13.4). Table 13.2 shows comparison between two algorithms, i.e., IBMS and IFPMS. From this table IFPMS algorithm works well and can find two roots like IBMS.

2. **Case 2:** Root of $f(x) = 4x^{10} - 5x^9 - 10x^2$
Figure 13.2 is curve of function $f(x) = 4x^{10} - 5x^9 - 10x^2$. This function has three roots. Exact roots of this function are −1.0142501623072, 0, and 1.4424309174598.

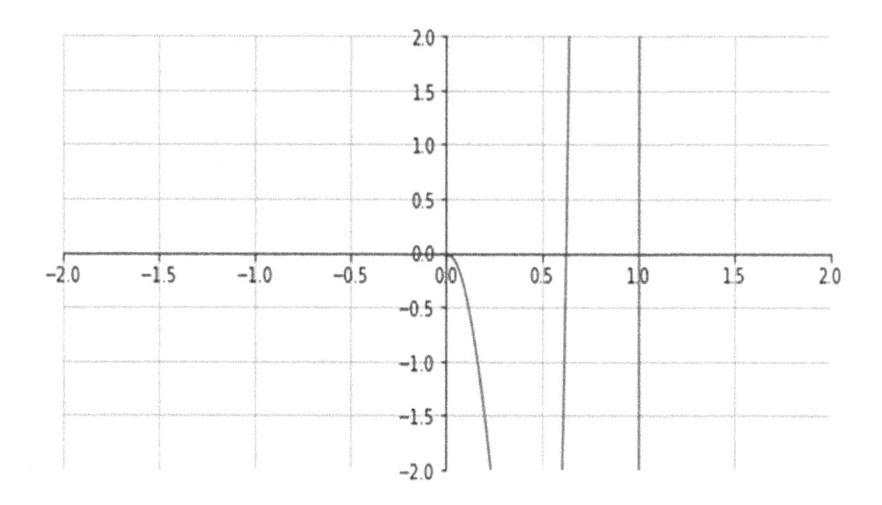

FIGURE 13.1 Curve of $f(y) = -A_1 P_{pr} + \frac{y+y^2+y^3-y^4}{(1-y)^3} - A_2 y^2 + A_3 y^{A_4}$.

TABLE 13.2 Comparison between IFPMS and IBMS for Roots of $f(y) = -A_1 P_{pr} + \frac{y+y^2+y^3-y^4}{(1-y)^3} - A_2 y^2 + A_3 y^{A_4}$

		IBMS		IFPMS	
Iteration	**Number of Roots**	**Root**	ε_r(%)	**Root**	ε_r(%)
0	1	0.62204	0.7168041166	0.626517	0.002234526304
	2	1.001752	0.1752	1.002925	0.2925
1	1	0.626121	0.06543969891	0.626531	0
	2	0.999847	0.0153	0.987273	1.2727
2	1	0.626026	0.08060255598	0.626528	0.0004788270652
	2	0.999263	0.0737	1.003461	0.3461
3	1	0.626314	0.03463515772	0.62653	0.0001596090217
	2	0.999939	0.0061	1.004067	0.4067
4	1	0.626468	0.01005536837	0.626531	0
	2	0.999971	0.0029	1.002738	0.2738
5	1	0.626476	0.008778496196	0.626531	0
	2	0.999675	0.0325	0.995163	0.4837
6	1	0.626565	0.005426706739	0.626531	0
	2	0.999887	0.0113	0.998113	0.1887
7	1	0.626566	0.005586315761	0.626531	0
	2	0.999997	0.0003	1.000168	0.0168
8	1	0.626556	0.003990225544	0.626531	0
	2	0.99997	0.003	1.004227	0.4227
9	1	0.626542	0.001755699239	0.626531	0
	2	1	0	0.999548	0.0452

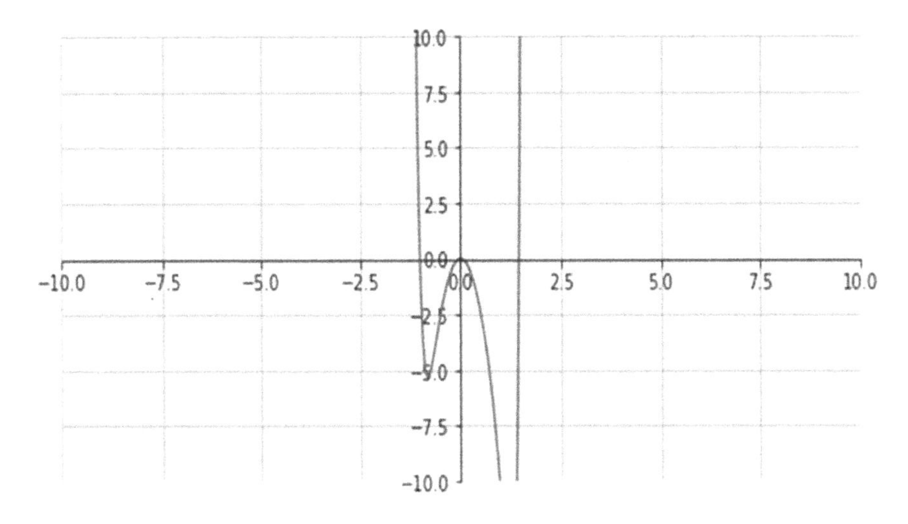

FIGURE 13.2 Curve of $f(x)=4x^{10}-5x^{9}-10x^{2}$.

For simulation of this case, interval $-10 \le root \le 10$ is chosen to search roots with IBMS and IFPMS algorithms with 1000 number of sampling and 10 iteration. Result for this simulation are shown in Table 13.3.

3. **Case 3:** Root of $f(x)=x\ cos(x)^2$
   For interval: $-10 \le root \le 10$, curve of $f(x)=x\ cos(x)^2$ is shown in Figure 13.3. This function, for this interval, has seven roots i.e. $-\frac{5\pi}{2}, -\frac{3\pi}{2}, -\frac{\pi}{2}, 0, \frac{\pi}{2}, \frac{3\pi}{2}, \frac{5\pi}{2}$.

   Simulation of this function with IFPMS and IBMS algorithm has been done with 1000 sampling and 10 iteration. Result of simulation is shown in Table 13.4. Table 13.4 consists of root and tangent root. Root is value of x which intersects the x-axis and tangent roots is value of the point offend the x-axis. In this case, function have one root and six tangent roots.

4. **Case 4:** Roots of $f(x)= e^{x-2}-x^{10}+3x^{9}$
   Curve of function $f(x)= e^{x-2}-x^{10}+3x^{9}$ is shown in Figure 13.4. This function has two exact root -0.646 and 3.

   In this simulation, interval $-10 \le root \le 10$ is used with 1000 sampling and 10 iterations. Result of simulation of IFPMS and IBMS is presented in Table 13.5.

TABLE 13.3 Comparison between IFPMS and IBMS for Roots of $f(x)=4x^{10}-5x^{9}-10x^{2}$

Iteration	Number of Roots	IBMS		IFPMS	
0		**Root**	$\varepsilon_r(\%)$	**Root**	$\varepsilon_r(\%)$
	1	−1.026989	1.255985768	−1.008530	0.5639794323
	2	1.443210	0.0540117749	1.441827	0.0418680335
		Tangent Root	$\varepsilon_{abs}(\%)$	**Tangent Root**	$\varepsilon_{abs}(\%)$
	3	–	–	−0.007312	0.007312
1		**Root**	$\varepsilon_r(\%)$	**Root**	$\varepsilon_r(\%)$
	1	−1.012893	0.1338094247	−1.014230	0.001987902783
	2	1.446034	0.2497923815	1.440983	0.1003803678
		Tangent Root	$\varepsilon_{abs}(\%)$	**Tangent Root**	$\varepsilon_{abs}(\%)$
	3	-0.002606	–	−0.005171	0.005171
2		**Root**	$\varepsilon_r(\%)$	**Root**	$\varepsilon_r(\%)$
	1	−1.012169	0.2051922084	−1.014233	0.001692117767
	2	1.441417	0.07029227171	1.442421	0.0006875518044
		Tangent Root	$\varepsilon_{abs}(\%)$	**Tangent Root**	$\varepsilon_{abs}(\%)$
	3	–	–	−0.09441	0.09441
3		**Root**	$\varepsilon_r(\%)$	**Root**	$\varepsilon_r(\%)$
	1	−1.014431	0.01782969326	−1.014249	0.0001145976844
	2	1.439588	0.1970921051	1.442431	0.00000572229831
		Tangent Root	$\varepsilon_{abs}(\%)$	**Tangent Root**	$\varepsilon_{abs}(\%)$
	3	−0.000477	0.000477	−0.015088	0.015088
4		**Root**	$\varepsilon_r(\%)$	**Root**	$\varepsilon_r(\%)$
	1	−1.013183	0.1052168732	−1.014246	0.0004103826999
	2	1.440753	0.1163256721	1.442422	0.0006182243941
		Tangent Root	$\varepsilon_{abs}(\%)$	**Tangent Root**	$\varepsilon_{abs}(\%)$
	3	−0.000122	0.000122	−0.00353	0.00353
5		**Root**	$\varepsilon_r(\%)$	**Root**	$\varepsilon_r(\%)$
	1	−1.014516	0.0262102687	−1.01425	0.00001600267921
	2	1.442702	0.01879345048	1.442431	0.00000572229831
		Tangent Root	$\varepsilon_{abs}(\%)$	**Tangent Root**	$\varepsilon_{abs}(\%)$
	3	−0.000337	0.000337	−0.006086	0.006086

(Continued)

TABLE 13.3 (Continued) Comparison between IFPMS and IBMS for Roots of $f(x) = 4x^{10} - 5x^9 - 10x^2$

Iteration	Number of Roots	IBMS		IFPMS	
6		**Root**	$\varepsilon_r(\%)$	**Root**	$\varepsilon_r(\%)$
	1	−1.014195	0.005438727964	−1.01425	0.00001600267921
	2	1.442252	0.01240388414	1.442431	0.00000572229831
		Tangent Root	$\varepsilon_{abs}(\%)$	**Tangent Root**	$\varepsilon_{abs}(\%)$
	3	−0.000266	0.000266	−0.01051	0.01051
7		**Root**	$\varepsilon_r(\%)$	**Root**	$\varepsilon_r(\%)$
	1	−1.014224	0.002579472814	−1.01425	0.00001600267921
	2	1.442204	0.01573159983	1.442431	0.00000572229831
		Tangent Root	$\varepsilon_{abs}(\%)$	**Tangent Root**	$\varepsilon_{abs}(\%)$
	3	−0.000162	0.000162	−0.00551	0.00551
8		**Root**	$\varepsilon_r(\%)$	**Root**	$\varepsilon_r(\%)$
	1	−1.014238	0.001199142741	−1.01425	0.00001600267921
	2	1.442458	0.001877562376	1.442431	0.00000572229831
		Tangent Root	$\varepsilon_{abs}(\%)$	**Tangent Root**	$\varepsilon_{abs}(\%)$
	3	−0.00022	0.00022	−0.001009	0.001009
9		**Root**	$\varepsilon_r(\%)$	**Root**	$\varepsilon_r(\%)$
	1	−1.01426	0.0009699473725	−1.01425	0.00001600267921
	2	1.442461	0.002085544606	1.442431	0.00000572229831
		Tangent Root	$\varepsilon_{abs}(\%)$	**Tangent Root**	$\varepsilon_{abs}(\%)$
	3	−0.000433	0.000433	−0.001688	0.001688

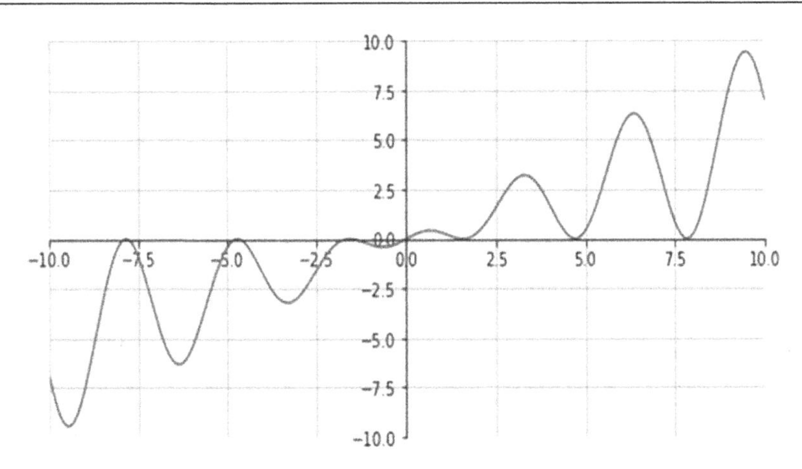

FIGURE 13.3 Curve of $f(x) = x\ cos(x)^2$.

TABLE 13.4 Comparison between IFPMS and IBMS for Roots of $f(x)=x\ cos(x)^2$

Iteration	Number of Roots	IBMS		IFPMS	
0		**Root**	$\varepsilon_{abs}(\%)$	**Root**	$\varepsilon_{abs}(\%)$
	1	0.017951	0.017951	0.000005	0.000005
		Tangent Root	$\varepsilon_r(\%)$	**Tangent Root**	$\varepsilon_r(\%)$
	2	−7.855885	0.07496815287	−7.856377	0.08123566879
	3	–	–	−4.72258	0.2670912951
	4	−1.582567	0.8004458599	−1.576642	0.4230573248
	5	1.570043	0.002738853503	1.554184	1.007388535
	6	–	–	4.710436	0.009256900212
	7	–	–	7.846436	0.04540127389
1		**Root**	$\varepsilon_{abs}(\%)$	**Root**	$\varepsilon_{abs}(\%)$
	1	0.000011	0.000011	0	0
		Tangent Root	$\varepsilon_r(\%)$	**Tangent Root**	$\varepsilon_r(\%)$
	1	-7.866761	0.2135159236	-7.866761	0.2135159236
	2	-4.71855	0.1815286624	-4.720302	0.2187261146
	3	-1.585633	0.9957324841	-1.582171	0.7752229299
	4	1.56488	0.3261146497	1.554373	0.9953503185
	5	4.68842	0.4581740977	4.707577	0.05144373673
	6	7.848897	0.01405095541	7.848897	0.01405095541
2		**Root**	$\varepsilon_{abs}(\%)$	**Root**	$\varepsilon_{abs}(\%)$
	1	-0.000054	0.000054	0	0
		Tangent Root	$\varepsilon_r(\%)$	**Tangent Root**	$\varepsilon_r(\%)$
	2	-7.856727	0.08569426752	-7.863932	0.177477707
	3	−4.720455	0.2219745223	−4.725504	0.3291719745
	4	−1.57511	0.325477707	−1.572425	0.1544585987
	5	1.569729	0.0172611465	1.566082	0.2495541401
	6	4.697829	0.2584076433	4.709536	0.009851380042
	7	7.845011	0.06355414013	7.853808	0.04850955414
3		**Root**	$\varepsilon_{abs}(\%)$	**Root**	$\varepsilon_{abs}(\%)$
	1	−0.000837	0.000837	0	0
		Tangent Root	$\varepsilon_r(\%)$	**Tangent Root**	$\varepsilon_r(\%)$
	2	−7.85629	0.08012738854	−7.855553	0.0707388535
	3	−4.71573	0.121656051	−4.716913	0.1467728238
	4	−1.571209	0.07700636943	−1.572764	0.1760509554
	5	1.565705	0.273566879	1.566697	0.2103821656
	6	4.70984	0.003397027601	4.702311	0.1632484076
	7	7.853176	0.04045859873	7.851959	0.02495541401

(Continued)

TABLE 13.4 (Continued) Comparison between IFPMS and IBMS for Roots of $f(x) = x\ cos(x)^2$

Iteration	Number of Roots	IBMS		IFPMS	
4		**Root**	$\varepsilon_{abs}(\%)$	**Root**	$\varepsilon_{abs}(\%)$
	1	0.001676	0.001676	0	0
		Tangent Root	$\varepsilon_r(\%)$	**Tangent Root**	$\varepsilon_r(\%)$
	2	−7.857315	0.09318471338	−7.86032	0.1314649682
	3	−4.712838	0.06025477707	−4.718101	0.1719957537
	4	−1.573786	0.2411464968	−1.574685	0.2984076433
	5	1.570037	0.002356687898	1.567623	0.1514012739
	6	4.71224	0.04755838641	4.712024	0.04297239915
	7	7.853027	0.03856050955	7.827562	0.2858343949
5		**Root**	$\varepsilon_{abs}(\%)$	**Root**	$\varepsilon_{abs}(\%)$
	1	−0.000028	0.000028	0	0
		Tangent Root	$\varepsilon_r(\%)$	**Tangent Root**	$\varepsilon_r(\%)$
	2	−7.855	0.06369426752	−7.857633	0.09723566879
	3	−4.714217	0.0895329087	−4.718481	0.1800636943
	4	−1.571869	0.119044586	−1.571433	0.09127388535
	5	1.570444	0.02828025478	1.570355	0.02261146497
	6	4.712139	0.04541401274	4.698497	0.2442250531
	7	7.853481	0.04434394904	7.844739	0.06701910828
6		**Root**	$\varepsilon_{abs}(\%)$	**Root**	$\varepsilon_{abs}(\%)$
	1	0.000317	0.000317	0	0
		Tangent Root	$\varepsilon_r(\%)$	**Tangent Root**	$\varepsilon_r(\%)$
	2	−7.854116	0.05243312102	−7.860028	0.1277452229
	3	−4.713292	0.06989384289	−4.715092	0.1081104034
	4	−1.571952	0.1243312102	−1.574318	0.2750318471
	5	1.569576	0.02700636943	1.549737	1.290636943
	6	4.711806	0.03834394904	4.711319	0.02800424628
	7	7.853287	0.04187261146	7.853862	0.04919745223
7		**Root**	$\varepsilon_{abs}(\%)$	**Root**	$\varepsilon_{abs}(\%)$
	1	0.000051	0.000051	0	0
		Tangent Root	$\varepsilon_r(\%)$	**Tangent Root**	$\varepsilon_r(\%)$
	2	−7.854272	0.05442038217	−7.865715	0.2001910828
	3	−4.712535	0.05382165605	−4.712731	0.05798301486
	4	−1.571668	0.1062420382	−1.57312	0.1987261146
	5	1.570326	0.02076433121	1.569214	0.05006369427
	6	4.712074	0.04403397028	4.711842	0.03910828025
	7	7.853774	0.04807643312	7.847296	0.03444585987

(Continued)

TABLE 13.4 (Continued) Comparison between IFPMS and IBMS for Roots of $f(x)=x\ cos(x)^2$

Iteration	Number of Roots	IBMS		IFPMS	
8		**Root**	$\varepsilon_{abs}(\%)$	**Root**	$\varepsilon_{abs}(\%)$
	1	0.00001	0.00001	0	0
		Tangent Root	$\varepsilon_r(\%)$	**Tangent Root**	$\varepsilon_r(\%)$
	2	−7.854561	0.05810191083	−7.854368	0.0556433121
	3	−4.712705	0.05743099788	−4.726835	0.3574309979
	4	−1.580372	0.6606369427	−1.574164	0.2652229299
	5	1.570272	0.01732484076	1.570339	0.02159235669
	6	4.71207	0.04394904459	4.710404	0.008577494692
	7	7.852213	0.0281910828	7.853902	0.04970700637
9		**Root**	$\varepsilon_{abs}(\%)$	**Root**	$\varepsilon_{abs}(\%)$
	1	0.000044	0.000044	0	0
		Tangent Root	$\varepsilon_r(\%)$	**Tangent Root**	$\varepsilon_r(\%)$
	2	−7.85867	0.1104458599	−7.854081	0.05198726115
	3	−4.716394	0.1357537155	−4.712567	0.05450106157
	4	−1.573569	0.2273248408	−1.5716487	0.1050127389
	5	1.570469	0.02987261146	1.570209	0.01331210191
	6	4.711907	0.04048832272	4.712209	0.04690021231
	7	7.853788	0.04825477707	7.850041	0.0005222929936

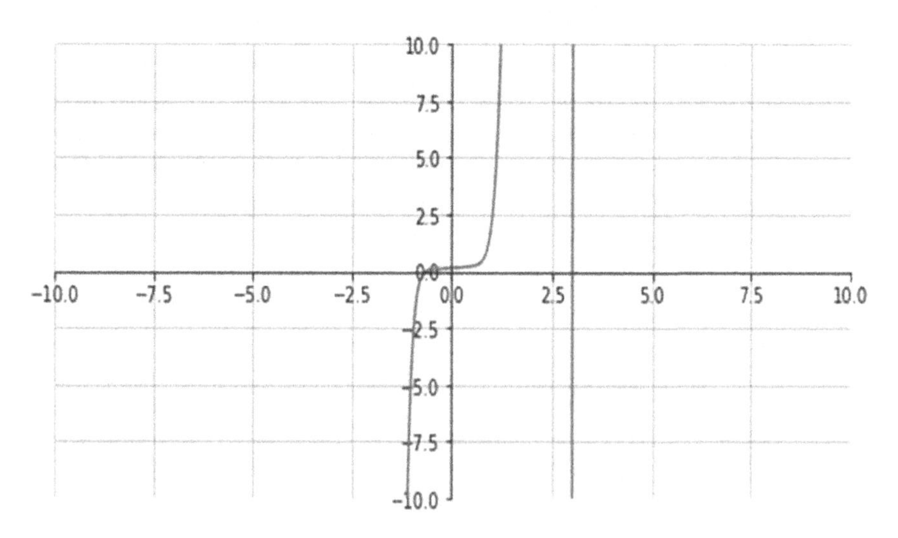

FIGURE 13.4 Curve of $f(x)=e^{x-2}-x^{10}+3x^9$.

TABLE 13.5 Comparison between IFPMS and IBMS for Roots of $f(x)=e^{x-2}-x^{10}+3x^{9}$

		IBMS		IFPMS	
Iteration	Number of Roots	Root	ε_r	Root	ε_r
0	1	−0.628969	2.636377709	−0.644519	0.2292569659
	2	2.999339	0.02203333333	2.99981	0.006333333333
1	1	−0.643333	0.4128482972	−0.644976	0.1585139319
	2	3.003603	0.1201	2.999849	0.005033333333
2	1	−0.647039	0.1608359133	−0.645538	0.07151702786
	2	2.994476	0.1841333333	3.000138	0.0046
3	1	−0.646262	0.04055727554	−0.645538	0.07151702786
	2	3.001189	0.03963333333	3.000138	0.0046
4	1	−0.6456	0.06191950464	−0.645538	0.07151702786
	2	3.000876	0.0292	3.000138	0.0046
5	1	−0.645983	0.002631578947	−0.645538	0.07151702786
	2	3.000336	0.0112	3.000138	0.0046
6	1	−0.645749	0.03885448916	−0.645538	0.07151702786
	2	2.9996	0.01333333333	3.000138	0.0046
7	1	−0.645506	0.07647058824	−0.645538	0.07151702786
	2	3.000068	0.002266666667	3.000138	0.0046
8	1	−0.64545	0.08513931889	−0.645538	0.07151702786
	2	3.000131	0.004366666667	3.000138	0.0046
9	1	−0.645562	0.06780185759	−0.645538	0.07151702786
	2	3.000129	0.0043	3.000138	0.0046

5. **Case 5:** Roots of $f(x)=\left(0.5^{2}+x^{2}+\frac{x}{19}\right)-18$

   Function $f(x)=\left(0.5^{2}+x^{2}+\frac{x}{19}\right)-18$ has two exact roots, i.e., −4.239 and 4.187. Curve of this function is shown in Figure 13.5.

   Interval for simulation is $-10 \leq root \leq 10$ with 1000 sampling and 10 iterations. Roots of this function for this simulation are shown in Table 13.6.

6. **Case 6:** Roots of $f(x)=x^{2}-1$

   Function $f(x)=x^{2}-1$ has two exact roots that are −1 and 1. Curve of this function is shown in Figure 13.6. For simulation this case is chosen with interval $-10 \leq root \leq 10$ and number of sampling is 1000 and number of iterations are 10.

   Table 13.7 presents result of this simulation. From this table it appears that the IFPMS algorithm is faster to get the exact root.

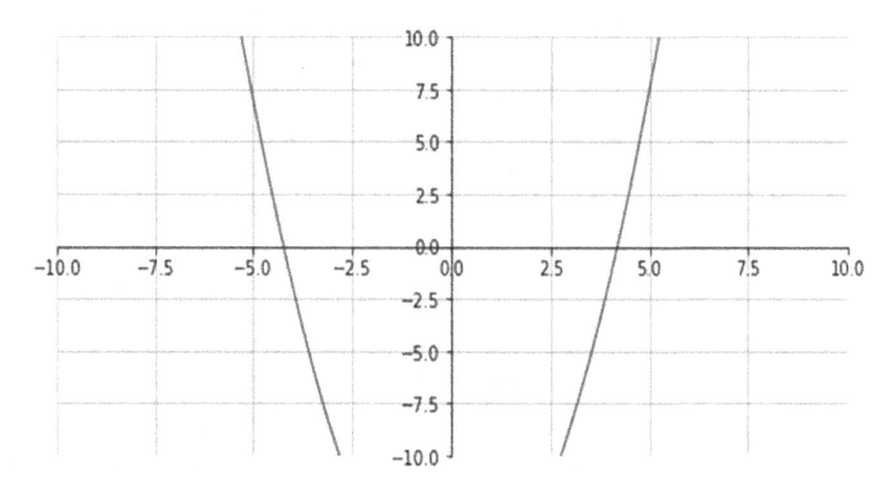

FIGURE 13.5 Curve of $f(x)=\left(0.5^2+x^2+\frac{x}{19}\right)-18$.

TABLE 13.6 Comparison between IFPMS and IBMS for Roots of $f(x)=\left(0.5^2+x^2+\frac{x}{19}\right)-18$

		IBMS		IFPMS	
Iteration	**Number of Roots**	**Root**	ε_r	**Root**	ε_r
0	1	−4.241123	0.05008256664	−4.23947	0.01108752064
	2	4.19	0.07165034631	4.186427	0.01368521615
1	1	−4.23993	0.02193913659	−4.239472	0.01113470158
	2	4.179074	0.189300215	4.186841	0.003797468354
2	1	−4.234515	0.1058032555	−4.239473	0.01115829205
	2	4.183754	0.07752567471	4.186841	0.003797468354
3	1	−4.238456	0.01283321538	−4.239473	0.01115829205
	2	4.189433	0.05810843086	4.186841	0.003797468354
4	1	−4.239157	0.003703703704	−4.239473	0.01115829205
	2	4.187113	0.002698829711	4.186841	0.003797468354
5	1	−4.240046	0.02467563105	−4.239473	0.01115829205
	2	4.18663	0.008836876045	4.186841	0.003797468354
6	1	−4.239575	0.01356451993	−4.239473	0.01115829205
	2	4.186881	0.002842130404	4.186841	0.003797468354
7	1	−4.23959	0.01391837698	−4.239473	0.01115829205
	2	4.186526	0.01132075472	4.186841	0.003797468354
8	1	−4.239447	0.01054493984	−4.239473	0.01115829205
	2	4.18684	0.003821351803	4.186841	0.003797468354
9	1	−4.239479	0.01129983487	−4.239473	0.01115829205
	2	4.186831	0.004036302842	4.186841	0.003797468354

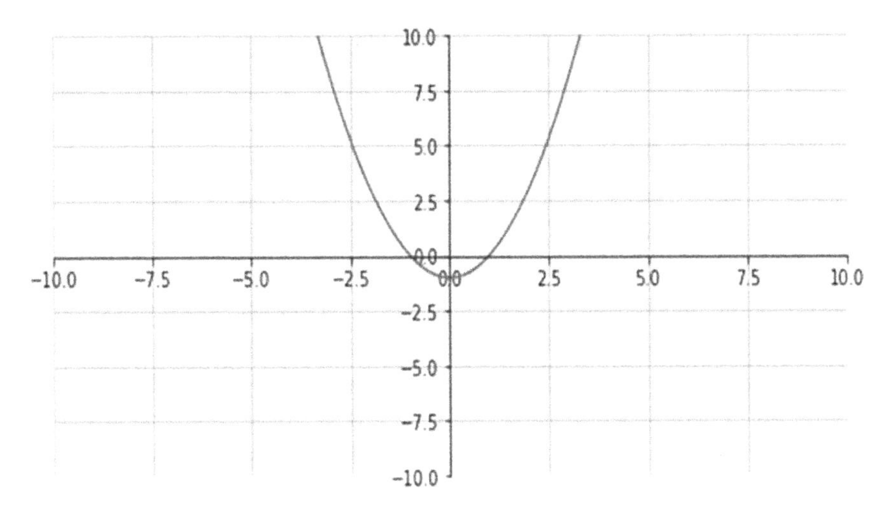

FIGURE 13.6 Curve of $f(x) = x^2 - 1$.

TABLE 13.7 Comparison between IFPMS and IBMS for Roots of $f(x) = x^2 - 1$

		IBMS		IFPMS	
Iteration	**Number of Roots**	**Root**	ε_r	**Root**	ε_r
0	1	−0.984087	1.5913	−0.999857	0.0143
	2	0.991339	0.8661	0.999924	0.0076
1	1	−1.002377	0.2377	−1	0
	2	1.000623	0.0623	1	0
2	1	−1.002909	0.2909	−0.999999	0.0001
	2	1.009223	0.9223	1	0
3	1	−1.000385	0.0385	−1	0
	2	1.000436	0.0436	1	0
4	1	−1.001484	0.1484	−1	0
	2	0.999854	0.0146	1	0
5	1	−0.999755	0.0245	−1	0
	2	1.000069	0.0069	1	0
6	1	−0.999857	0.0143	−1	0
	2	0.999985	0.0015	1	0
7	1	−0.999924	0.0076	−1	0
	2	1.000043	0.0043	1	0
8	1	−0.999903	0.0097	−1	0
	2	0.999853	0.0147	1	0
9	1	−1.00001	0.001	−1	0
	2	0.999995	0.0005	1	0

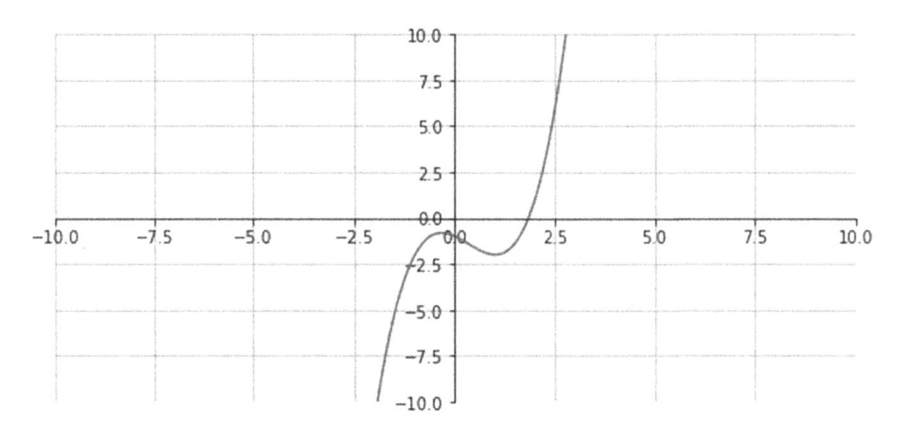

FIGURE 13.7 Curve of $f(x)=x^3-x^2-x-1$.

7. **Case 7:** Roots of $f(x)=x^3-x^2-x-1$
   For this case, IFPMS and IBMS algorithms are used to search roots of $f(x)=x^3-x^2-x-1$. Exact root of the function is 1.839 and the curve of this function is shown in Figure 13.7. Interval of $-10 \le root \le 10$, with 1000 number of sampling and 10 as number of iterations is chosen for this simulation.

   Table 13.8 presents the results of simulation. Like other use case, the IFPMS methods have lower error compared to the exact root.

TABLE 13.8 Comparison between IFPMS and IBMS for Roots of $f(x)=x^3-x^2-x-1$

	IBMS		IFPMS	
Iteration	**Root**	ε_r	**Root**	ε_r
0	1.816278	1.250960729	1.839154	0.007217743489
1	1.853412	0.7679740509	1.839286	0.00004104852047
2	1.843975	0.2548947296	1.839287	0.0000133203808
3	1.84076	0.08009871196	1.839287	0.0000133203808
4	1.83491	0.2379593605	1.839287	0.0000133203808
5	1.839699	0.02241330771	1.839287	0.0000133203808
6	1.839294	0.0003939026897	1.839287	0.0000133203808
7	1.839541	0.0138230213	1.839287	0.0000133203808
8	1.839379	0.005015259298	1.839287	0.0000133203808
9	1.839278	0.0004759997307	1.839287	0.0000133203808

13.5 RESULT AND DISCUSSION

After we compare the IBMS and the IFPMS algorithm with some use case functions in Section 3.2, it can be observed that both IFPMS and IBMS can give a very small error, i.e., below 0.1%, with a certain number of iterations. In this section, some figures are presented to see the results of IFPMS algorithm compared to IBMS algorithm. Figure 13.8 is plot of roots for function (4). This function has two roots. From this figure, it appears that IFPMS algorithm has more precision in every iteration for root 1, but that IBMS algorithm has more precision in every iteration for root 2.

Root of function $f(x)=e^{x-2}-x^{10}+3x^{9}$is shown in Figure 13.8. It shows that the resulting root values of the two algorithms (IFPMS and IBMS) are close together in each iteration.

Figure 13.9 shows root of function x^3-x^2-x-1 for 10 iterations. From this figure we can see that IFPMS algorithm also has more precision to exact root in each iteration.

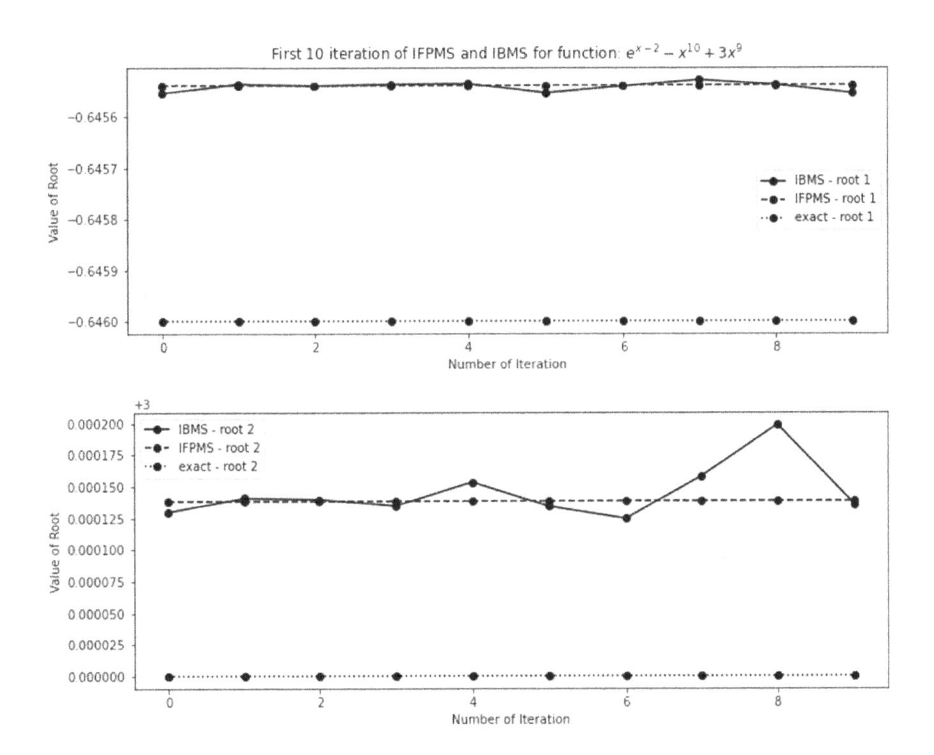

FIGURE 13.8 Roots of function $f(x)=e^{x-2}-x^{10}+3x^{9}$.

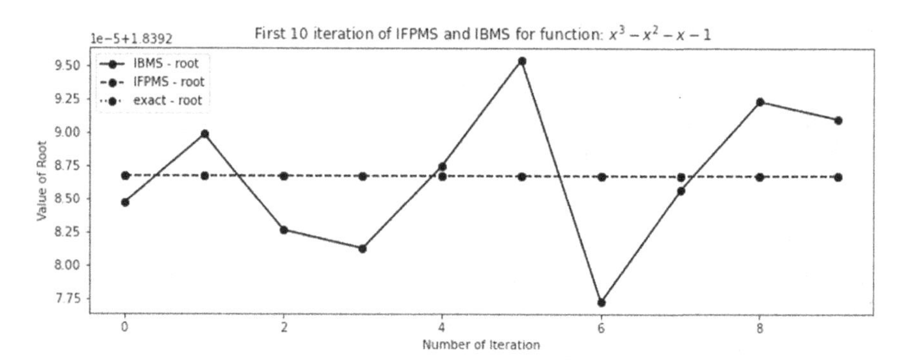

FIGURE 13.9 Roots of function $x^3 - x^2 - x - 1$.

However, for an environment with small number or no iteration at all the IFPMS algorithm result often gives lower error compared to the IBMS algorithm, meanwhile the IBMS algorithm may not get all roots of the function if the iteration number is too small. The error from root finding using IFPMS algorithm is more constant, i.e., no change or very small change than the IBMS algorithm. This means that the proposed IFPMS algorithm gives a better approximation of roots with relatively little to no iteration.

13.6 SUMMARY

In this chapter, we propose a new numerical method for root finding on nonlinear function called IFPMS algorithm, based on IBMS algorithm by Reza et al. The IFPMS algorithm integrates Monte Carlo and false position method to find every root of a nonlinear function inside a certain range. The finished IFPMS algorithm shows that it can get a more accurate approximation of exact root with relatively small number of iterations compared to the IBMS algorithm.

REFERENCES

1. M. M. A. Vahid & B. B. Shila., (2016, August 12). The Approximate Calculation of the Roots of Algebraic Equation Through Monte Carlo Method. *International Journal of Mathematics and Computational Science, Vol. 2, No. 3.* https://www.researchgate.net/publication/306558591_The_Approximate_Calculation_of_the_Roots_of_Algebraic_Equation_Through_Monte_Carlo_Method
2. E. Reza., M. Mohsen., M. Mashaallah., & A. G. Reza., (2021, August 25). A New Method for Rooting Nonlinear Equations Based on the Bisection Method. *MethodsX, Volume 8.* https://doi.org/10.1016/j.mex.2021.101502

3. K. P. Kingdom & B. N. Oriji., (2012, August). A New Computerized Approach to Z-Factor Determination. *Transnational Journal of Science and Technology, Vol. 2, No. 7.* https://www.researchgate.net/publication/261707648_A_NEW_COMPUTERIZED_APPROACH_TO_Z-FACTOR_DETERMINATION
4. S.C. Chapra and R.P. Canale, *Numerical Methods for Engineers*, New York: Mc Graw Hill, 2010.
5. L. B. Peter., (2001, February). A Brief Introduction to Monte Carlo Simulation. *Clinical Pharmacokinetics, Volume 40.* https://www.researchgate.net/publication/12094719_A_Brief_Introduction_to_Monte_Carlo_Simulation

CHAPTER 14

Computing of Anxiety or Depression Symptoms Indicators Using Lagrange Exponential Modified Euler Method

Mohammad Khatim Hasan, Jumat Sulaiman, and Samsul Arifin Abdul Karim

14.1 INTRODUCTION

Anxiety (Stein & Craske, 2017) and/or depression (Weinberger et al., 2018) are significant public health problems in many countries worldwide. Mental problems are among the leading cause of disability (Kessler et al., 2005). The problem also increases the risk of other disorders like alcoholism (Grillon et al., 2019) and drug abuse. These disorders can cause misery and poor health (Chisholm et al., 2016).

Anxiety disorders are the dominant mental health issue in this era of globalization in the United States. Despite its rarity, depression, also known as major depressive disorder (MDD), affects over 16 million individuals. The United States experiences these diseases in approximately 18% of its population, or over 40 million adults annually (Weinberger et al., 2018). In 2020, the World Health Organization (WHO) estimated that the COVID-19 epidemic caused a 27.6% increase in depression and 25.5% growth in anxiety disorders worldwide. Anxiety disorders are not

DOI: 10.1201/9781003400387-14

limited to worrying about difficult or unexpected events. More common concerns may arise, such as our health and potential worries about school, work, or relationships. Such worries can cause lasting concerns and eventually impact daily life. Anxiety, fear, worry, irritability (depression), physical restlessness, or feeling on edge are the primary symptoms of ongoing anxiety. Patients may experience a sense of fear, doom, panic, sleep difficulties, and persistent tiredness. Some patients even had suicidal thoughts during the COVID-19 pandemic (WHO, 2022).

The condition of depression is marked by enduring sadness and loss of enthusiasm. Various emotional and physical issues can arise from clinical depression or MDD which affect our emotions, thoughts, and behavior. In addition, depression can have negative effects on well-being and functioning in one's work, school, and family. People tend to have frequent episodes. Symptoms of these episodes may manifest throughout the day, almost daily, and can include feelings of sadness, tears, emptiness, or hopelessness. In addition, they may experience angry outbursts, irritability, or frustration, even in small matters. Most common activities, such as leisure pursuits, hobbies, or sports, cease to engage them. Each year, suicide rates reach almost 0.8 million. The fourth highest cause of death for 15–19-year-olds is depression, as identified by WHO in 2021. The WHO has estimated that depression affected over 300 million individuals worldwide by 2021. The diseases that cause disability are also significant. Therefore, depression causes a significant burden of disease. Depression affects the adult population at around 5% prevalence in different cultures and 20% in milder forms. Adults who are middle-aged are at the highest risk. Worldwide depression increased by 18% between 2005 and 2015 (Weinberger et al., 2018).

To gather information on the social and economic effects of COVID-19 on American households, the US Census Bureau partnered with five federal agencies to initiate the Household Pulse Survey to evaluate the pandemic's impact on employment status, consumer spending, food security, housing, education disruptions, and measures of physical and mental well-being. The survey was created to provide precise and weekly estimates (United States Census Bureau, 2023).

Differential equations, such as ordinary differential equations, are seldom used to model problems in science and engineering (Hasan et al., 2011). Many ordinary differential equations are nonlinear and cannot be solved analytically. Therefore, approximation methods such as numerical methods are beneficial. The time-step size restriction is a common feature of conventional finite difference methods, but it can lead to undesirable solution behavior. Several researchers, beginning with Mickens (1993), suggested nonstandard finite difference methods (NSFD) to maintain certain essential features of the nonlinear ODE. After that, many researchers continued to develop modified methods based on NSFD methods and applied them to various problems in science and engineering, such as Ebola virus transmission (Anguelov et al., 2020) and epidemiological models (Anguelov et al., 2014). Researchers apply the NSFD to different standard methods, including the second-order explicit Euler method (Gupta et al., 2020), a modified nonstandard theta method (Gupta et al., 2021), a nonstandard trimean method (Othman & Hasan, 2017), a harmonic-polygon modified Euler method (Yusop et al., 2017a), a harmonic polygon Euler method (Yusop et al., 2017b), a nonstandard harmonic Euler method on a stiff problem (Yusop & Hasan, 2015), a nonstandard Gauss–Seidel with exponential space step size, a hyperbolic step size on MHD convection in a porous cavity (Yaacob et al., 2022, p. 64), and nonstandard successive over-under-relaxation with sinus and hyperbolic sinus step size (Yaacob & Hasan, 2015). The NSFD was recently utilized by Hasan et al. (2022b, p. 114) to simulate macroeconomic interaction. The Nelder–Mead and least-square methods were employed to align the observed data with the dynamic macroeconomic model. Hasan et al. (2022a, p. 128) utilized a similar approach to fit crude oil price data to various ordinary differential models. All the studies show that nonstandard schemes produce awe-inspiring simulation results.

This study used a numerical method to anticipate analyzing indicators of anxiety or depression symptoms. We utilize mathematical methods to analyze because mathematics gives ways to simulate and understand patterns, quantify relationships, and forecast the future. This chapter applies Lagrange and differentiation to generate an ordinary differential equation that approximately models the rate of anxiety or depression symptoms frequency. Then we propose an exponential modified Euler method to simulate the occurrence frequency of the symptoms.

14.2 MATERIAL

This chapter analyzes data on anxiety or depression symptoms of US citizens during the COVID-19 pandemic from US Census Bureau website. However, we only take ten consecutive monthly data, as given in Table 14.1.

TABLE 14.1 Anxiety or Depression Indicators of Ten Consecutive Months

M1	M2	M3	M4	M5	M6	M7	M8	M9	M10
41.1	56.9	43.9	43.2	41.2	38.4	41.7	36.9	34.6	33.4

14.3 MODEL CONSTRUCTION

The first step of this analysis is to construct an approximate polynomial model for data in Table 14.1. This chapter uses the Lagrange interpolating polynomial method to construct an approximate polynomial. The method can construct, at most, a polynomial of degree n for a data-set that consists of $n + 1$ data points. Many researchers have used the method to solve various problems such as transportation (Roy, 2015), mobility prediction (Li et al., 2016), reversible data hiding schemes (Jana, 2017), multi-secret sharing schemes (Cheraghi, 2014), agriculture (Celik, 2018a), income generation (Kira, 2019), milk production (Celik, 2018b), heat conduction (Prasopchingchana, 2017), and incompressible fluid (Prasopchingchana, 2021).

The general formulation of Lagrange interpolating polynomial is given by

$$P_n(t)=\sum_{j=0}^{k} L_j(t)f(t) \text{ where } L_j(t)=\prod_{\substack{m=0\\ m\neq j}}^{k}\frac{t-t_m}{t_j-t_m}.$$

From the indicator of symptoms data, we construct $L_j(x)$ for $j=0,1,2,9$.

$$L_0(t)=\frac{(t-t_1)(t-t_2)(t-t_3)(t-t)(t-t_5)(t-t_6)(t-t_7)(t-t_8)(t-t_9)}{(t_0-t_1)(t_0-t_2)(t_0-t_3)(t_0-t_4)(t_0-t_5)(t_0-t_6)(t_0-t_7)(t_0-t_8)(t_0-t_9)}.$$

Replacing known observed data inside the formula, we get

$$L_0(t)=\frac{(t-2)(t-3)(t-4)(t-5)(t-6)(t-7)(t-8)(t-9)(t-10)}{(1-2)(1-3)(1-4)(1-5)(1-6)(1-7)(1-8)(1-9)(1-10)}.$$

Thus, we arrive at

$$L_0(t)=10-19.289683t+15.855754t^2-7.3185736t^3+2.0975694t^4-0.3882523t^5$$
$$+0.0465278t^6-0.0034888t^7+0.0001488t^8-0.0000028t^9.$$

Next, we calculate

$$L_1(t)=\frac{(t-t_0)(t-t_2)(t-t_3)(t-t_4)(t-t_5)(t-t_6)(t-t_7)(t-t_8)(t-t_9)}{(t_1-t_0)(t_1-t_2)(t_1-t_3)(t_1-t_4)(t_1-t_5)(t_1-t_6)(t_1-t_7)(t_1-t_8)(t_1-t_9)}.$$

Replacing known observed data inside the formula, we get

$$L_1(t) = \frac{(t-1)(t-3)(t-4)(t-5)(t-6)(t-7)(t-8)(t-9)(t-10)}{(2-1)(2-3)(2-4)(2-5)(2-6)(2-7)(2-8)(2-9)(2-10)}.$$

Thus, we get

$$L_1(t) = -45 + 109.30357t - 103.50268t^2 + 52.533135t^3 - 16.106076t^4 + 3.1331597t^5 - 0.3899306t^6 + 0.0301091t^7 - 0.0013145t^8 + 0.0000248t^9.$$

Next, we calculate

$$L_2(t) = \frac{(t-t_0)(t-t_1)(t-t_3)(t-t_4)(t-t_5)(t-t_6)(t-t_7)(t-t_8)(t-t_9)}{(t_2-t_0)(t_2-t_1)(t_2-t_3)(t_2-t_4)(t_2-t_5)(t_2-t_6)(t_2-t_7)(t_2-t_8)(t_2-t_9)}.$$

Replacing known observed data inside the formula, we get

$$L_2(t) = \frac{(t-1)(t-2)(t-4)(t-5)(t-6)(t-7)(t-8)(t-9)(t-10)}{(3-1)(3-2)(3-4)(3-5)(3-6)(3-7)(3-8)(3-9)(3-10)}.$$

Thus, we get

$$L_2(t) = 120 - 311.47619t + 317.91984t^2 - 172.11865t^3 + 55.620833t^4 - 11.289583t^5 + 1.4541667t^6 - 0.1154762t^7 + 0.0051587t^8 - 0.0000992t^9.$$

Next, we calculate

$$L_3(t) = \frac{(t-t_0)(t-t_1)(t-t_2)(t-t_4)(t-t_5)(t-t_6)(t-t_7)(t-t_8)(t-t_9)}{(t_3-t_0)(t_3-t_1)(t_3-t_2)(t_3-t_4)(t_3-t_5)(t_3-t_6)(t_3-t_7)(t_3-t_8)(t_3-t_9)}.$$

Replacing known observed data inside the formula, we get

$$L_3(t) = \frac{(t-1)(t-2)(t-3)(t-5)(t-6)(t-7)(t-8)(t-9)(t-10)}{(4-1)(4-2)(4-3)(4-5)(3-6)(3-7)(3-8)(3-9)(3-10)}.$$

Thus, we get

$$L_3(t) = -210 + 562.58333t - 597.40833t^2 + 337.3088t^3 - 113.41181t^4 + 23.849306t^5 - 3.1680556t^6 + 0.2583333t^7 - 0.0118056t^8 + 0.0002315t^9.$$

Next, we calculate

$$L_4(t)=\frac{(t-t_0)(t-t_1)(t-t_2)(t-t_3)(t-t_5)(t-t_6)(t-t_7)(t-t_8)(t-t_9)}{(t_4-t_0)(t_4-t_1)(t_4-t_2)(t_4-t_3)(t_4-t_5)(t_4-t_6)(t_4-t_7)(t_4-t_8)(t_4-t_9)}.$$

Replacing known observed data inside the formula, we get

$$L_4(t)=\frac{(t-1)(t-2)(t-3)(t-4)(t-6)(t-7)(t-8)(t-9)(t-10)}{(5-1)(5-2)(5-3)(5-4)(5-6)(5-7)(5-8)(5-9)(5-10)}.$$

Thus, we get

$$L_4(t)=252-687.7t+748.125t^2-434.36806t^3+150.41319t^4-32.560069t^5$$
$$+4.4444444t^6-0.3715278t^7+0.0173611t^8-0.0003472t^9.$$

Next, we calculate

$$L_5(t)=\frac{(t-t_0)(t-t_1)(t-t_2)(t-t_3)(t-t_4)(t-t_6)(t-t_7)(t-t_8)(t-t_9)}{(t_5-t_0)(t_5-t_1)(t_5-t_2)(t_5-t_3)(t_5-t_4)(t_5-t_6)(t_5-t_7)(t_5-t_8)(t_5-t_9)}.$$

Replacing known observed data inside the formula, we get

$$L_5(t)=\frac{(t-1)(t-2)(t-3)(t-4)(t-5)(t-7)(t-8)(t-9)(t-10)}{(6-1)(6-2)(6-3)(6-4)(6-5)(6-7)(6-8)(6-9)(6-10)}.$$

Thus, we get

$$L_5(t)=-210+580.08333t-641.37361t^2+379.76528t^3-134.44479t^4$$
$$+29.794792t^5-4.1645833t^6+0.35625t^7-0.0170139t^8+0.0003472t^9.$$

Next, we calculate

$$L_6(t)=\frac{(t-t_0)(t-t_1)(t-t_2)(t-t_3)(t-t_4)(t-t_5)(t-t_7)(t-t_8)(t-t_9)}{(t_6-t_0)(t_6-t_1)(t_6-t_2)(t_6-t_3)(t_6-t_4)(t_6-t_5)(t_6-t_7)(t_6-t_8)(t_6-t_9)}.$$

Replacing known observed data inside the formula, we get

$$L_6(t)=\frac{(t-1)(t-2)(t-3)(t-4)(t-5)(t-6)(t-8)(t-9)(t-10)}{(7-1)(7-2)(7-3)(7-4)(7-5)(7-6)(7-8)(7-9)(7-10)}.$$

Thus, we get

$$L_6(x) = 120 - 334.33333t + 373.98333t^2 - 224.66574t^3 + 80.898611t^4 - 18.272917t^5 + 2.6069444t^6 - 0.2277778t^7 + 0.0111111t^8 - 0.0002315t^9.$$

Next, we calculate

$$L_7(t) = \frac{(t-t_0)(t-t_1)(t-t_2)(t-t_3)(t-t_4)(t-t_5)(t-t_6)(t-t_8)(t-t_9)}{(t_7-t_0)(t_7-t_1)(t_7-t_2)(t_7-t_3)(t_7-t_4)(t_7-t_5)(t_7-t_6)(t_7-t_8)(t_7-t_9)}.$$

Replacing known observed data inside the formula, we get

$$L_7(t) = \frac{(t-1)(t-2)(t-3)(t-4)(t-5)(t-6)(t-7)(t-9)(t-10)}{(8-1)(8-2)(8-3)(8-4)(8-5)(8-6)(8-7)(8-9)(8-10)}.$$

Thus, we get

$$L_7(t) = -45 + 126.17857t - 142.38214t^2 + 86.486706t^3 - 31.561806t^4 + 7.2409722t^5 - 1.0513889t^6 + 0.0936508t^7 - 0.0046627t^8 + 0.0000992t^9.$$

Next, we calculate

$$L_8(t) = \frac{(t-t_0)(t-t_1)(t-t_2)(t-t_3)(t-t_4)(t-t_5)(t-t_6)(t-t_7)(t-t_9)}{(t_8-t_0)(t_8-t_1)(t_8-t_2)(t_8-t_3)(t_8-t_4)(t_8-t_5)(t_8-t_6)(t_8-t_7)(t_8-t_9)}.$$

Replacing known observed data inside the formula, we get

$$L_8(t) = \frac{(t-1)(t-2)(t-3)(t-4)(t-5)(t-6)(t-7)(t-8)(t-10)}{(9-1)(9-2)(9-3)(9-4)(9-5)(9-6)(9-7)(9-8)(9-10)}.$$

Thus, we get

$$L_8(t) = 10 - 28.178571t + 32.014484t^2 - 19.617163t^3 + 7.2364583t^4 - 1.6817708t^5 + 0.2479167t^6 - 0.0224702t^7 + 0.0011409t^8 - 0.0000248t^9.$$

And lastly, we calculate

$$L_9(t) = \frac{(t-t_0)(t-t_1)(t-t_2)(t-t_3)(t-t_4)(t-t_5)(t-t_6)(t-t_7)(t-t_8)}{(t_9-t_0)(t_9-t_1)(t_9-t_2)(t_9-t_3)(t_9-t_4)(t_9-t_5)(t_9-t_6)(t_9-t_7)(t_9-t_8)}.$$

Replacing known observed data inside the formula, we get

$$L_9(t) = \frac{(t-1)(t-2)(t-3)(t-4)(t-5)(t-6)(t-7)(t-8)(t-9)}{(10-1)(10-2)(10-3)(10-4)(10-5)(10-6)(10-7)(10-8)(10-9)}.$$

Thus, we get

$$L_9(t) = -1 + 2.8289683t - 3.2316468t^2 + 1.9942681t^3 - 0.7421875t^4 + 0.1743634t^5 - 0.0260417t^6 + 0.0023975t^7 - 0.000124t^8 + 0.0000028t^9.$$

Then, we can start constructing the approximate polynomial by replacing all the $L_j(t)$, $j = 0, 1, \ldots$, nine that we have calculated inside Equation (14.1).

$$y(t) \approx P_9(t) = \sum_{j=0}^{9} L_j(t) f(t). \tag{14.1}$$

Yielding Equation (14.2).

$$y(t) \approx 21 - 167.87925t + 445.98152t^2 - 398.25119t^3 + 180.91641t^4 - 47.512234t^5 + 7.5157292t^6 - 0.706541t^7 + 0.0363467t^8 - 0.0007876t^9. \tag{14.2}$$

Then, we derivate Equation (14.2) to get the approximate ordinary differential equation displayed in Equation (14.3).

$$\frac{dy}{dt} \approx f(x) = -167.87925 + 891.96304t - 1194.7536t^2 + 723.66562t^3 - 237.56117t^4 + 45.094375t^5 - 4.945787t^6 + 0.2907738t^7 - 0.0070883t^8. \tag{14.3}$$

14.4 GENERATE PREDICTION DATA

We present three numerical approaches for determining the symptoms of anxiety or depression. The newly proposed method is compared using the two existing numerical methods as control. These control methods include the fourth-order Runge–Kutta method and the modified Euler method. The fourth-order Runge–Kutta method was chosen since it is one of the most accurate and popular methods frequently applied to solve various science and engineering problems. While the modified Euler method, which is sometimes also known as the second-order Runge–Kutta method, was chosen since this is the primary method that we modified to produce the new method proposed in this chapter.

14.5 THE FOURTH-ORDER RUNGE–KUTTA METHOD

The fourth-order Runge–Kutta method is a numerical technique consisting of five equations as given in Equation (14.4) with $f(x,y)$ given by Equation (14.3).

$$\begin{aligned}
k_1 &= f(t_i, y_i) \\
k_2 &= f\left(t_i + \frac{h}{2}, y_i + \frac{k_1}{2}\right) \\
k_3 &= f\left(t_i + \frac{h}{2}, y_i + \frac{k_2}{2}\right) \\
k_4 &= f(t_i + h, y_i + k_3) \\
y_{i+1} &= y_i + \frac{h}{6}(k_1 + 2k_2 + 2k_3 + k_4), i = 0,1,2\ldots,n-1
\end{aligned} \tag{14.4}$$

Researchers have utilized the Runge–Kutta method for solving various problems such as COVID-19 cases (Iskandar & Tiong, 2022), the concentration of bacteria (Gowri et al., 2017), transient analysis (Henry et al., 2019), dynamic force identification (Lai et al., 2017), determine deviation in the weapon arm vehicle (Nugraha, 2015), *Lü* chaotic system (Mehdi & Kareem, 2017), epidemic model (Kovalnogov et al., 2020), initial value problem (Islam, 2015), and system of first-order ordinary differential equations (Abraha, 2020).

14.6 MODIFIED EULER METHOD

The formula of the modified Euler method to generate the prediction data is given by Equation (14.5) with $f(x,y)$ was given by Equation (14.3).

$$y_{i+1} = y_i + hf\left(t_i + \frac{h}{2}, y_i + \frac{h}{2} f(t_i, y_i)\right) \tag{14.5}$$

Researchers have utilized the modified Euler method for solving various problems such as a system of first-order ordinary differential equations (Abraha, 2020), COVID-19 (Nazir et al., 2021), and tension leg problem (TLP) (Tabeshpour et al., 2013).

14.7 THE EXPONENTIAL MODIFIED EULER METHOD

The exponential modified Euler method formula to generate the prediction data is given by Equation (14.6) with $f(x,y)$ was given by Equation (14.3). In this proposed method, we substitute the h in the outer part of the Equation (14.3) with an exponential type of function, $f(h)=\frac{1-e^{-\frac{h}{4}}}{\frac{1}{4}}$. The resulting exponential modified Euler formula is given in Equation (14.6).

$$y_{i+1}=y_i+\left(\frac{1-e^{-\frac{h}{4}}}{\frac{1}{4}}\right)f\left(t_i+\frac{h}{2},y_i+\frac{h}{2}f\left(t_i,y_i\right)\right) \tag{14.6}$$

14.8 STABILITY AND CONSISTENCY ANALYSIS

The stability analysis for the exponential modified Euler method can be obtained by using Dahlquist's test problem (Corless et al., 2019).

$$\frac{dy}{dt}=\lambda y_i;y(t_0)=y_0,\lambda\in\mathbb{C} \tag{14.7}$$

To check the stability of the proposed method, we substitute Equation (14.7) into Equation (14.6), and we obtain $y_{i+1}=y_i+H(h)\lambda y_i+0.5H\left(h^2\right)\lambda^2 y_i$, where $z=H(h)\lambda$ with $H(h)=\frac{1-e^{-\frac{h}{4}}}{\frac{1}{4}}$. Hence, the stability region of the proposed method is the region given by $|G(z)|=\left|\frac{y_{i+1}}{y_i}\right|=\left|1+z+\frac{z^2}{2}\right|\leq 1$. The stability region plot using octave software is given in Figure 14.1.

The consistency of the initial value problem can be checked by taking the limit of the EME formula for $h\to 0$. Let us rewrite Equation (14.6) as $y_{i+1}=y_i+H(h)Y\left(t_i,y_i,h\right)$. Then take the limit of the term with the h parameter (third term). Taking the h approaching zero will impact $H(h)$ approaching h and the $Y\left(t_i,y_i,h\right)$ approaching $f\left(x_i,y_i\right)$. Thus,

$$\lim_{h\to 0}H(h)Y\left(t_i,y_i,h\right)=hf\left(x_i,y_i\right)$$

which shows that the proposed method is consistent.

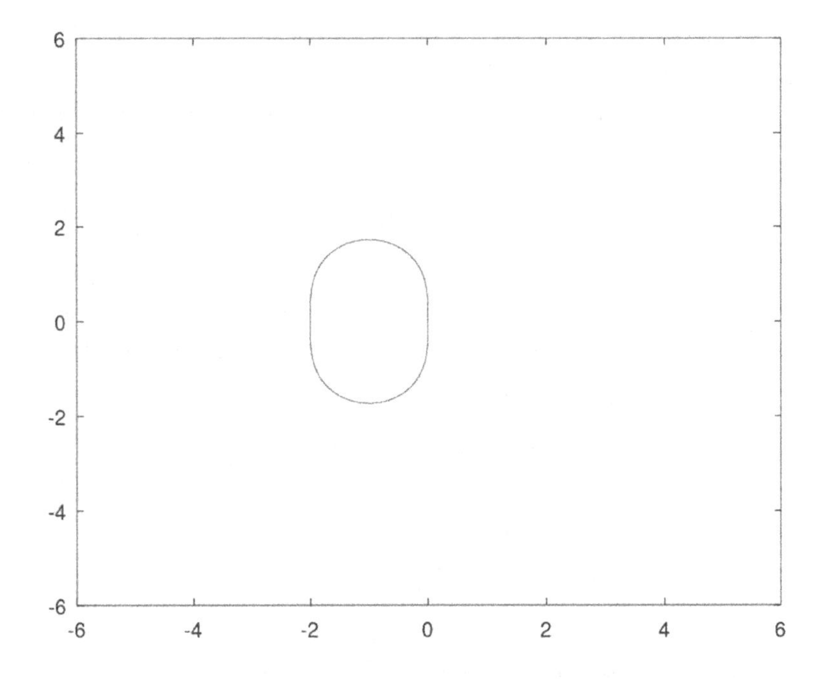

FIGURE 14.1 Stability region for exponential modified Euler method.

Algorithm 14.1: Algorithm for Lagrange Exponential Modified Euler

1. Start Coding
2. Gather data from the user
3. Apply Lagrange (L) to Construct Polynomial, $P(x)$.
4. Find derivation for Polynomial, $P'(x)$.
5. Gather initial data from the user
6. Discretize solution domain
7. Start timing
8. Calculate the denominator exponential function
9. Calculate predicted observation using Exponential Modified Euler (EME)
10. Calculate APE and MAPE
11. Stop timing
12. Display result
13. End Coding

14.9 RESULTS AND DISCUSSIONS

We analyze the performance of the modified Euler, RK4, and exponential modified Euler by computing the indicators of anxiety, and depression symptoms approximation occurred in ten time periods by using five sizes of time steps which are $h = 0.1, 0.05, 0.01, 0.005, 0.001$. The results output was fascinating since the exponential modified Euler method shows an outstanding accuracy even though the method is second order while RK4 is fourth order. The algorithm for the exponential modified Euler method is given in Algorithm 14.1. The results are shown in tables (Table 14.2–Table 14.6) and figures (Figure 14.2–Figure 14.8).

TABLE 14.2 Comparison of Observed Data and Prediction by Numerical Methods with h = 0.1

Data	Observed	LME	LRK4	LEME
1	41.1	41.1	41.1	41.1
2	56.9	56.87691	56.89989	56.68133
3	43.9	43.85915	43.8996	43.82495
4	43.2	43.1674	43.19876	43.14177
5	41.2	41.1638	41.19612	41.16301
6	38.4	38.3506	38.38739	38.38468
7	41.7	41.63102	41.65977	41.62444
8	36.9	36.749	36.78042	36.80294
9	34.6	34.2223	34.27448	34.30756
10	33.4	32.78262	32.58715	32.88572

TABLE 14.3 Comparison of Observed Data and Prediction by Numerical Methods with h = 0.05

Data	Observed	LME	LRK4	LEME
1	41.1	41.1	41.1	41.1
2	56.9	56.89419	56.89992	56.79588
3	43.9	43.88953	43.89963	43.87217
4	43.2	43.19096	43.19879	43.17794
5	41.2	41.18808	41.19614	41.18753
6	38.4	38.37823	38.38741	38.39517
7	41.7	41.65263	41.6598	41.64919
8	36.9	36.7726	36.78045	36.79954
9	34.6	34.2615	34.27453	34.30406
10	33.4	32.63621	32.58728	32.68889

TABLE 14.4 Comparison of Observed Data and Prediction by Numerical Methods with h = 0.01

Data	Observed	LME	LRK4	LEME
1	41.1	41.1	41.1	41.1
2	56.9	56.89969	56.89992	56.87996
3	43.9	43.89923	43.89963	43.89573
4	43.2	43.19847	43.19879	43.19585
5	41.2	41.19582	41.19615	41.1957
6	38.4	38.38705	38.38741	38.39044
7	41.7	41.65952	41.6598	41.65882
8	36.9	36.78013	36.78045	36.78553
9	34.6	34.27401	34.27453	34.28253
10	33.4	32.58925	32.58729	32.59988

TABLE 14.5 Comparison of Observed Data and Prediction by Numerical Methods with h = 0.005

Data	Observed	ME	RK4	EME
1	41.1	41.1	41.1	41.1
2	56.9	56.89986	56.89992	56.88999
3	43.9	43.89953	43.89963	43.89778
4	43.2	43.19871	43.19879	43.1974
5	41.2	41.19607	41.19615	41.19601
6	38.4	38.38732	38.38741	38.38902
7	41.7	41.65973	41.6598	41.65938
8	36.9	36.78037	36.78045	36.78307
9	34.6	34.2744	34.27453	34.27866
10	33.4	32.58778	32.58729	32.5931

TABLE 14.6 Comparison of Observed Data and Prediction by Numerical Methods with h = 0.005

Data	Observed	ME	RK4	EME
1	41.1	41.1	41.1	41.1
2	56.9	56.89992	56.89992	56.89794
3	43.9	43.89963	43.89963	43.89928
4	43.2	43.19878	43.19879	43.19852
5	41.2	41.19614	41.19615	41.19613
6	38.4	38.38741	38.38741	38.38775
7	41.7	41.6598	41.6598	41.65973
8	36.9	36.78044	36.78045	36.78098
9	34.6	34.27452	34.27453	34.27538
10	33.4	32.58731	32.58729	32.58837

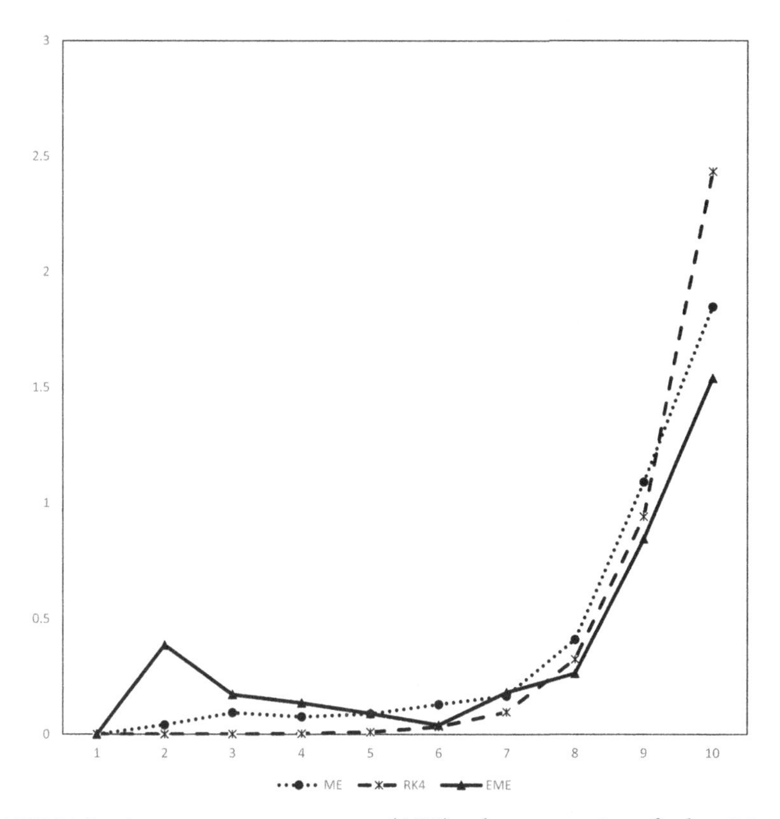

FIGURE 14.2 Average percentage error (APE) value comparison for h = 0.1.

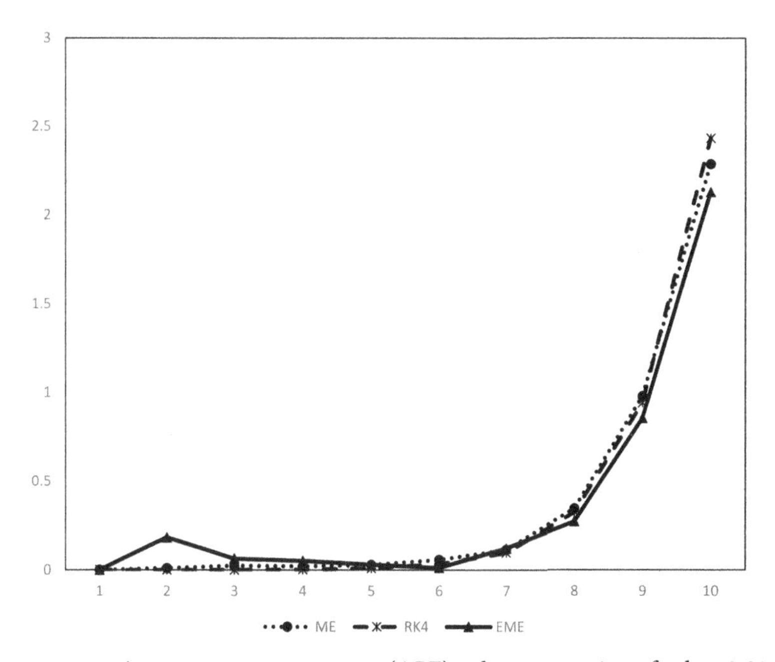

FIGURE 14.3 Average percentage error (APE) value comparison for h = 0.05.

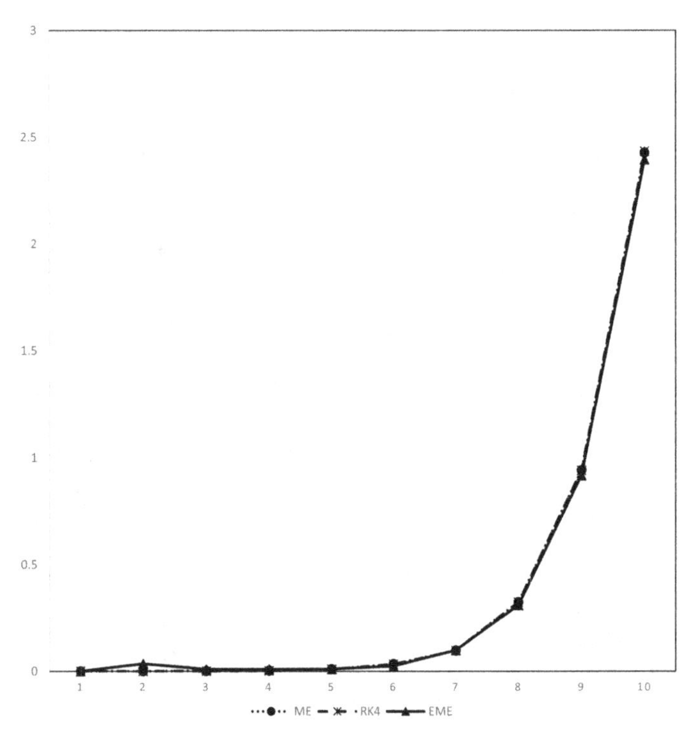

FIGURE 14.4 Average percentage error (APE) value comparison for h = 0.01.

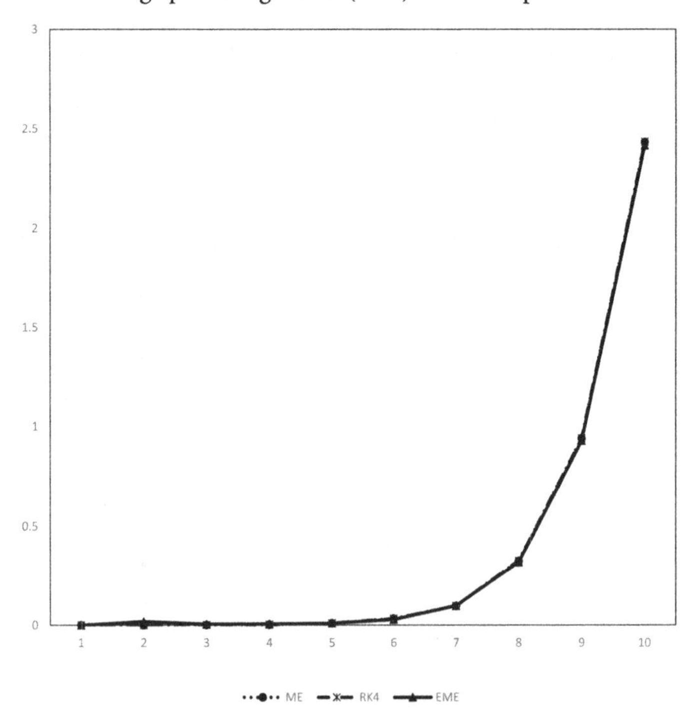

FIGURE 14.5 Average percentage error (APE) value comparison for h = 0.005.

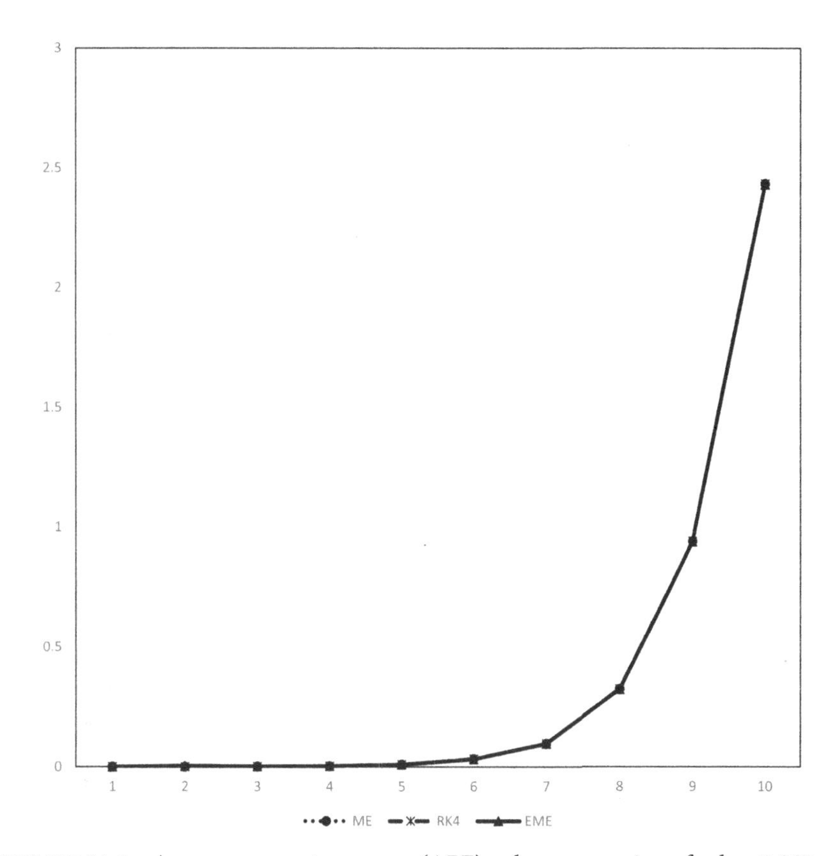

FIGURE 14.6 Average percentage error (APE) value comparison for h = 0.001.

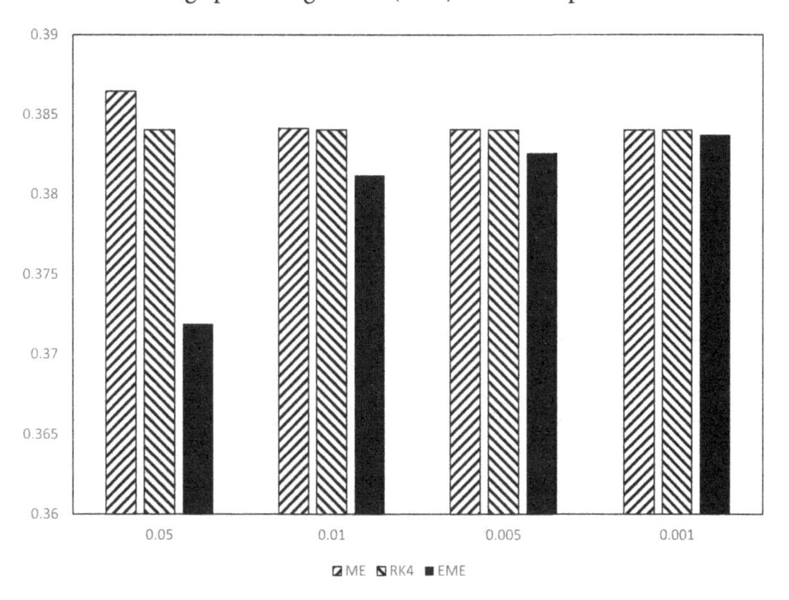

FIGURE 14.7 Mean average percentage error (MAPE) value comparison for various step sizes.

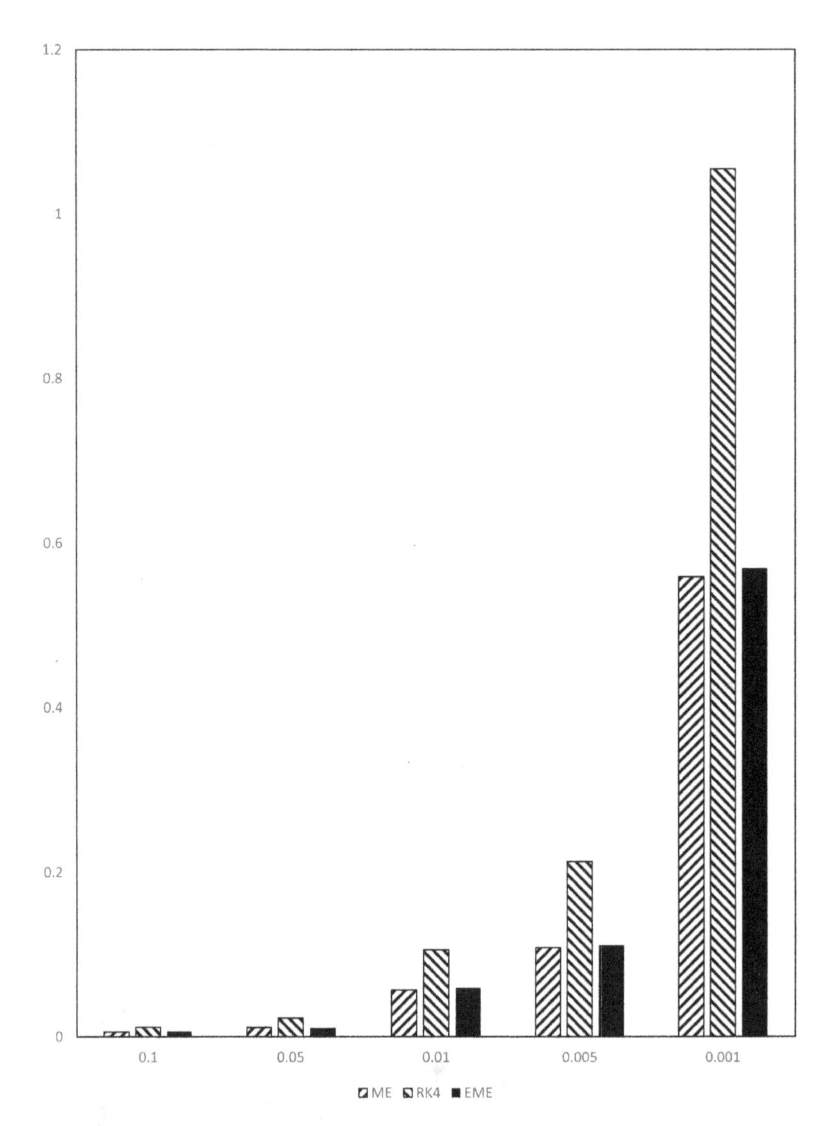

FIGURE 14.8 Computational time comparison for various step sizes.

14.10 CONCLUSION

It is worth noting that the values of h affect the accuracy and efficiency of the computation. Theory always shows that the smaller the step size, the more accurate the computation, but it will also be more computationally extensive. However, findings in this chapter show that only modified Euler method shows those characteristics. While fourth-order Runge–Kutta method shows above given characteristic until $h = 0.01$, and after that, the accuracy looks saturated. In contrast, exponential modified Euler method shows opposite of that characteristic. The exponential modified Euler predicted the highest accuracy when using $h = 0.1$. The accuracy is reduced when the h becomes smaller. The exponential function of h may cause the explanation of this phenomenon to be used instead of only the step size h in the denominator formulation. The findings also show that the exponential modified Euler method with any size of h is more accurate than the modified Euler and fourth-order Runge–Kutta methods, even though the exponential modified Euler is of order two. The method also computed faster than fourth-order Runge–Kutta method by 77.09%–108.49% and only slower than modified Euler by 1.87%–10.99%. With these findings, the exponential modified Euler is the best numerical method compared to the modified Euler and fourth-order Runge–Kutta methods.

ACKNOWLEDGMENT

We acknowledge the Malaysian Ministry of Higher Education FRGS Grants FRGS/1/2021/ICT06/UKM/02/2 for sponsoring this research.

REFERENCES

Abraha, J. D. (2020). Comparison of Numerical Methods for System of First Order Ordinary Differential Equations. *Pure and Applied Mathematics Journal*, *9*(2), 32. https://doi.org/10.11648/j.pamj.20200902.11

Anguelov, R., Berge, T., Chapwanya, M., Djoko, J. K., Kama, P., Lubuma, J. M. -S., & Terefe, Y. (2020). Nonstandard finite difference method revisited and application to the Ebola virus disease transmission dynamics. *Journal of Difference Equations and Applications*, *26*(6), 818–854. https://doi.org/10.1080/10236198.2020.1792892.

Anguelov, R., Dumont, Y., Lubuma, J. M. -S, & Shillor, M. (2014). Dynamically consistent nonstandard finite difference schemes for epidemiological models. *Journal of Computational and Applied Mathematics*, 255, 161–182. https://doi.org/10.1016/j.cam.2013.04.042

Celik, S. (2018a). Lagrange interpolating polynomial and an application in agriculture. *International Journal of Latest Engineering and Management Research*, *3*(7), 81–87. http://www.ijlemr.com/papers/volume3-issue7/12-IJLEMR-33264.pdf

Celik, S. (2018b). Using Lagrange interpolation to determine the milk production amount by the number of milked animals. *American Journal of Engineering Research*, *7*(8), 264–271. ZZF0708264271.pdf (ajer.org)

Cheraghi, A. (2014). Sharing Several Secrets Based on Lagrange's Interpolation Formula and Cipher Feedback Mode. *International Journal of Nonlinear Analysis and Applications*, *2*(5), 60–66. DOI: 10.22075/ijnaa.2014.127

Chisholm, D., Sweeny, K., Sheehan, P., Rasmussen, B., Smit, F., Cuijpers, P., & Saxena, S. (2016). Scaling-up treatment of depression and anxiety: A global return on investment analysis. *Lancet Psychiatry*, *3*, 415–24. https://doi.org/10.1016/S2215-0366(16)30024-4

Corless, R. M., Kaya, C. Y., Moir, R. H. C. (2019). Optimal residuals and the Dahlquist test problem. *Numerical Algorithms, 81*(4), 1253–1274. https://link.springer.com/article/10.1007/s11075-018-0624-x

Gowri, P., Priyadharsini, S., & Maheswari, T. (2017). A case study on Runge-Kutta 4th order differential equations and its application. *Imperial Journal of Interdisciplinary Research, 3*(2), 134–139. https://www.researchgate.net/publication/320124650_A_CaseStudy_onRunge_Kutta_4_th_Order_Differential_Equations_and_Its_application

Grillon, C., Robinson, O. J., & Cornwell, B. et al. (2019). Modeling anxiety in healthy humans: A key intermediate bridge between basic and clinical sciences. *Neuropsychopharmacology, 44,* 1999–2010. https://doi.org/10.1038/s41386-019-0445-1

Gupta, M., Slezak, J. M., Alalhareth, F. K., Roy, S. & Kojouharov, H. V. (2020). *Second-order nonstandard explicit Euler method,* AIP Conference Proceedings, 2302(1), 110003. https://doi.org/10.1063/5.0033534

Gupta, M., Slezak, J.M., Alalhareth, F. K., Roy, S. & Kojouharov, H.V. (2021). Second-order modified nonstandard Runge-Kutta and theta methods for one-dimensional autonomous differential equations. *International Journal of Applications and Applied Mathematics, 16*(2), 788–803. https://digitalcommons.pvamu.edu/cgi/viewcontent.cgi?article=1877&context=aam

Hasan, M. K., Othman, N. A., & Idrus, B. (2022b). Data-driven Macroeconomic model analysis using nonstandard trimean algorithm in S. A. Abdul Karim (ed.), *Intelligent Systems Modeling and Simulation II, Studies in Systems, Decision and Control 444,* pp. 113–126. Springer Nature. https://link.springer.com/chapter/10.1007/978-3-031-04028-3_9

Hasan, M. K., Sulaiman, J., & Abdul Karim, S. A. (2022a). Data-driven ordinary differential equations model for predicting Missing Data and Forecasting Crude Oil Prices in S. A. Abdul Karim (ed.), *Intelligent Systems Modeling and Simulation II, Studies in Systems, Decision and Control 444,* pp. 127–143. Springer Nature. https://link.springer.com/chapter/10.1007/978-3-031-04028-3_10

Hasan, M. K., Sulaiman, J., Karim, S. A. A., & Othman, M. (2011). Development of some numerical methods applying complexity reduction approach for solving scientific problems. *Journal of Applied Sciences, 11*(7), 1255–1260. https://doi.org/10.3923/jas.2011.1255.1260

Henry, O., Albert, B., & Justice, I. (2019). Application of numerical methods in transient analysis. *International Journal of Science and Research, 8*(5), 2108–2112. https://www.ijsr.net/archive/v8i5/ART20198178.pdf

Iskandar, D., & Tiong, O. C. (2022). *The application of the Runge-Kutta Fourth Order Method in SIR Model for simulation of COVID-19 Cases.* Proceedings of Science and Mathematics, 10, 61–70. https://science.utm.my/procscimath/wp-content/uploads/sites/605/2022/10/61-70-Danial-Iskandar_Ong-Chee-Tiong.pdf

Islam, M. A. (2015). A comparative study on numerical solutions of initial value problems (IVP) for ordinary differential equations (ODE) with Euler and Runge-Kutta methods. *American Journal of Computational Mathematics, 5,* 393–404. http://dx.doi.org/10.4236/ajcm.2015.53034

Jana, B. (2017). Reversible data hiding scheme using sub-sampled image exploiting lagrange's interpolating polynomial. *Multimedia Tools and Applications*, *77*(7), 8805–8821. https://doi.org/10.1007/s11042-017-4775-x.

Kessler, R. C., Chiu, W. T., Demler, O., Merikangas, K. R., & Walters, E. E. (2005). Prevalence, severity, and comorbidity of 12-month DSM-IV disorders in the National Comorbidity Survey Replication. *Arch Gen Psychiatry*, 62, 590–592. doi: 10.1001/archpsyc.62.7.709

Kira, J. (2019). Application of Lagrange polynomial for interpolating income generation from certain students' fee structure. *Mathematical Theory and Modeling*, *9*(5), 55–64. Doi: 10.7176/MTM

Kovalnogov, V. N., Simos, T. E., Tsitouras, C. (2020). Runge-Kutta pairs suited for SIR-type epidemic model. *Mathematical Methods in the Applied Sciences*, *44*, 5210–5216. https://doi.org/10.1002/mma.7104

Lai, T., Yi, T.-H., Li, H.-N., & Fu, X. (2017). An explicit fourth order Runge-Kutta method for dynamic force identification. *International Journal of Structural Stability and Dynamics*, *17*(10), 1750120 (21 pages). https://doi.org/10.1142/S0219455417501206

Li, B. Zhang, H. & Lu, H. (2016). *User mobility prediction based on Lagrange's Interpolation in Ultra-dense Networks.* IEEE 27th Annual International Symposium on Personal, Indoor, and Mobile Radio Communications (PIMRC), Valencia, Spain, pp. 1–6. Doi: 10.1109/PIMRC.2016.7794984.

Mehdi, S. A., Kareem, R. S. (2017). Using fourth-order Runge-Kutta to solve Lü Chaotic System. *American Journal of Engineering Research*, *6*(1), 72–77. https://www.ajer.org/papers/v6(01)/K06017277.pdf

Mickens, R. (1993). Nonstandard Finite Difference Models of Differential Equations, *World Scientific*. Singapore. https://doi.org/10.1142/2081

Nazir, G., Zeb, A., Shah, K., Saeed, T., Ali Khan, R., & Ullah Khan, S. I. (2021). Study of COVID-19 mathematical model of fractional order via modified Euler method. *Alexandria Engineering Journal*, *60*, 5287–5296. https://doi.org/10.1016/j.aej.2021.04.032

Nugraha, A. S. (2015). The Selection of time step in Runge-Kutta fourth order to determine deviation in the weapon arm vehicle. *Energy Procedia*, *68*(2015), 363–369. https://doi.org/10.1016/j.egypro.2015.03.267

Othman, N. A., & Hasan, M. K. (2017). New Hybrid two-step method for simulating Lotka-Volterra model. *Pertanika Journal of Science and Technology*, *25*(56), 115–124. http://www.pertanika.upm.edu.my/resources/files/Pertanika%20PAPERS/JST%20Vol.%2025%20(S)%20Jun.%202017%20(View%20Full%20Journal).pdf

Prasopchingchana, U. (2017). *Modification of the Lagrange Interpolating Polynomial Scheme for Using with the Finite Difference Method.* MATEC Web of Conferences 136, 01005 (2017), 2017 2nd International Conference on Design, Mechanical, and Material Engineering (D2ME 2017). Doi: 10.1051/matecconf/201713601005.

Prasopchingchana, U. (2021). A method of the Lagrange polynomial interpolation with weighting factors for reducing computational time. *Journal of Mechanical Engineering*, *18*(1), 1–19, 2021. https://ir.uitm.edu.my/id/eprint/47613/1/47613.pdf.

Roy, S. K. (2015). Lagrange's Interpolating Polynomial Approach to Solve Multi-choice Transportation Problem. *International Journal of Applied and Computational Mathematics, 1*, 639–649. Doi: 10.1007/s40819-015-0041-y.

Stein, M. B., & Craske, M. G. (2017). Treating anxiety in 2017: Optimizing care to improve outcomes. *Journal of American Medical Association*, 318(3), 235–236. doi:10.1001/jama.2017.6996

Tabeshpour, M.R., Golafshani, A.A., and Seif, M.S. (2013). Stability of the modified Euler method for nonlinear dynamic analysis of TLP. *International Journal of Maritime Technology, 1*(1), 23–34. https://ijmt.ir/article-1-155-en.html

United States Census Bureau https://www.census.gov

Weinberger, A. H., Gbedemah, M., Martinez, A. M., Nash, D., Galea, S., & Goodwin, R. D. (2018). Trends in depression prevalence in the USA from 2005 to 2015: Widening disparities in vulnerable groups. *Psychological Medicine, 48*(8),1308–1315. doi:10.1017/S0033291717002781.

WHO. (2021). https://www-who-int.ezproxy.uio.no/news-room/fact-sheets/detail/suicide.

World Health Organization (WHO). (2022). Mental health and COVID-19: Early evidence of the pandemic's impact. Scientific Brief, 2 March 2022. 11 pages. https://iris.who.int/bitstream/handle/10665/352189/WHO-2019-nCoV-Sci-Brief-Mental-health-2022.1-eng.pdf?sequence=1&isAllowed=y

Yaacob, Z., & Hasan, M. K. (2015). *Nonstandard finite difference schemes for natural convection in an inclined porous rectangular cavity.* [Paper presentation]. ICEEI 2015: Bridging the Knowledge between Academic, Industry, and Community, Denpasar, Indonesia. Pp. 665–669, doi: 10.1109/ICEEI.2015.7352582.

Yaacob, Z., Hasan, M. K., & Idrus, B. (2022). Efficient Numerical Schemes for MHD Convection in Porous Cavity in S. A. Abdul Karim, A. Shafie (eds.), *Towards Intelligent Systems Modeling and Simulation: With Applications in Energy, Epidemiology and Risk Assessment, Studies in Systems, Decision and Control 383*, pp. 63–78. Springer Nature. https://link.springer.com/chapter/10.1007/978-3-030-79606-8_3

Yusop, N.M.M, & Hasan, M.K. (2015). Development of a new harmonic Euler using a nonstandard finite difference technique for solving stiff problems. *Jurnal Teknologi, 77*(20), 19–24. https://doi.org/10.11113/jt.v77.6546

Yusop, N.M.M, Hasan, M.K., Wook, M., Amran, M.F.M., & Ahmad, S.R. (2017a). Comparison new algorithm modified Euler based on harmonic-polygon approach for solving an ordinary differential equation. *Journal of Telecommunication, Electronic and Computer Engineering, 9*(2–11), 29–32. https://core.ac.uk/reader/235221710

Yusop, N.M.M, Hasan, M.K., Wook, M., Amran, M.F.M., & Ahmad, S.R. (2017b). *A new Euler scheme based on harmonic-polygon approach for solving first-order ordinary differential equations.* AIP Conference Proceedings 1891, 020102. https://doi.org/10.1063/1.5005435

CHAPTER 15

Hybridization of Simulated Kalman Filter and Minimization of Metabolic Adjustment for Succinate and Lactate Production

Nurul Syifa Kamarolzaman, Habibollah Haron, Yee Wen Choon, Nurul Athirah Nasarudin, Muhammad Akmal Remli, and Mohd Saberi Mohamad

15.1 INTRODUCTION

Recent advancements in techniques of molecular biology, experimental methodologies, and mathematical instruments have contributed to the growth of interest in the use of metabolic engineering. The gene knockout technique is a vital tool in molecular biology and finds widespread application in industrial metabolic engineering. This method is commonly used to analyze the functions of specific genes, protein structures, and biochemical production. By selecting a particular gene that encodes a certain protein and deleting or knocking it out, certain reactions can be prevented, affecting the production of compounds. In addition, researchers can infer

DOI: 10.1201/9781003400387-15

the functions of genes with unknown roles by comparing knockout genes with normal ones. However, this technique has limitations when dealing with genes that have irrelevant functions.

Escherichia coli is a microorganism commonly utilized for DNA and protein synthesis, with applications in various living organisms. It serves as a versatile host for foreign DNAs in laboratory procedures, and it can adapt to both aerobic and anaerobic conditions, making it a pioneering model for biotechnology. In the field of biotechnology, *E. coli* plays a vital role in facilitating the production of high yields of succinate and lactate for commercial purposes. However, when certain genes involved in the production of succinate are disrupted in combination, it hinders the improvement of succinate production.

Succinate is widely utilized as a foundational material in various chemical industries, including pharmaceuticals, agriculture, food products, and more. Similarly, lactate production is of great interest in biotechnology, specifically for the manufacturing of polylactate (PLA)-based plastic materials which are in high demand. However, the conventional methods employed to produce succinate and lactate have limitations, resulting in low yields compared to their theoretical maximum.

Minimization of metabolic adjustment (MOMA) is a modeling technique introduced by Segre et al. in 2012. MOMA is utilized to model and simulate mutant fluxes by sampling the solution space derived from wild-type Flux Balance Analysis (FBA), enabling more accurate predictions of metabolic phenotypes in gene knockout bacteria. By employing quadratic programming (QP), MOMA identifies a point in the flux space closest to the wild-type solution while adhering to the gene deletion constraint. Moreover, as a computational method for analyzing genetically modified bacteria, MOMA aids in understanding the mechanisms of cell adaptation to gene loss through regulatory and evolutionary optimization (Tang et al., 2015).

Prokaryotic organisms are widely utilized in various industrial applications that are driven by the increasing demand for biochemical compounds in fields such as automotive engineering, nonwovens, furniture, and food. However, challenges arise from the complex regulatory cellular and metabolic networks, coupled with a lack of effective modeling and optimization methods. Conventional approaches to increasing succinate and lactate production suffer from high processing costs, lengthy computing times, and the fact that succinate and lactate are considered minor products. Furthermore, the intricate nature of metabolic networks makes it difficult to identify the effects of genetic modifications on targeted phenotypes.

Many algorithms have been employed for enhancing the production of succinate and lactate, but they have not yielded high product yields due to inherent limitations. In addition, manipulating the central metabolic pathways in *E. coli* for increased production can result in slow growth and reduced metabolic capabilities. Furthermore, certain algorithms outperform simulated Kalman filter (SKF) in specific composition functions but fail to provide fitness values for common algorithms.

This chapter addresses the limitations found in previous research and introduces a combination of SKF and MOMA (SKFMOMA) as a solution for identifying gene knockout strategies that maximize the production of succinate and lactate in *E. coli*. The aim is to achieve improved results in this study.

15.2 MATERIALS AND METHODS

15.2.1 Parameter Settings of SKFMOMA

There are some parameters involved in optimizing the performance of SKFMOMA. The target reactions are declared as *targetRxn* in the MATLAB® where "*targetRxn* = *EX_succ(e)*" indicates as succinate, while "*targetRxn* = *EX_lac-D(e)*" represents lactate.

The number of population created in this algorithm must be higher than the number of iteration which is 100. Each population iterates once in each agent which basically summarize that the number of it iterates is 100 times. These agents can evaluate the fitness value of each population and update the best solution to date. These values indicate the best outcomes based on the SKFMOMA trial and error method.

Furthermore, SKFMOMA is running 50 times that it has 100 agents or can be said to have 100 iterations in each run with different number of gene knockouts, which is triple to quintuple. Table 15.1 shows the parameter setting for SKFMOMA. MOMA evaluates the population fitness value in each agent. The growth rate of an organism is fundamental in obtaining high yield of targeted biochemical product. Moreover, the growth rate of *Lactiplantibacillus plantarum* is only 0.1 after 48 hours under anaerobic conditions in the experiment by Teusink et al. (2009). Thus, it is significant to predict the growth rate of a microorganism using constraint-based model in order to get a high yield of desired biochemical products. At an early stage, a mutant cell is able to produce lots of demanded productions but the growth rate might drop to 0 after a period of time, this is meaningless because biochemical productions cannot occur if the cell is dead.

TABLE 15.1 Parameter Setting for SKFMOMA

Parameter	Value
Number of Population	100
Number of Maximum Iterations	1
Number of Agents	100
Noise Covariance Estimate, P(0)	500
Measurement Noise, R	0.5
Process Noise, Q	0.5
Number of Runs (Cycle)	50
Target reaction	Succinate→ *EX_succ(e)* *Lactate*→ *EX_lac-D(e)*
Number of Gene Knockouts	3, 4 and 5

Therefore, a value of 0.1 is set for growth rate in this research to ensure the cell must be alive before gene knockout is applied and every population with growth rate less than 0.1 unit per hour is removed.

15.2.2 A Hybrid of SKFMOMA

The development of SKFMOMA involves three primary stages, outlined as follows:

i. Initialization;

ii. Evaluating fitness using MOMA and assuming the true value; and

iii. Prediction, measurement, and estimation.

15.2.2.1 MOMA (Minimization of Metabolic Adjustment)

The objective functions in MOMA are production and growth rate of *E. coli* as shown in the second step of the flowchart at Figure 15.1. The fitness of each population for gene knockout is calculated by MOMA. This means that the point closest to the wild type represents the knockout gene, enabling the improvement of succinate and lactate production. Upon the creation of a new population, fitness calculations are conducted to evaluate the growth rate and production rates of lactate and succinate. Subsequently, these rates are employed to determine the most favorable solution among the population.

15.2.2.2 Initialization

SKFMOMA begins by generating a population at random with a matrix of 95 × 100. Ninety-five reactions are involved in this research. To guarantee

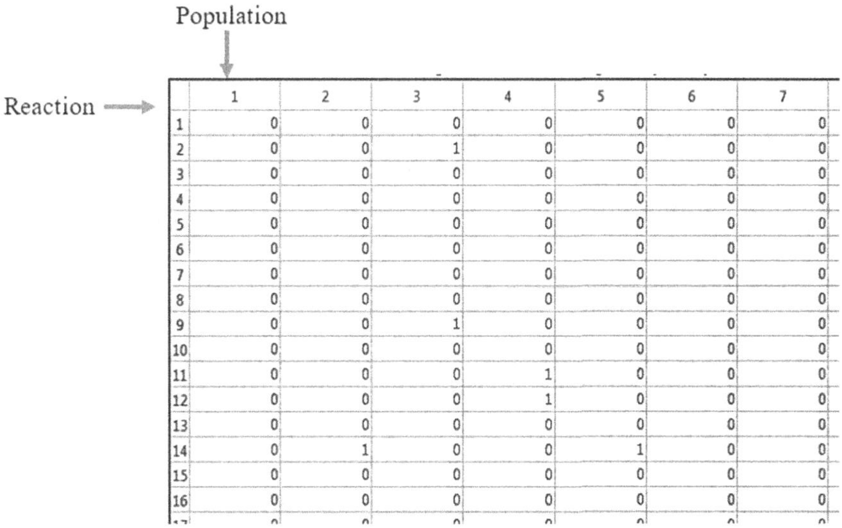

Population

Reaction

	1	2	3	4	5	6	7
1	0	0	0	0	0	0	0
2	0	0	1	0	0	0	0
3	0	0	0	0	0	0	0
4	0	0	0	0	0	0	0
5	0	0	0	0	0	0	0
6	0	0	0	0	0	0	0
7	0	0	0	0	0	0	0
8	0	0	0	0	0	0	0
9	0	0	1	0	0	0	0
10	0	0	0	0	0	0	0
11	0	0	0	1	0	0	0
12	0	0	0	1	0	0	0
13	0	0	0	0	0	0	0
14	0	1	0	0	1	0	0
15	0	0	0	0	0	0	0
16	0	0	0	0	0	0	0

FIGURE 15.1 Initial population of SKFMOMA generated in MATLAB®.

that SKFMOMA selects each reaction, it is necessary to have a population size greater than the number of reactions. Hundred populations are created in the matrix. Subsequently, this matrix is randomly populated with values of 0 and 1. A value of 0 signifies that the reaction is not eligible for knockout, whereas a value of 1 denotes that the reaction can be knocked out. Figure 15.1 shows the initial population of SKFMOMA in MATLAB. The rows and columns are reactions and populations respectively. Each reaction is carried out by a specific gene. Besides that, the agent's estimate state, *X(0)* is initialized in the search space by using the command *(rand(agentNo, dimNo))*200-100* where *agentNo* represents for number of agent (100 agents) while *dimNo* represents for number of dimension (5 dimensions). The noise covariance estimate, *P(0)*; the measurement noise, *R*; and the process noise, *Q* are set at 500, 0.5, and 0.5, respectively as initial value. The specified values are adjusted to meet the necessary criteria for achieving optimal outcomes.

15.2.2.3 MOMA Fitness Evaluation and Assume True Value

Assume that *maxIteration* represents the number of outer iterations, which is set to 1, and *agentNo* represents the number of agents in the inner iteration, which is 100. The process begins by calculating the fitness of each approximate state through agents utilizing MOMA. In this context, each agent functions as an individual Kalman filter. The MOMA approach serves not only to forecast the growth optimization process but also to predict the metabolic phenotype by minimizing the gaps in the flux space caused by perturbations.

After initializing the primary parameters, the list of reactions eligible for knockout is provided as input to MOMA. MOMA utilizes this information to evaluate the objective function based on the predicted metabolic flux distribution of the wild-type organism. The method proceeds by calculating the growth rate for each knockout reaction, observing if the cells continue to survive even after the reaction is removed. A growth rate exceeding 0.1 indicates cell survival following gene knockout. The MOMA fitness value calculation is then used to determine the extent of improvement in the production rate. For significant improvement, the value should be greater than -1e-3. Values below this threshold indicate negligible enhancement resulting from the knockout reactions.

Equations 15.1 and 15.2 explain the Euclidean distance between the maximal growth point attained through FBA in the wild-type condition and the vector in the mutant flux space.

$$D(w,x)=\sqrt{\sum_{i=1}^{N}(w_i-x_i)^2} \tag{15.1}$$

$$f(x)=L.x+\frac{1}{2}x^TQx \tag{15.2}$$

where L = vector with length of N

$Q = n \times n$ matrix

N = length of vector x

w = optimal growth vector by FBA

x = vector in mutant flux space

x^T = transpose of x

w is also known as wild-type or unperturbed state flux distribution and x also represents flux distribution on gene deletion that is to be solved for. While MOMA is being processed, FBA is concurrently utilized to compute the optimal flux distribution of the wild type using a predefined linear objective function. MOMA searches for the point that is closest to the optimal FBA of the wild type. After fitness calculation, the true value, x_{true} is assumed. This true value represents the best solution at that moment and will be updated if better solution is found. During each iteration, the agents within the population enhance their estimation by incorporating feedback obtained from a simulated measurement process.

15.2.2.4 Predict, Measure, and Estimate

In this particular stage, SKF utilizes two sets of Kalman equations: the "predict" equations and the "estimate" equations. The "predict" equations are employed to compute a priori estimates for the subsequent step, while the "estimate" equations are used to calculate a posteriori estimates. Time update equations are utilized to forecast the state and error covariance estimates based on the current estimates for the upcoming step. These estimates are commonly known as a priori estimates:

$$X(t|t-1) = X(t-1) \tag{15.3}$$

$$P(t|t-1) \leftarrow P(t-1) + Q \tag{15.4}$$

Equation 15.3 consists of time-update equations that forecast the state and error covariance estimates for the subsequent step, utilizing the current estimates as a basis. The measurement step acts as feedback for the estimation process. The calculation considers random positions based on the predicted state estimation, taking into account every position. Equation 15.5 simulates the calculated position of each individual agent.

$$Z_i = X_i(t\,|\,t-1) + \sin(rand(2\pi)) \times abs(X_i(t\,|\,t-1) - x_{true} \tag{15.5}$$

Finally, in the last step, the estimation is carried out, involving the calculation of the Kalman gain, *K(t)*, through the utilization of Equation 15.6.

$$K(t) = \frac{P(t|t-1)}{P(t|t-1) + R} \tag{15.6}$$

Subsequently, the measurement-update equations are utilized to generate posteriori estimates based on the prior estimates, as demonstrated in the following formulation:

$$X_i(t) = X_i(t|t-1) + K(t) \times (Z_i(t) - X_i(t|t-1) \tag{15.7}$$

$$P(t) = (1 - K(t)) \times P(t|t-1) \tag{15.8}$$

SKFMOMA algorithm

Require: noPopulation ← 100, agentNo ←100, maxIteration ←1, P(t-1) ←500, Q ←0.5, R ←0.5
Initialize population of agents' state estimate **X**(t-1) randomly over the search space
while not stopping condition **do**
Evaluate fitness for each agent using MOMA
Assume true value: $x_{true} \leftarrow x_{min}$
for all agents **do**
procedure PREDICT

$$X(t|t-1) = X(t-1)$$
$$P(t|t-1) \leftarrow P(t-1) + Q$$

end procedure
procedure MEASURE

$$Z_i = X_i(t|t-1) + \sin(rand(2\pi)) \times abs(X_i(t|t-1) - x_{true}$$

end procedure
procedure ESTIMATE

$$K(t) = \frac{P(t|t-1)}{P(t|t-1) + R}$$
$$X_i(t) = X_i(t|t-1) + K(t) \times (Z_i(t) - X_i(t|t-1)$$
$$P(t) = (1 - K(t)) \times P(t|t-1)$$

end procedure
end for
end while
Re-evaluate fitness for each agents
return $x_{true} \leftarrow x_{min}$ (best solution)

FIGURE 15.2 Pseudocode of SKFMOMA algorithm.

The feedback obtained from the measured position, influenced by the Kalman gain value, guides the agent in approximating the optimal position. The iteration process persists until the maximum number of iterations is reached. Figure 15.2 shows the pseudocode of SKFMOMA algorithm.

15.2.2.5 Termination

After 1 maximum iteration along with 100 agents has been completed, the best solution for the list of reactions that need to be knocked out is generated. This best solution represents the highest fitness value within the population. As a result, the algorithm is terminated.

15.2.3 Comparison of SKF and SKFMOMA

Figures 15.3 and 15.4 show the differences between previous work (SKF) and proposed method (SKFMOMA).

15.3 DATASET AND EXPERIMENTAL SETUP

The dataset used for this chapter is the *E. coli* metabolic model, which is a subset of the iAF1260. This model represents a highly comprehensive and up-to-date metabolic reconstruction of *E. coli*, offering the potential for new discoveries. It is available in SBML format and can be freely

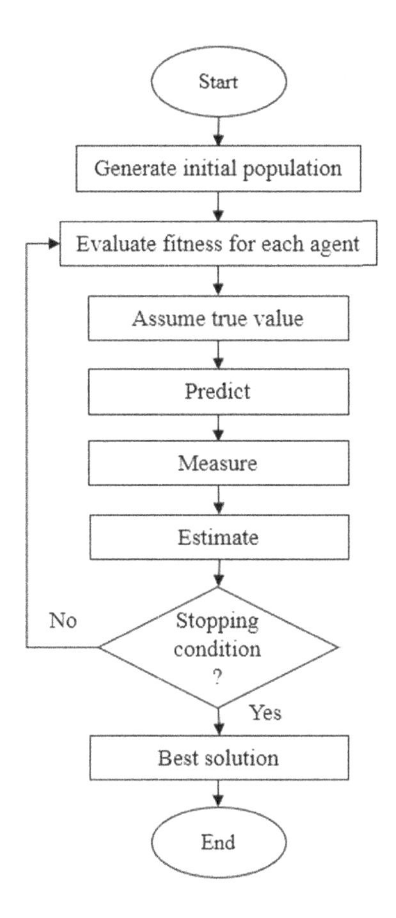

FIGURE 15.3 Flowchart of previous work (SKF).

downloaded from various sources such as BioModels Database, KEGG Database, and System Biology Research Group. Furthermore, the iAF1260 reconstruction serves as a foundation for other *E. coli* metabolic reconstructions and related phenotypes.

The SBML format extension file is saved as an XML file and can be read in MATLAB by executing the corresponding command; *model = readCbModel ('ecolicoremodel.xml')*. Table 15.2 shows the details of the model.

15.4 EXPERIMENTAL RESULTS AND DISCUSSION

After conducting 50 runs of SKFMOMA, three sets of knockout lists were identified that resulted in the highest production of succinate. Table 15.3 shows the maximum production of succinate by triple to quintuple the number of gene knockouts.

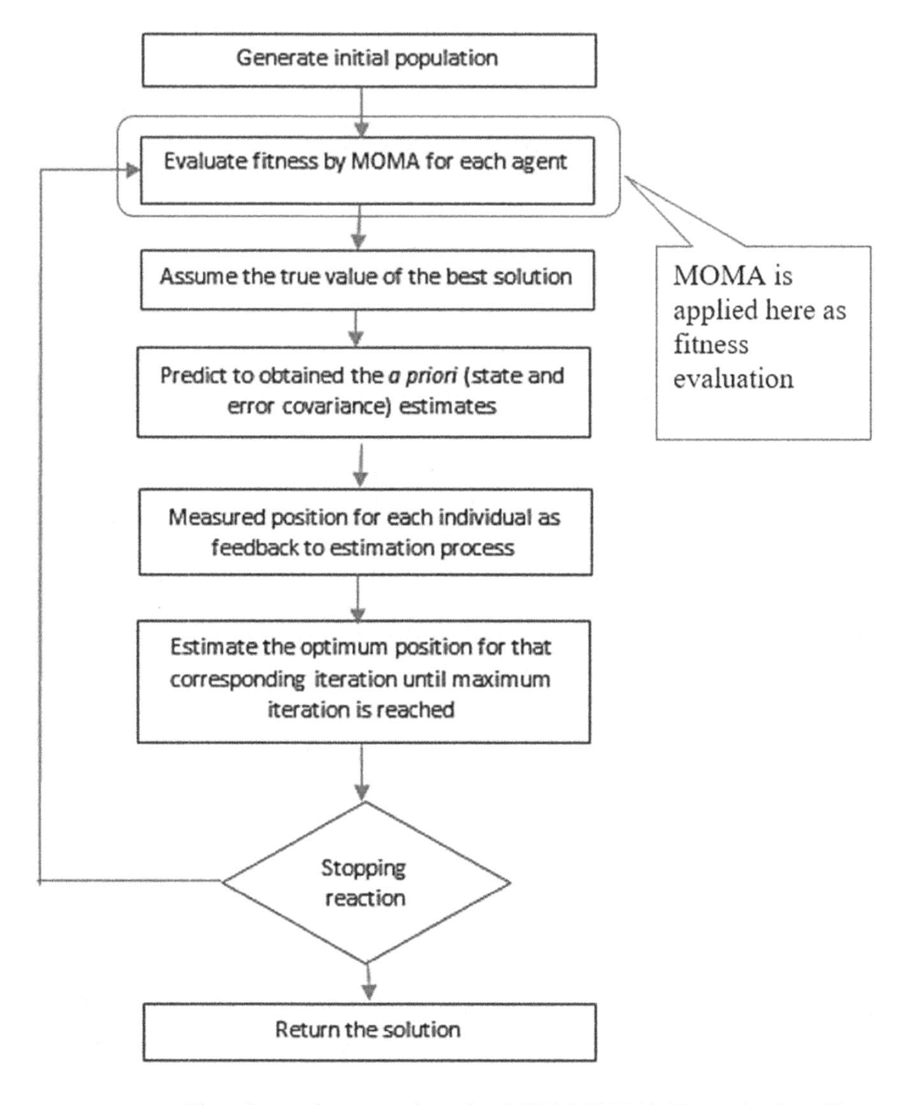

FIGURE 15.4 Flowchart of proposed method (SKFMOMA). The red colored box indicates the modified parts.

TABLE 15.2 Details of *i*AF1260 Model

Dataset	Number of Reactions	Number of Metabolites	Number of Genes	Database (URL)
*E. coli i*AF1260	95	72	137	System Biology Research Group (https://systemsbiology.ucsd.edu/Downloads/EcoliCore)

TABLE 15.3 Knockouts List for Succinate in *E. coli* *i*AF1260 Model

Number of Gene Knockouts	Gene Knockouts	Suggestions for Gene Deletion	Succinate Production Rate (mmol gDW^{-1} hr^{-1})	Growth Rate (hr^{-1})
3 (Triple)	FUM	*fumA, fumB, fumC*	6.7228	0.4312
	GND	*Gnd*		
	ME2	*maeB*		
4 (Quadruple)	FUM	*fumA, fumB, fumC*	**7.7987**	0.2863
	PDH	*aceE, aceF, lpd*		
	THD2	*pntB, pntA*		
	TKT1	*tktA, tktB*		
5 (Quintuple)	ACALDt	*s0001*	6.7266	0.4308
	FUM	*fumA, fumB, fumC*		
	GLNabc	*glnP, glnQ, glnH*		
	PGL	*Pgl*		
	THD2	*pntB, pntA*		

Note: Bold font represents the best result.

As seen in the findings in Table 15.3, the *fum*, *gnd*, and *maeB* gene knockouts increased succinate production to 6.7228 $mmol\ gDW^{-1}h^{-1}$. The growth rate of the *E. coli* is 0.4312 h^{-1} that suggests that *E. coli* cell still survived, although three genes have been deleted. For this set of knockouts, the growth rate is greater than others. Fumarase catalyzes the interconversion between fumarate and malate, which is just a by-product in the study by Park & Gunsalus (1995). The gene that encodes phosphogluconate dehydrogenase and *maeB* gene that encodes the malic enzyme or *NADP* must also be removed to improve the production of succinate.

The second group of knockouts is the *fum*, *pdh*, *tkt*, and *pnt* genes, which increase the production of succinate to 7.7987 $mmol\ gDW^{-1}h^{-1}$. The growth rate of 0.2863 h^{-1} was decreased compared to the triple gene knockout. As mentioned above, *fum* knockout can increase the production of succinate. Pyruvate dehydrogenase catalyzed *acetyl-CoA* formed by *pdh*. Acetate and ethanol can be restricted and NADH can be completely used in the TCA cycle, thus increasing succinate (Vemuri et al., 2002). *Pnt* encodes genes for NAD(P) transhydrogenase while *tkt* encodes genes for transketolase, all of which needed to be eliminated in order to achieve substantially succinate production.

For the last collection of the list of knockouts, *s0001*, *fum*, *pgl*, *gln*, and *pnt* raise the production of succinate to 6.7266 $mmol\ gDW^{-1}h^{-1}$ lower than the above. *S0001* is a reversible transport of acetaldehyde that is not required for the processing of succinate and is better removed. The growth

TABLE 15.4 Knockouts List for Lactate in *E. coli* *i*AF1260 Model

Number of Gene Knockouts	Gene Knockouts	Suggestions for Gene Deletion	Lactate Production Rate (mmol gDW^{-1} hr^{-1})	Growth Rate (hr^{-1})
3 (Triple)	FORt2	*focA, focB*	5.4100	0.1728
	NADH16	*nuoK, nuoH, nuoF, nuoB, nuoN, nuoL, nuoA, nuoE, nuoJ, nuoI, nuoG, nuoC, nuoM*		
	FUM	*fumA, fumB, fumC*		
4 (Quadruple)	G6PDH2r	*Zwf*	**8.5690**	0.1283
	AKGt2r	*kgtP*		
	FORt2	*focA, focB*		
	NADH16	*nuoK, nuoH, nuoF, nuoB, nuoN, nuoL, nuoA, nuoE, nuoJ, nuoI, nuoG, nuoC, nuoM*		
5 (Quintuple)	NADH16	*nuoK, nuoH, nuoF, nuoB, nuoN, nuoL, nuoA, nuoE, nuoJ, nuoI, nuoG, nuoC, nuoM*	5.6957	0.1347
	FORt2	*focA, focB*		
	GLUSy	*gltD, gltB*		
	PPS	*ppsA*		
	MDH	*Mdh*		

Note: Bold font represents the best result.

rate is 0.4308 h^{-1} better than that set above, which is however lower than the first triple gene knockout. In order to increase the succinate production rate genes *pgl* and *gln* are encoding for 6-phosphogluconolactonase and *GLNabc* had to be removed. Table 15.4 indicates the maximum production of lactate by triple until quintuple gene knockouts.

Among the results presented in Table 15.4, the first set of gene knockouts includes the *foc*, *nuo*, and *fum* genes, which led to a production rate of 5.4100 $mmol.g^{-1}$ h^{-1} and a growth rate of 0.1728 h^{-1}. Although the production rate for triple gene knockouts is low, it exhibits the highest growth rate compared to other knockouts. The *nuo* gene is responsible for encoding NADH dehydrogenase, an enzyme involved in NADH oxidation in both directions. The absence of NADH oxidation pathways results in a loss of aerobic growth potential in central carbon metabolism, potentially limiting the citric cycle. Eliminating the *fum* gene increases lactate production rate, along with the *foc* gene, which encodes a proton symport transporter that facilitates lactate production.

In the second group of knockouts, the *zwf, kgtP, foc*, and nuo genes were knocked out, leading to a significant improvement in lactate production rate, reaching 8.5690 $mmol.g^{-1}$ h^{-1}, and a growth rate of 0.1283 h^{-1}. This knockout combination achieves the highest production rate compared to other knockouts. Similar to the previous case, the *nuo* gene restricts the citric cycle by catalyzing NADH to NAD+ in both directions. Moreover, reducing G6PDH2r, catalyzed by the *zwf* gene, plays a vital role in increasing lactate production. The elimination of the *kgtP* gene, responsible for reversible transport of 2-oxoglutarate, along with the *foc* gene, enhances the likelihood of increasing lactate production.

In the last set of gene knockouts, which includes *nuo, foc, glt, ppsA*, and *mdh* genes, the lactate production rate reaches 5.6957 $mmol.g^{-1}$ h^{-1}, while the growth rate is 0.1347 h^{-1}. Knocking out the *foc* gene disables the transport of formate via proton symport, leading to a significant improvement in lactate production. In addition, removing the *glt* gene, which is involved in generating NADPH and glutamate synthase, can increase the lactate yield by eliminating competing pathways. The *ppsA* gene encodes *E. coli* malate dehydrogenase, and the *mdh* gene encodes phosphoenolpyruvate synthase. In an anaerobic environment, lactate secretion is notably improved with the inactivation of the *mdh* and *ppsA* genes.

Table 15.5 indicates the mean and standard deviations of succinate and lactate production after 50 SKFMOMA runs by triple until quintuple gene knockouts, respectively. Unfortunately, the lactate mean is low. This rarely happens and can be influenced by a limited number of reactions or the need to adjust a new dataset with a wide search area. In general, the standard deviation is below 0.1, indicating a high level of similarity to the mean value. This suggests that the obtained results are relatively consistent and exhibit minimal divergence.

Furthermore, the assessment of both valid solution accuracy and optimum solution accuracy was conducted for triple to quintuple gene

TABLE 15.5 The Growth Rate of Mean and Standard Deviation in Succinate and Lactate for Triple to Quintuple Gene Knockout with 50 Runs of SKFMOMA

	Succinate		**Lactate**	
	Growth Rate			
Metabolite/Number of Gene Knockouts	**Mean**	**Standard Deviation**	**Mean**	**Standard Deviation**
Triple	0.4654	0.1411	0.1309	0.1010
Quadruple	0.4779	0.1179	0.1204	0.0396
Quintuple	0.4541	0.1445	0.1415	0.1118

knockouts. The evaluation involved performing 50 runs of SKFMOMA to determine the accuracy of valid solutions and the accuracy of optimum solutions. All solutions validly proved that SKFMOMA is capable of producing valid solutions. Results for optimal solution are all 100% suggesting that all solutions achieved are optimal solutions due to no scaling problems and unbounded problems in the SKFMOMA algorithm.

After that the gene knockout of the proposed method is explained on the basis of a wet laboratory experiment. This can be illustrated by the list of gene knockouts from SKFMOMA that could be applied to a real wet laboratory experiment. The comparison of the findings for the proposed method with the previous gene knockout analysis is also explained in the next subsection.

Table 15.6 displays the experimental results of the proposed method, SKFMOMA with OptKnock and MOMAKnock for succinate production.

TABLE 15.6 Experimental Result of OptKnock, MOMAKnock, and SKFMOMA for Succinate

Targeted Reaction	**Succinate**			
Method/Features	OptKnock (Burgard et al., 2003)	MOMAKnock (Ren et al., 2013)	SKFMOMA	
Production (*mmol gDW*$^{-1}$ *h*$^{-1}$)	6.21	5.02	**7.7987**	**6.7266**
Number Gene Knockouts	Quadruple	Quadruple	**Quadruple**	**Quintuple**
List of Knockout	PYK, ACKr, PTAr, GLCpts	SUCDi, ACKr, MTHFD, TKT1	**FUM, PDH, THD2, TKT1**	**FUM, ACALDt, GLNabc, PGL, THD2**
Reactions	adp + pep → atp + pyr	q8 + succ → fum + q8h2	**fum + h2o ⇌ mal-L**	**fum + h2o ⇌ mal-L**
	Actp + adp ⇌ ac + atp	ac + atp → actp + adp	**coa + nad + pyr → accoa + co2 + nadh**	**acald ⇌ acald**
	accoa + pi ⇌ actp + coa	h2o + methf → 10fthf+ h	**nadh + nadp + 2h → 2h + nad + nadph**	**atp + h2o + gln-L→ adp + gln-L + h + pi**
	glc + pep → g6p + pyr	5p + xu5p-d → g3p + s7p		**6pgl + h2o → 6pgc + h**
				nadh + nadp + 2h → 2h + nad + nadph

Note: Bold font represents the best result.

TABLE 15.7 Experimental Result of OptKnock and SKFMOMA for Lactate

Target Reaction	Lactate	
Method/Features	**OptKnock (Burgard et al., 2003)**	**SKFMOMA**
Production (*mmol* g^{-1} h^{-1})	5.58	8.5690
Number Gene Knockouts	Double	**Quadruple**
List of Knockout	PTAr or PFK, ALDH2	**NADH16, AKGt2r, FORt2, G6PDH2r**
Reactions	accoa + pi ⇌ actp + coa or atp + f6p → adp + fdp + h accoa + 2nadh ⇌ coa + eth + 2 nad	**h + nadh + q8 → 3 h + nad + q8h2** **akg + h ⇌ akg + h** **for + h → for + h** **g6p + nadp ⇌ 6pgl + h + nadph**

Note: Bold font represents the best result.

Four gene knockouts led to the highest production rates in OptKnock and MOMAKnock, 6.21 *mmol* gDW^{-1} h^{-1} and 5.02 *mmol* gDW^{-1} h^{-1}, respectively. SKFMOMA has acquired outstanding performance with 7.7987 *mmol* gDW^{-1} h^{-1} for 4 gene knockouts and 5 gene knockouts.

Table 15.7 displays the experimental findings for SKFMOMA and OptKnock for lactate production. The obtained results have demonstrated promising performance in identifying the optimal gene knockout for enhancing lactate production through the utilization of SKFMOMA. OptKnock is only 5.58 *mmol* gDW^{-1} h^{-1} with double gene knockout, while SKFMOMA is 8.5690 *mmol* gDW^{-1} h^{-1} with triple gene knockout.

15.5 CONCLUSION AND FUTURE WORKS

In recent times, metabolic engineering has gained significant attention across various fields, particularly in genetic modification, aiming to achieve higher production levels of desired biochemical substances. These substances have applications in diverse areas such as accessible enzymes, food production, energy resources, environmental conditions, and healthcare. Drawing inspiration from the Kalman filter Predict-Measure-Estimation process, SKF is a novel approach that aims to estimate the optimal solution effectively. On the other hand, MOMA is a precise tool used to predict the metabolic characteristics of gene knockout bacteria. Hence, this research study presents a novel combination SKF and MOMA as an approach to improve and optimize modeling techniques for augmenting the production of succinate and lactate through gene knockout strategies. Moreover, the chapter highlights the potential advantages of utilizing these techniques in the commercial production of biochemicals. There are some enhancements mentioned that can be introduced to enhance the efficiency of the hybrid method for future studies. First, some modifications may be made to enhance the efficiency of the proposed method. Second, it is proposed that regulatory on/off metabolic (ROOM) be introduced in SKF. By introducing ROOM in SKF, a comparison can be made with SKFMOMA so that improved hybrid approaches and findings can be produced. Third, an increase in the number of iterations and agents for the proposed method could result in more reliable results for the target biochemical compound, along with an increase in the number of agents by tuning some of the relevant parameters. Alternative dataset models, such as fungi and yeast, offer the potential to achieve significant yields of the desired biochemical substance. These models have the capacity to generate unexpectedly high production rates, which are crucial to meeting consumer demand.

ACKNOWLEDGEMENT

We would like to thank the Fundamental Research Grant Scheme-Malaysia Research Star Award (FRGS-MRSA) from Malaysia Ministry of Higher Education for their support to this research (no grant: R/FRGS/A0800/01655A/003/2020/00720).

REFERENCES

Burgard, A. P., Pharkya, P., & Maranas, C. D. (2003). 'Optknock: A Bilevel Programming Framework For Identifying Gene Knockout Strategies For Microbial Strain Optimization'. *Biotechnology and Bioengineering*, 84(6), 647–657.

Kauffman, K. J., Prakash, P., & Edwards, J. S. (2003). 'Advances in Flux Balance Analysis'. *Current Opinion in Biotechnology*, 14(5), 491–496.

Park, S. J., & Gunsalus, R. P. (1995). 'Oxygen, Iron, Carbon, and Superoxide Control of the Fumarase *fum*A and *fum*C Genes of *Escherichia coli*: Role of the *arcA*, *fnr*, and *soxR* Gene Products'. *Journal of Bacteriology*, 177(21), 6255–6262.

Ren, S., Zeng, B., & Qian, X. (2013). 'Adaptive Bi-Level Programming for Optimal Gene Knockouts for Targeted Overproduction Under Phenotypic Constraints'. *BMC Bioinformatics*, 14(Suppl 2), S17.

Tang, P. W. et al. (2015). 'Optimising the Production of Succinate and Lactate in *E. coli* using a Hybrid of Artificial Bee Colony Algorithm and Minimisation of Metabolic Adjustment'. *Journal of Bioscience and Bioengineering*, 119(3), 363–368.

Teusink, B., Wiersma, A., Jacobs, L., Notebaart, R. A., & Smid, E. J. (2009). Understanding the Adaptive Growth Strategy of *Lactobacillus plantarum* by *In Silico* Optimization. *PLoS Computational Biology*. 5(6): doi:10.1371/journal.pcbi.1000410

Vemuri, G. N., Eiteman, M. A., & Altman, E. (2002). 'Succinate Production in Dual-Phase *Escherichia coli* Fermentations Depends on the Time of Transition from Aerobic to Anaerobic Conditions'. *Journal of Industrial Microbiology & Biotechnology*, 28(6), 325–332.

CHAPTER 16

Intelligent Air Conditioning Systems

Enhancing Energy Efficiency and Indoor Air Quality through IoT and Air Conditioning Unit Using Machine Learning

Chockalingam Aravind Vaithilingam,
Samsul Ariffin Abdul Karim, and Amutha Prabha

16.1 INTRODUCTION

Figure 16.1 demonstrates that the cooling or air conditioning systems use a lot of energy (45.78%), and that system component failure, poor maintenance, and decreasing fluid levels usually occur due to maintenance requirements for HVAC systems. A number of literature works have independently attempted to addressed and document these issues with help of modern computational tools. Furthermore, from several studies it is inferred that the indoor air quality (IAQ) is typically 2–5 times more bad than outside air quality. There are two main core causes for these two issues. First off, poor air conditioning system upkeep and cleaning have resulted in poor air conditioning performance, which raises power usage. For instance, a clean air filter can reduce the energy use of the air conditioner by 15%. In addition, a dust-clogged air filter reduces the evaporator coil's ability to effectively control the system's humidity level. Second, no monitoring system exists to give consumers useful information about the location of air conditioning parts in need of repair and maintenance.

DOI: 10.1201/9781003400387-16

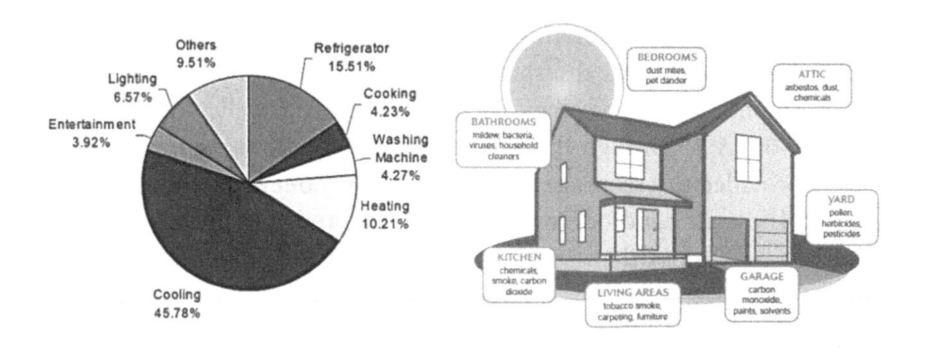

FIGURE 16.1 Typical electrical consumption and indoor air quality.

The monitoring system of an air conditioning unit also prevents consumers from knowing the carbon dioxide and humidity levels. Therefore, there is a chance to address these problems and their underlying causes by developing and implementing an integrated Internet of Things (IoT) system that can gather and analyze data and then provide consumers with air conditioning unit insights(Xu, 2022). Figure 16.2 shows the conceptual system proposed with the above approach.

The system is made up of networked objects that can exchange data and communicate with one another. With technologies like radio-frequency identification (RFID) and wireless sensor networks (WSN), the IoT enables things to communicate with one another without the need for human intervention. The system first gathers and sends data on the air conditioning parts to a cloud server. Data analysis and modeling is done on the cloud server. The planned IoT system warns customers through mobile app notification if it detects system irregularities, such as the compressor

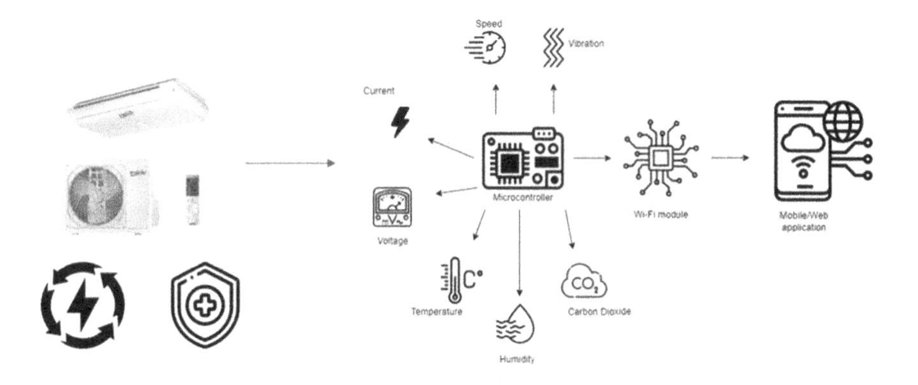

FIGURE 16.2 Integrated intelligent air conditioning system.

of the air conditioner vibrating excessively or detecting that the carbon dioxide level is above the threshold. Customers learn which air conditioning unit components need service and maintenance as well as receive alerts about poor IAQ. Through the mobile app, air conditioning unit insights will be provided to encourage consumers to carry out maintenance to save energy. These air conditioning units influence and contribute to the creation of a technology that significantly improves IAQ in the settings where customers reside, whether at home or at work. Since customers spend most of their time inside, the environment must be healthy and conducive to regular breathing. In addition, the air shouldn't be prone to the growth of mold and fungi, which might happen due to changes in the room's humidity and temperature.

16.2 APPROACH USED IN BUILDING THE SYSTEM

Regardless of whether the system in question is an air handling unit (AHU), variable refrigerant volume (VRV), or single/multi-split system, this component can be found in all air conditioning systems. The next stage is to identify the variables that affect the system's energy usage and quantify those variables using instruments and digital sensors. The sensor then uploads the data it has collected to the cloud. This fact is amply demonstrated by the study's experimental design, shown in Figure 16.3. The user interface would allow the end user to view the processed and visualized data and this interface may resemble a web or mobile application.

After identifying the issues with the air conditioning equipment based on the conceptualization stage, pertinent sensors are investigated and chosen.

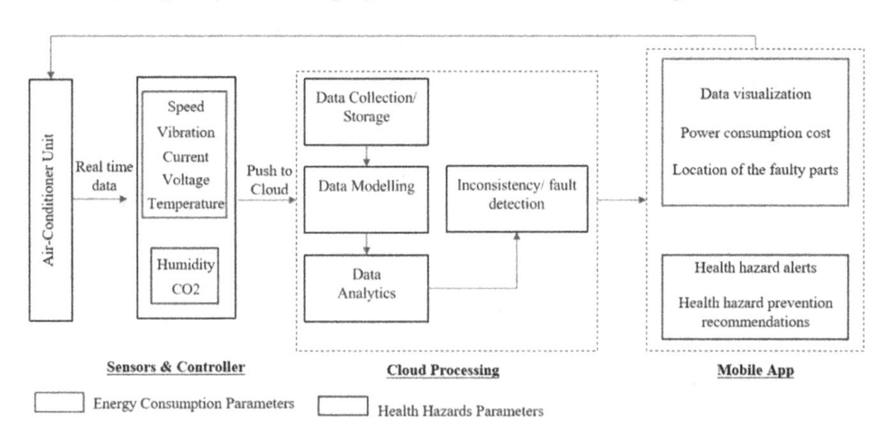

FIGURE 16.3 Stages of the integration of air conditioning unit towards intelligent system.

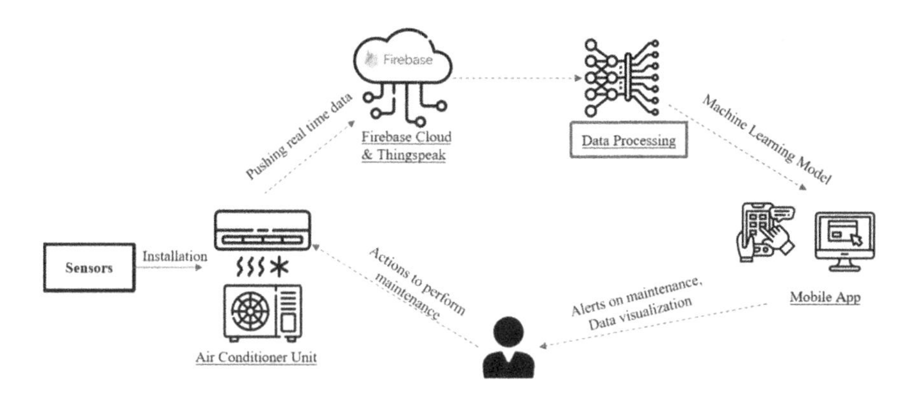

FIGURE 16.4 Overarching IoT integrated system.

The microcontroller for the sensor is coupled with the code for Firebase and ThingSpeak. The air conditioner is then turned on, allowing the sensors to collect data and transmit it to the servers of ThingSpeak and Firebase. The sensor data is regularly posted into the Firebase database of all cloud-based sensor data. By exporting the data to a spreadsheet, the patterns in the data are also examined. The data that is uploaded to Firebase consists of data that is collected over different dates. The initial data collected is for building the intelligent system for use as shown in Figure 16.4. The data stored is referred for fault detection during maintenance for the available date and time of occurrence as in Figure 16.5.

People can upload, visualize, and analyze numerous datapoints in real time using the visual analytics platform ThingSpeak's cloud service. Because the data is obtained visually, a second cloud service is being used. The data may be simply downloaded from this location and then instantly uploaded into MATLAB® for analysis and visualization using their program as can be seen in Figure 16.6. In addition, the graphs shown on their website can be instantly updated and directly uploaded to a website or mobile application. Using machine learning businesses may utilize RapidMiner's comprehensive and user-friendly machine learning technique to harness the power of data and gain insightful information. Through its powerful algorithms, automated model generation, and integration capabilities, RapidMiner assists businesses in today's data-driven world in making data-driven decisions, resolving complex problems, and gaining a competitive advantage. The data is split into two sets: training data and test data, after it has been taken from ThingSpeak. As more training data would enable for more observations to be made, more accurate results would

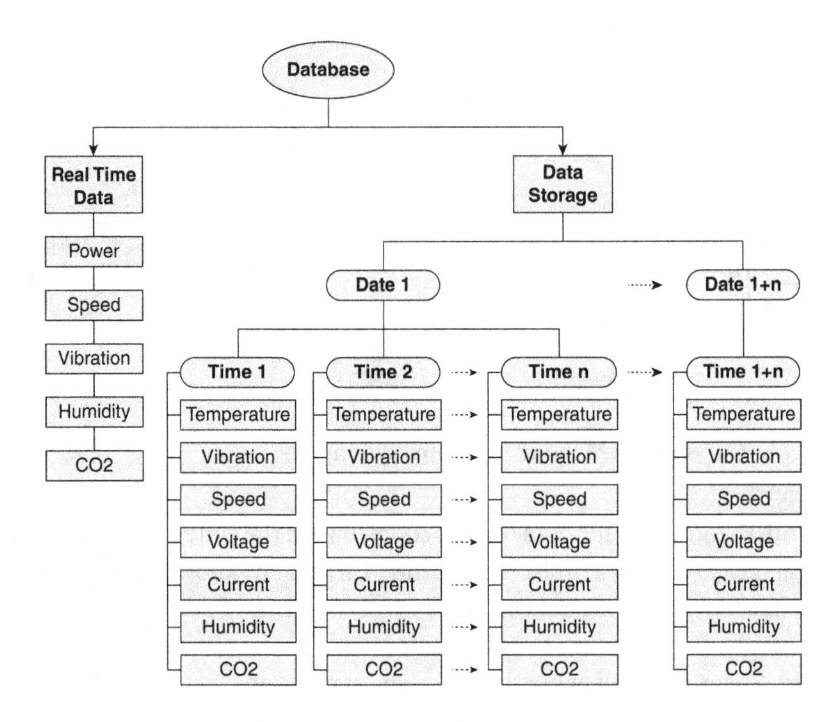

FIGURE 16.5 Overview of cloud storage.

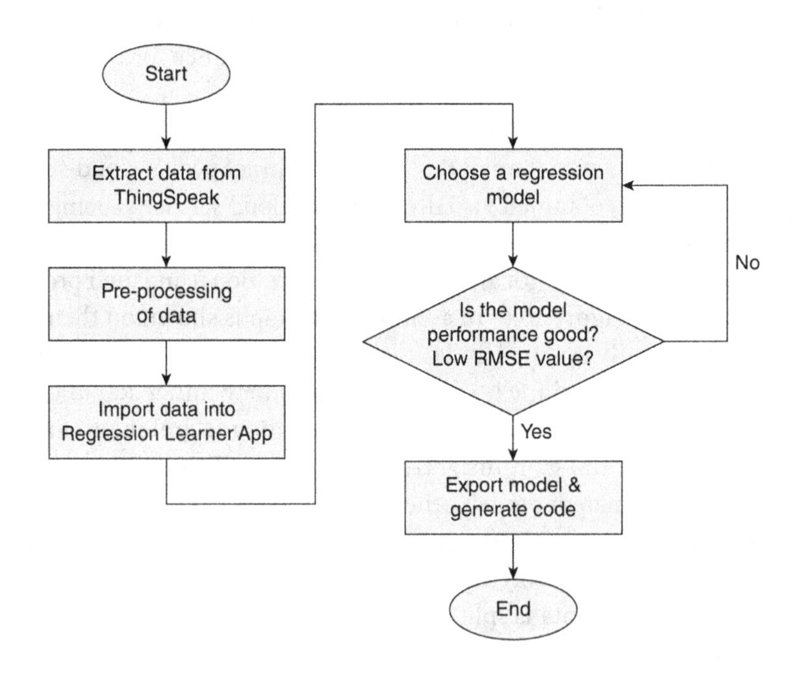

FIGURE 16.6 Approach to machine learning (MATLAB®).

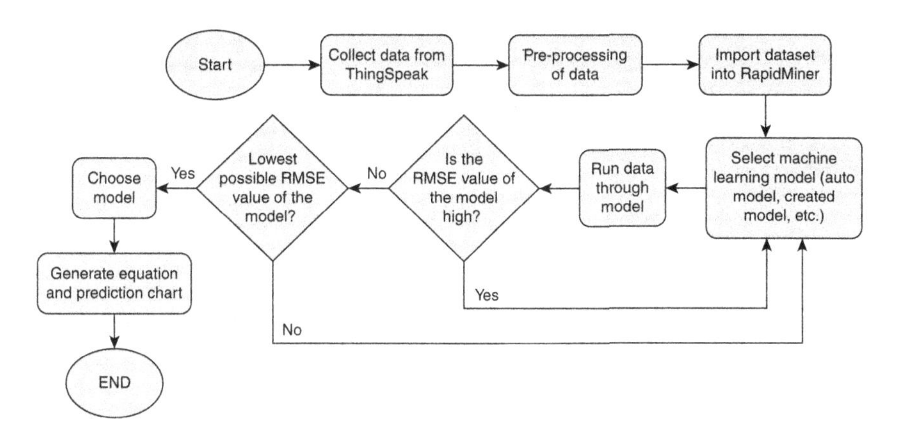

FIGURE 16.7 Approach to machine learning (RapidMiner).

follow from the separation of this data into 70% and 30%, respectively. The next step is to enter this data into the MATLAB Regression Learner App to evaluate the findings and produce the equation required for machine learning as in Figure 16.7.

16.3 TESTING AND PERFORMANCE MEASURES

When the data is selected in the app, the response can be chosen and the predictors that will be used to determine the response can be selected. The linear regression model is used to perform linear regression analysis based on the selected parameters for the target variable, which in this case is power as shown in Figure 16.8. The model is run in the chosen machine

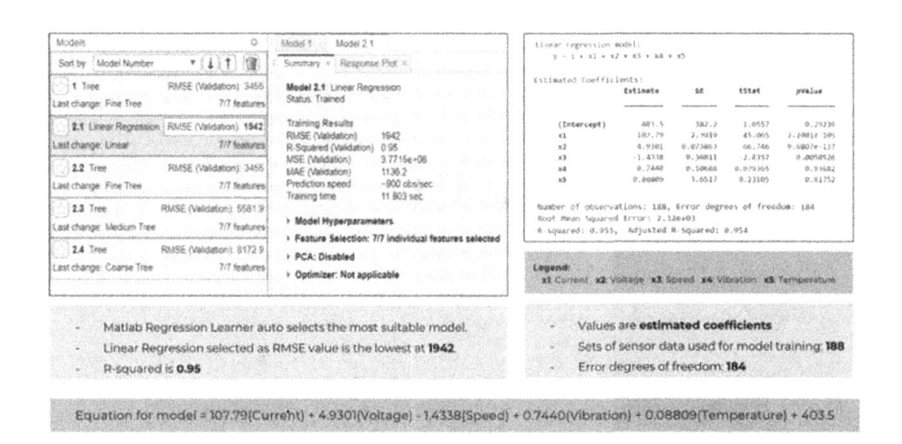

FIGURE 16.8 Regression equation.

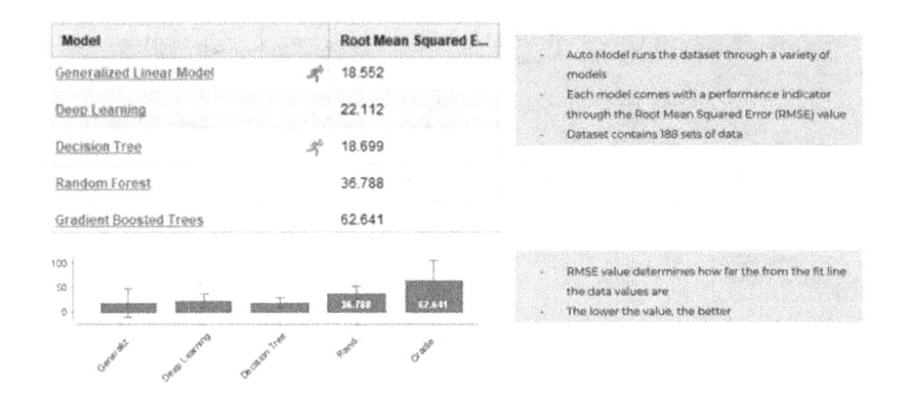

FIGURE 16.9 Outcome from the RapidMiner.

learning model for the root mean square error (RMSE) value to iterate till the system optimizes to generate the generic equation. Higher R-squared values represent smaller differences between the observed data and the fitted values. The number of independent bits of information (degrees of freedom) used to produce the estimate affects the R-squared value, which is a measure of the strength of the relationship between variables and how well the data fit a regression line (Figure 16.9).

After several iterations and validations, the optimized linear regression model equation is achieved as in Figure 16.10. The Flutter framework and the dart programming language were used to build the mobile

Attribute	Coefficient	Std. Error	Std. Coefficient	Tolerance	t-Stat	p-Value	Code
Current	107.802	2.379	0.674	0.845	45.307	0	****
Voltage	4.932	0.073	1.018	0.951	67.481	0	****
Speed	-1.050	0.365	-0.040	0.982	-2.877	0.004	***
(Intercept)	440.444	345.545	?	?	1.275	0.204	

LinearRegression

```
  107.802 * Current
+ 4.932 * Voltage
- 1.050 * Speed
+ 440.444
```

PerformanceVector

```
PerformanceVector:
root_mean_squared_error: 9.167 +/- 0.000
squared_correlation: 0.946
```

- Optimized linear regression model gave the best RMSE value of 9.167
- Same dataset of 188 sets of data were used
- The squared correlation value is 0.946, which means its above 0.75 and indicates that this is a very reliable model. The closer the squared correlation value is to 1, the better the model's explanation in the variance of the data points.

Equation → Model = 107.802(Current) + 4.932(Voltage) - 1.050(Speed)+ 440.444

FIGURE 16.10 Linear regression from optimized model.

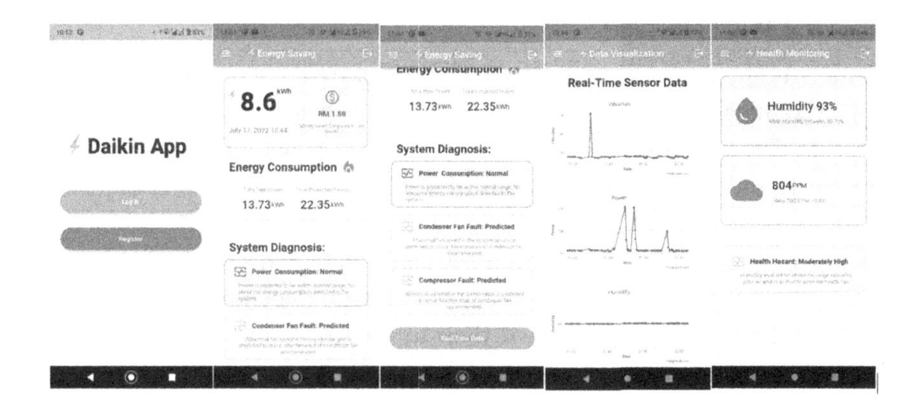

FIGURE 16.11 Overview of mobile app design through intelligent system.

application. In Figure 16.11, the alpha UI design of the app screen illustrates the fundamental layout for the app screen flow. Two buttons on the app's launch screen will direct users to sign up or log in. Users' email addresses and passwords are connected to the cloud-based Firebase authentication service for user authentication and registration. The user is directed to the energy-saving screen after logging in, where they may examine the energy and money they have saved. In addition, the energy-saving screen would compare the total energy used by the air conditioner to the total energy expected to be used. The mobile application helps to coordinate with service assistance in order to create an integrated system between manufacturer service and consumer service, as shown in Figure 16.12.

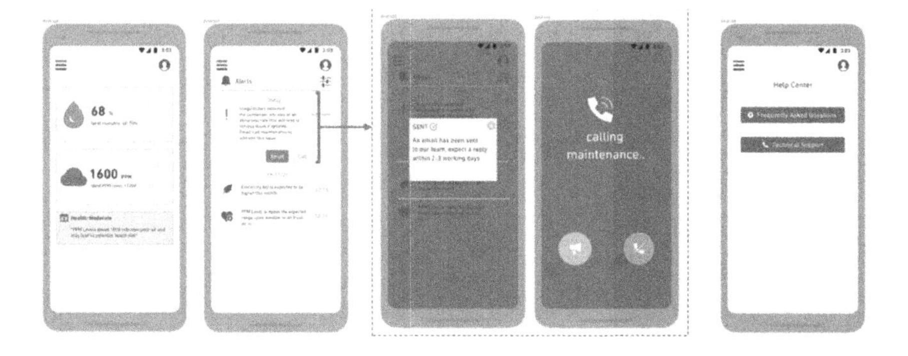

FIGURE 16.12 System support make it integrated.

16.4 CONCLUSIONS

Concerns about energy usage and health risks are related to air conditioning systems and these can be addressed with help of machine learning aspects. The system is trained to forecast power equations using machine learning, and the output equations of MATLAB and fast miner are compared. The predictive power equation can be calculated by the mobile application using the most recent updated data. The predictive power equation will compare to the actual power and, depending on the comparison, inform the consumer if their air conditioning systems need to be serviced. In addition to predictive capability, an algorithm has been created for the air conditioner that tracks vibration, humidity, CO_2, and temperature readings and alerts the user to any changes via an alert message on the app. To motivate users to do maintenance when it is necessary, the app calculates the amount of money saved. The method uploads a sensor from the air conditioner to the cloud, where it is analyzed before being pulled to a mobile app. Utilizing an air conditioning unit in a study to learn how to estimate power use and alert customers to the need for repair will motivate customers to call for service.

ACKNOWLEDGMENT

The authors would like to express their gratitude to Taylor's University in Malaysia, Daikin Research and Development Sungai Buloh Malaysia for the test unit sponsor, and the project lead by Naushua for managing the experimental design there.

REFERENCES

Xu, D. (2022). *Learning efficient dynamic controller for HVAC system*. Mobile Information Systems. https://www.hindawi.com/journals/misy/2022/4157511/

CHAPTER 17

A Comprehensive Analysis of Air Quality Data

A Case Study Approach with the OpenAir Package in R

Murugan Thangiah, Chockalingam Aravind Vaithilingam, Kang Jia Cheng, Faysal Hossain, Rainaf Akif, Atul Dsouza Victor Francis, and Youssef Karim Ahmed Abdelkader

17.1 INTRODUCTION

17.1.1 Background

The World Health Organization (WHO) has estimated that 99% of people worldwide continue to breathe unhealthy air that exceeds WHO air quality standards (Al Mandhari et al., 2022). For instance, pollutants with diameters equal to or less than 10 μm (PM_{10}) or 2.5 μm ($PM_{2.5}$) are capable of penetrating deep into the lungs and entering the bloodstream, causing cardiovascular and cerebrovascular issues (stroke) (Lin et al., 2022). In addition, NO_2 can cause respiratory conditions, particularly asthma, which may result in hospital admissions, ER visits, and respiratory symptoms like coughing, wheezing, or difficulty in breathing.

The air tends to change from time to time. Correspondingly, research on air quality typically only describes a phenomenon at the time of conducting (in terms of location and study period), leading to limited data

DOI: 10.1201/9781003400387-17

for the present air quality. Data on good air quality are measured down to the tiniest scale, for instance, the PM_{10} hourly concentration for Delhi, leading to a more accurate interpretation of the data. This small measurement scale, however, tends to increase the amount of data and lengthen the analysis process. As a result, this analysis requires a tool or application to examine the massive amounts of air quality data, but even Microsoft Excel has its limitations in this regard.

Big data analysis tools that use computer programming languages are available, which can more accurately predict spatial and temporal pollutants and interpret ambient air quality. Examples of computer programming languages for scientific computation, including air quality modeling, include FORTRAN, C++, and R. R is a language dedicated to statistical calculation that provides more in-depth analysis. Also, R has developed an OpenAir model or OpenAir package to analyze air quality data; many functions are designed specifically for performing air quality monitoring analysis.

17.1.2 Study Area and Purpose

Delhi is one of the fastest-growing economic centers of South Asia, and in 2018, it witnessed a total of 11.2 million motor vehicles on the road (Gulia et al., 2022). This number poses significant environmental and health threats in Delhi. Carbon monoxide (CO), nitrogen oxides (NO_x), and particulate matter (PM) are the primary pollutants released from the transportation automobiles (Dutta and Jinsart, 2022). In addition, secondary air pollutants like ozone (O_3) have seen a resurgence in levels.

Many of Delhi's air quality monitoring stations collect large amounts of data on the city's air quality. To better understand the air quality issues, more in-depth analysis of the pollutant concentrations and meteorological data is critical. This study will look at different applications of the open-source OpenAir model, including, temporal variation (pollutant fluctuations over time), and correlations of meteorological factors. With increasing levels of air pollution, we are in dire need of a model that monitors and predicts the air quality ($PM_{2.5}$, PM_{10}, O_3, NO_2, SO_2, CO) over a period of time. This chapter proposes a scientific contribution to this challenge.

17.1.3 Types of Air Pollutants

a. *Particulate matter ($PM_{2.5}$, PM_{10})*: The main sources of particulate matter in Delhi are vehicular emissions from heavy motor diesel vehicles, kerb-side dust, thermal power plants, and industrial and domestic combustion processes. More harmful to human health than PM_{10} is a substance known as respirable suspended particle matter ($PM_{2.5}$). The maximum allowable amount of $PM_{2.5}$ pollution is 60 μg/m^3, yet in all of Delhi's neighbourhoods, the level is 300 μg/m^3 or more.

b. *Nitrogen oxide (NO_x)*: Nitrogen oxides are typically produced in vehicle exhaust during industrial combustion operations. Because of transportation, NO_x levels are highest in metropolitan areas. It is a crucial component of photochemical smog that covers the urban atmosphere in a haze-like blanket. The negative consequences include a variety of respiratory issues in both children and adults.

c. *Sulphur dioxide (SO_2)*: This gas is mostly produced through the combustion of fossil fuels, particularly in thermal power plants. This pollutant impairs lung function and causes acid rain.

d. *Ozone (O_3)*: Created by a chemical reaction between nitrogen dioxide and volatile organic molecules in the presence of sunshine, ozone levels are typically greater in summer. Ozone at ground level also aids in the development of photochemical smog.

e. *Carbon monoxide (CO)*: CO is a poisonous air pollutant created when carbon-containing fuels are partially burned. One of the main causes is the acceleration of moving and idle vehicles.

17.2 RELATED WORKS

Delhi, being a densely populated and highly industrialized city, faces significant challenges in controlling carbon emissions. Several sources contribute to carbon emissions in Delhi, including vehicular emissions, industrial activities, biomass burning, and construction. Understanding these sources is crucial for designing effective strategies to mitigate carbon emissions and improve air quality in the city.

17.2.1 Vehicular Emissions

Air pollution is the term used to describe the tainting of the atmosphere by gases, particles, and other pollutants. There are many sources of air pollution, including the burning of fossil fuels, pollution from factories and other businesses, deforestation, mining operations, and automobile emissions, among others (Kumar et al., 2021). One of the primary sources of carbon emissions in Delhi is vehicular traffic. The city has a large number of vehicles, including personal cars, motorcycles, buses, and trucks, which contribute significantly to carbon emissions. Combustion of fossil fuels, such as gasoline and diesel, in vehicles releases carbon dioxide (CO_2), nitrogen oxides (NO_x), and particulate matter (PM), leading to air pollution. The growth in private vehicle ownership and inadequate public transport infrastructure contribute to the high levels of vehicular emissions in the city.

According to several studies (Chen et al., 2020b; Guo et al., 2019; Amann et al., 2017; Marrapu et al., 2014), the National Capital Region (NCR)'s (the region that includes Delhi) and adjacent states' anthropogenic local sources are the primary causes of air pollution in Delhi. In order to reduce Delhi's $PM_{2.5}$ pollution, a coordinated emissions reduction strategy is necessary. In Delhi, a major source of $PM_{2.5}$ pollution is the local on-road transportation industry. According to research on source apportionment based on measurements of $PM_{2.5}$ at locations in Delhi, automotive emissions contribute between 17% and 30% of the total seasonal $PM_{2.5}$ pollution, with the post-monsoon and winter seasons accounting for the majority of this contribution (Mogno, et al., 2023; TERI & ARAI, 2018). According to Gajbhiye et al. (2023), Delhi has been facing increased levels of air pollution, with on-road vehicle emissions playing a significant role in the pollutant pool. About 72% of Delhi's total pollution load is due to automobile emissions. With a decadal growth rate of 47% and 17.1 million urban residents as of the 2011 Census, Delhi has seen a sharp expansion in its urban population. In order to accommodate the growing demand for transportation, this population growth finally resulted in higher energy consumption and accelerated motorization.

17.2.2 Industrial Activities

Delhi has a significant concentration of industries, including manufacturing plants, power generation units, and construction sites. These industries emit substantial amounts of carbon dioxide through combustion processes, such as the burning of fossil fuels for energy generation or other industrial processes. Industrial emissions also include other greenhouse gases like methane (CH_4) and nitrous oxide (N_2O) released from chemical reactions and industrial waste management. The combustion of coal, oil, and natural gas for power generation and industrial processes in Delhi contributes to carbon emissions and air pollution.

According to the research study by Appannagari (2017), massive amounts of contaminants, such as ions of chlorine, sulphate, bicarbonate, nitrate, sodium, magnesium, and phosphate, are released through sewage effluents into rivers and lakes as a result of growing industrial expansion, poisoning the water. The environment is negatively impacted in a number of ways by the emission of various gases, smokes, ashes, and other particles from factory chimneys. Coal and petroleum fuel combustion have modified the atmosphere's natural gaseous composition by raising the atmospheric concentration of CO_2. Because CO_2 intensifies the greenhouse effects of the atmosphere by allowing solar radiation to pass through the atmosphere and reach the earth's surface while blocking outgoing long-wave terrestrial radiation from escaping to space, an increase in the atmosphere's CO_2 content could change the global radiation and heat balance by increasing the amount of sensible heat in the atmosphere. Because less UV solar radiation is absorbed when the ozone layer is depleted, the earth's surface temperature rises significantly.

According to assessments done by the Central Pollution Control Board (CPCB), industrial clusters that produce excessive amounts of air, water, and soil pollution are both inside and beyond the nation's capital. Gurgaon-Delhi-Meerut Industrial Region, one of the most important economic sectors of the country, includes Delhi as one of its key industrial zones. Industrial areas occupy 51.81 km^2 of the National Capital Territory of Delhi. While the number of registered industries has been growing exponentially along with PM_{10}, NO_2 (nitrogen dioxide) and CO (carbon monoxide) have been growing at polynomial rates in Delhi (Parveen et al., 2021). The Anand Parbat, Naraina, Okhla, and Wazirpur industrial zones are part of Delhi's Najafgarh Drain Basin, which is the second most polluted cluster in India. Its air and water qualities are classified as "critical" and its soil as "severe" for their hazardous content. Industrial pollution contributes 18.6% to the poor air quality in the Delhi-NCR, which has up to 3,182 industrial facilities (Chatterji, 2021).

17.2.3 Biomass Burning

Biomass burning, particularly during the winter months, significantly contributes to carbon emissions in Delhi. Biomass fuels, such as wood, crop residues, and biomass pellets, are commonly used for cooking and heating purposes, especially in rural areas and slums. Incomplete combustion of these biomass fuels releases carbon monoxide, volatile organic compounds (VOCs), and fine particulate matter (PM2.5) into the atmosphere. The smoke and pollutants from biomass burning contribute to the overall carbon emissions and deteriorate air quality in Delhi.

It is well recognized that biomass burning (including wildfires, controlled burns, and agricultural burns) significantly worsens local and regional air quality by releasing a lot of pollutants into the atmosphere. Agricultural residue burning releases particulate matter and trace gases into the atmosphere such as sulfur dioxide (SO_2), methane (CH_4), carbon monoxide (CO), reactive nitrogen species (e.g., NH_3, NO_x, N_2O), and hydrocarbons (Bray et al., 2019). At both the local and regional levels, these trace gases can also aid in the development of $PM_{2.5}$ or secondary fine particulate matter.

Black carbon (BC) over Delhi is primarily produced by the burning of fossil fuels and biomass, with industry emissions, home biomass burning, and waste incineration contributing at relatively lesser rates (Bikkina et al., 2019; Goel et al., 2021). In order to evaluate the $PM_{2.5}$ build-up in Delhi throughout the dry season (Oct-May), Chowdhury et al. (2019) analyzed sixteen years' worth of satellite data. They discovered that open biomass burning could raise $PM_{2.5}$ concentrations by up to 9%. Due to the burning of crop residue in Punjab and Haryana, Jethva et al. (2019) found a concomitant increase of 60% in the mass concentration of $PM_{2.5}$ over Delhi.

Dumka et al. (2018) estimate that, globally, the burning of open biomass accounts for 40% of fossil fuel emissions, 20% of biofuel emissions, and 20% of BC. These proportions notably varied between urban, rural, and remote areas, with vehicle emissions predominating in urban settings and biomass/biofuel burning predominating in rural settings. Outdoor fires are another local source of air pollution, mostly from burning in agricultural areas, that contribute to the deterioration of air quality in addition to urban sources. Fires are generally caused by residue burning, which is most prevalent in northern India in April–May (pre-monsoon) and October to November (post-monsoon) months, which corresponds to burning after the wheat and rice harvests, respectively.

17.2.4 Construction and Dust

Construction activities generate significant amounts of dust and particulate matter, which contribute to carbon emissions in Delhi. The city's rapid urbanization and construction boom lead to increased excavation, transportation of construction materials, and land clearing, all resulting in dust emissions. Dust particles, especially those from construction sites and unpaved roads, contain carbonaceous material and contribute to the overall carbon emissions and air pollution in the city.

According to projections, India's construction and real estate sectors will grow from a predicted 120 billion USD in 2017 to a market size of 1 trillion USD by 2040, accounting for 13% of the country's GDP (IBEF, 2021). But it is also a given that the construction sector has a significant detrimental effect on both human health and the environment. Construction dust, which is released during such activities, is a substantial contributor to ambient air pollution. Construction dust contains small, solid particles that hang in the air and pose a greater threat to human health because of their size and morphological structure (Alshetty & Nagendra, 2022).

The activities and machinery used at the construction sites may release particles with diameters ranging from a few nanometers to 100 μm. According to Yan et al. (2019), construction activity increases the particle content in the immediate surroundings by 16–40%. In addition, the total PM_{10} emissions resulting from construction activities were estimated to be as high as 7.7 and 6.27 tons per day for Bangalore and Mumbai, respectively (Gargava and Rajagopalan, 2016). This accounts for approximately 15–20% of contribution in any urban area. According to a study published by Guttikunda and Calori (2013), construction operations in Delhi City are expected to produce 30 tons of PM_{10} per day, or 9% of the city's emissions.

According to Alshetty & Nagendra (2022), a combination of carcinogenic small particles from vehicle exhaust, brakes, tires, and road surface wear contribute to the harmful nature of road dust. Moreover, during construction activities, there is a large increase in the movement of heavy-duty vehicles. As a result, it can add to the PM emissions that must be measured. These findings clarify that one of the main causes of elevated PM concentrations in the area around a construction zone is the resuspension of road dust. It is logical to expect that in the near future exposure to construction dust will cause severe health and environmental problems.

17.2.5 Waste Management

Improper waste management practices, including open burning of waste and inadequate waste disposal infrastructure, contribute to carbon emissions in Delhi. The decomposition of organic waste in landfills produces methane, a potent greenhouse gas that contributes to carbon emissions. Open burning of waste, including plastic and other non-biodegradable materials, releases toxic gases and carbonaceous particles into the atmosphere, worsening air quality and increasing carbon emissions.

The amount of solid trash produced annually in Indian cities has climbed from 6 million tons in 1947 to 90 million tons in 2009, and this amount is expected to reach 300 million tons by 2047. The significant rise in municipal solid waste (MSW) production is the result of urban dwellers' changing lifestyles, eating habits, and standards of living (Mukherji et al., 2016). One of the main causes of urban air pollution in Indian cities is the open burning of MSW in neighborhoods and landfills, which is caused by the lack of proper waste-management infrastructure and services (Luthra et al., 2023). According to UNEP (n.d.), the terms "open waste burning" and "spontaneous combustion at landfills and dumpsites" refer to both intentional burning by municipal authorities and waste pickers (to make room at dumpsites, make scavenging easier, and serve as a heat source) and spontaneous combustion.

Throughout the post-monsoon, winter, and pre-monsoon seasons, industrial areas had high concentrations of PM_{10} and $PM_{2.5}$, whereas during the monsoon season, pollution levels drastically decreased. Although NO_2, O_3 and CO levels in many industrial regions were within the acceptable range throughout the monsoon and pre-monsoon seasons, they were in the moderate range during the winter and post-monsoon seasons. The present study aims to analyze the concentration of air pollutants (PM_{10}, $PM_{2.5}$, NO_2, O_3, and CO) through OpenAir package which is available using R package.

17.3 METHODOLOGY

In the realm of air quality management, the OpenAir model has emerged as a valuable tool for analyzing, interpreting, and comprehending air pollution data. By employing various statistical analyses specifically tailored for air quality modeling, such as linear regression, *p-value* decision-making, and coefficient determination, the OpenAir model offers a comprehensive framework for understanding the complexities of air pollution and devising effective strategies to improve air quality.

Within the OpenAir model, each function serves a distinct utility purpose, contributing to a holistic understanding of air pollution dynamics. In this research, we focus on three specific model functions: TheilSen, timeVariation, and scatterPlot. These functions play a crucial role in analyzing pollution concentration trends, temporal variations of pollutants, and the linear correlation between two variables, respectively.

The TheilSen function allows us to delve into pollution concentration trends, providing insights into the long-term patterns and changes in air quality. By examining the data through this lens, we can discern the trajectory of air pollution levels over time. In the case of this study, the results obtained from the TheilSen function indicate a concerning trend of deteriorating air quality from December 2020 to January 2023. This finding underscores the urgency of implementing effective air quality management strategies to mitigate the adverse impacts on public health and the environment.

Furthermore, the timeVariation function enables us to analyze the temporal variations of pollutants. By scrutinizing the fluctuations in pollution levels over different periods, we gain a deeper understanding of the factors influencing air quality dynamics. This function allows us to identify patterns, seasonal variations, and potential sources of pollution, aiding in the formulation of targeted interventions to address specific pollution hotspots or timeframes.

The scatterPlot function facilitates the examination of the linear correlation between two variables. By plotting the relationship between different parameters, we can uncover potential cause-and-effect relationships or identify variables that may influence air pollution levels. This function serves as a valuable tool for exploring the complex interplay of factors contributing to air pollution, enabling policymakers and researchers to make informed decisions based on empirical evidence.

The results obtained from the three functions of the OpenAir model collectively highlight the pressing need for enhanced air quality management in Delhi. As air pollutants continue to pose a significant threat to public health and environmental well-being, the utilization of OpenAir models will undoubtedly increase in the coming years. By harnessing the power of this model and expanding its application to encompass longer monitoring periods and a broader range of air pollutants, we can obtain more accurate and comprehensive air quality data. This, in turn, will enable policymakers to devise targeted interventions and policies to combat air pollution effectively.

In summary, the OpenAir model represents a valuable tool in the realm of air quality management. Through its various functions, such as TheilSen, timeVariation, and scatterPlot, this model empowers researchers and policymakers to analyze air pollution data, identify trends, understand temporal variations, and explore correlations between variables. The findings from this research underscore the deteriorating air quality in Delhi and emphasize the need for increased utilization of OpenAir models to obtain more accurate and comprehensive air quality data. By doing so, we can devise effective strategies to combat air pollution and safeguard the health and well-being of communities.

17.4 DATA ANALYSIS

An OpenAir model can monitor air quality data while handling atmospheric conditions, such as wind speed and direction (Agustine et al., 2017). It can assess model performance and analyze pollutant characteristics, source emissions, and trend estimates. The advantage of the OpenAir model is in its ability to manipulate or interpolate data, analyze statistical data, and produce and display high-quality graphics. Trend estimates analysis is the main topic of this study.

To ensure the package's availability, the OpenAir model should be downloaded first in R software on the official website (Carslaw et al., 2012). Once downloaded, the OpenAir model package can be activated in R software by typing "library (OpenAir)." The air quality data to be analyzed is fed as input from computer files or imported from monitoring stations across Delhi. After that, the datasets are processed and imported in the OpenAir model (software R) and are presented in comma-separated value (CSV) format, as one of the Microsoft Excel extension files.

The OpenAir model is an air quality modeling that has a function to stimulate the mathematical formula into the computer program. This model is a tool for statistically analyzing semi-empirical mathematical relationships between air pollutant concentration and other factors that may affect it. Some fundamental analyses in the OpenAir model include linear regression, decision-making with *p-values*, and coefficient of determination.

17.4.1 Linear Regression

Linear regression is a statistical method for developing a relationship model between the dependent and independent variables (Kazi et al., 2023). The model's coefficient represents the assumed parameter value for the actual condition. However, the regression model coefficients are an average value that may occur in the variable Y (dependent variable) corresponding to a value of X (independent variable). There are two types of regression coefficients: intercept (point intersection with the Y axis) and slope (line gradient). In statistics, the slope value is the average increase or decrease in variable Y for each unit increase in variable X.

17.4.2 Decision-Making with the *p-Value*

Statistics use sample data to infer the overall condition of the population. As a result, the potential for error in making a population decision is also relatively high. Nonetheless, the statistical concept seeks to minimize error as much as possible. A test criterion is required to determine whether H_0 is rejected or accepted. The *p-value* is the most used test criterion in a computer program (Bates et al., 2023).

P-value provides two pieces of information at once: the reason for rejecting the null hypothesis (H_0) and the probability of occurrence mentioned in H_0 (assuming H_0 is accurate). The definition of *p-value* is the slimmest level of meaning at which the result of a statistical test can still be meaningful. Furthermore, it can also be interpreted as the magnitude of the possibility of making a mistake when deciding whether to reject H_0. Generally, the *p-value* is compared to the significance level (α).

17.4.3 Coefficient of Determination

The coefficient of determination (R^2) is the amount of diversity (information) in the *Y* variable that the regression model can provide. R^2 has a value ranging from 0 to 1. When the value of R^2 is multiplied by 100%, the percentage of diversity (information) within *Y* (dependent variable) that is influenced by *X* (independent variable) is calculated. The higher the R^2 value, the better the regression model.

17.5 RESULTS AND DISCUSSION

Based on the outcomes of air quality studies in Delhi that used OpenAir model application, this study only discusses three functions of the OpenAir model as shown in Table 17.1. TheilSen for trend analysis, timeVariation for temporal variations, and scatterPlot for linear correlation analysis are the functions used in this research study for data analysis.

17.5.1 The TheilSen Function

This function helps understand concentration changes (trend) over time and for comparing with an air quality standard. It produces a slope value as the trend percentage and concentration value in the unit per period change. The positive slope of the linear regression line represents the value of the increasing trend, and the negative slope represents the value of the decreasing trend.

TABLE 17.1: The *OpenAir* Model Analysis Functions Used in This Study

No.	Function	Purpose
1	TheilSen	Calculate TheilSen slope estimates and uncertainties
2	scatterPlot	Traditional scatter plots with enhanced options
3	timeVariation	Diurnal, day of week and monthly variations

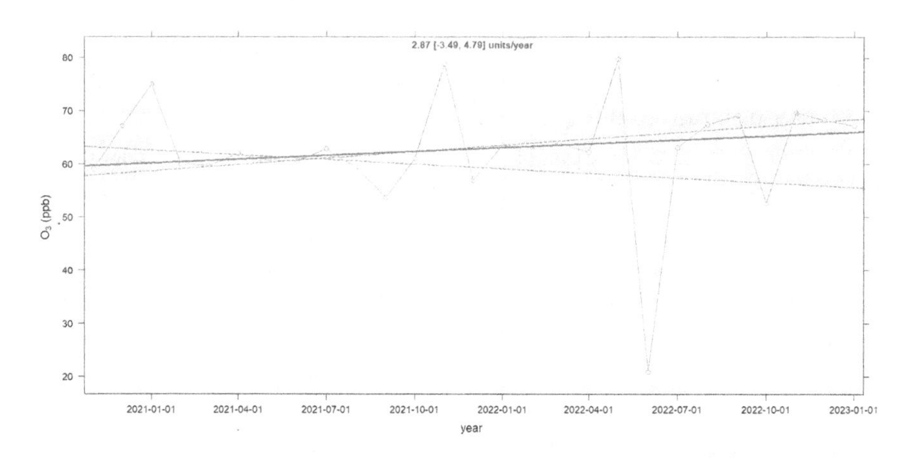

FIGURE 17.1 Trend of ozone (O_3) concentration to ambient air.

TheilSen function analyzes pollutant concentration trends using linear regression and the Mann–Kendall method, with a 95% confidence interval and a 5% significance level (α). The confidence interval and significance level will vary to account for the study's constraints. Whether the trend concentration change tends to be significant or not over the time is also observed. In this instance, an alternative hypothesis (H_a) formulates correspondingly as the trend changes by pollutant concentration.

Figure 17.1 shows the trends in ozone (O_3) in Delhi. The plot shows the deseasonalized monthly mean concentrations of O_3. The solid red line shows the trend estimate and the dashed red lines show the 95% confidence intervals for the trend based on resampling methods. The overall trend (as shown at the top left) is 2.87 (ppb) per year; the 95% confidence intervals in the slope from −3.49 to 4.79 ppb/year. In this situation, Delhi must take additional steps to reduce O_3 emissions.

17.5.2 The timeVariation Function

When using the timeVariation function, it is possible to visualize the temporal variation of pollutant concentration in a line graph. The output of this function is an image made up of four-line graphs based on different time scales, including hourly, daily, hourly, and monthly time. Based on a 95% confidence interval, this function analyzes data. The advantage of this function is that it can plot more than one pollutant simultaneously.

Knowing a pollutant's temporal variation over time allows one to predict when its concentration will be at its lowest or highest, on which days of the week, or in which months throughout the years.

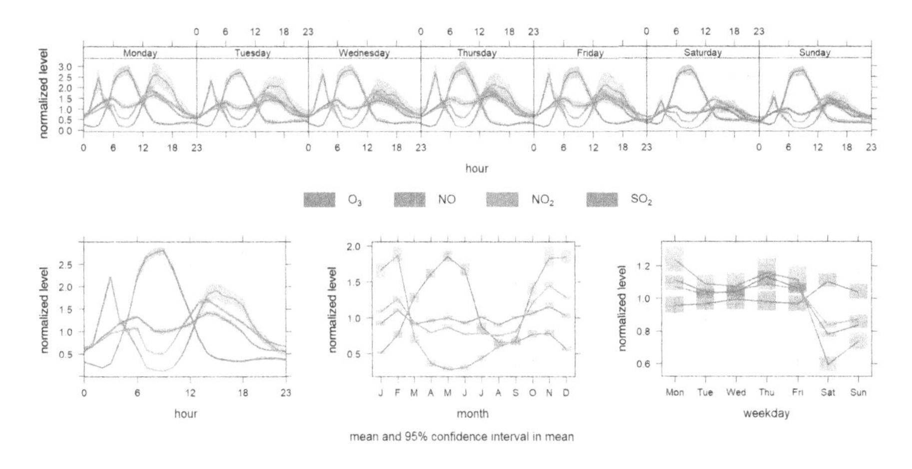

FIGURE 17.2 Normalized concentrations of o_3, NO, NO_2, and SO_2 concentrations.

Figure 17.2 suggests a peak during the morning rush hour from 7:00 to 9:00. The difference line also shows a more significant difference in pollutant emissions between weekdays and weekends. Also, given that the number of cars at this location is roughly constant throughout the day, the variation could result from the emissions of other vehicle types.

17.5.3 The ScatterPlot Function

Linear regression helps to determine whether a relationship exists between the dependent variable (pollutant concentration) and the independent variable (meteorological factor). It will produce the coefficient of determination (R^2), which can help to interpret the correlation/relationship result. If a relationship does exist between two variables (meteorological factors and pollutant concentration), the value of the relationship will either be positive or negative.

Positive values can be identified by the slope being an upwardly sloped curve (positive slope); negative values can be determined by the gradient being a downwardly sloped curve (negative slope). A positive relationship's direction indicates a proportional relationship, such as an increase in the condition/value of a meteorological factor followed by an increase in pollutant concentration. A negative relationship shows a reversed relationship that increases the condition/value of the meteorological facet, followed by a decrease in pollutant concentration.

In Figure 17.3, the number of occurrences in each bin is color-coded (but not on a linear scale). It is now possible to see where most of the data

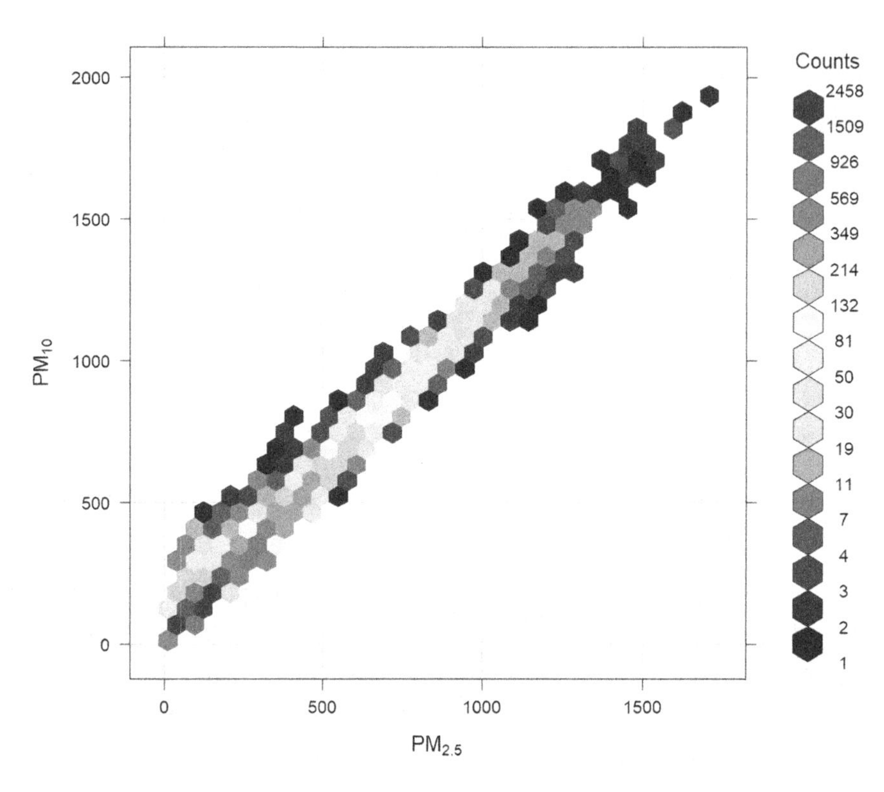

FIGURE 17.3 Scatter plot of hourly $PM_{2.5}$ vs. PM_{10} using hexagonal binding.

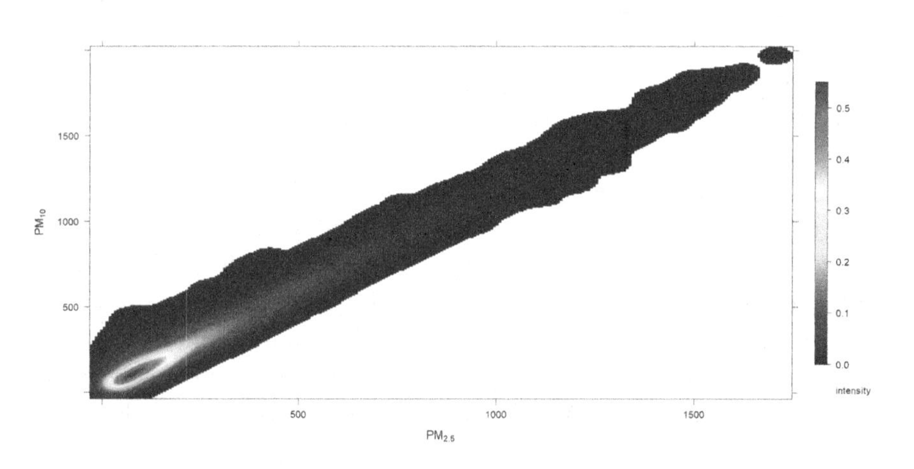

FIGURE 17.4 A scatter plot of hourly $PM_{2.5}$ vs. PM_{10} levels using a kernel density estimate.

lies in a clearer correlation picture between $PM_{2.5}$ and PM_{10}. On the other hand, the "intensity" measures how many dots of $PM_{2.5}$ and PM_{10} concentration exist in a unit area provides a clearer indication of the relationship between $PM_{2.5}$ and PM_{10}, as shown in Figure 17.4.

17.6 CONCLUSION

The OpenAir model is a tool to analyze, interpret, and comprehend air pollution data to improve air quality management. This model uses statistical analyses designed for air quality models, such as linear regression, *p-value* decision-making, and coefficient determination, all of which correlate to various functions. This research only looks at three OpenAir model functions: TheilSen for analyzing pollution concentration trends, timeVariation for analyzing pollutant temporal variations, and scatterPlot for analyzing the linear correlation between two variables. The results from the three functions show deteriorating air quality from 2020 to 2023. As air pollutants and monitoring periods lengthen, the use of OpenAir models in Delhi will soon increase to conclude more accurate air quality data.

17.7 CODE AND DATA AVAILABILITY

For OpenAir package, all development is carried out using Github for version control. Users can access all the code that is used to analyze, interpret, and understand air pollution data in OpenAir at https://github.com/davidcarslaw/openair. Data are typically regular time series and air quality measurement, meteorological data, and dispersion model output can be analyzed. The package is described in Carslaw and Ropkins (https://cran.r-project.org/web/packages/openair/index.html, 2012). The dataset used in this research study is downloaded from the https://www.kaggle.com/datasets/deepaksirohiwal/delhi-air-quality.

The software "Comprehensive R Archive Network" which is an open-source tool is available at https://cran.r-project.org/mirrors.html

REFERENCES

Agustine, I., Yulinawati, H., Suswantoro, E., & Gunawan, D. (2017). Application of open air model (R package) to analyze air pollution data. *Indonesian Journal of Urban and Environmental Technology*, 94–109. https://doi.org/10.25105/urbanenvirotech.v1i1.2430

Al-Mandhari, A., Al-Yousf, A., Malkawi, M., & El-Adawy, M. (2022). "Our planet, our health": saving lives, promoting health and attaining well-being by protecting the planet-the Eastern Mediterranean perspectives. *Eastern Mediterranean Health Journal*, *28*(4), 247–248.

Alshetty, D., & Nagendra, S. S. (2022). Impact of vehicular movement on road dust resuspension and spatiotemporal distribution of particulate matter during construction activities. *Atmospheric Pollution Research, 13*(1), 101256.

Amann, M., Purohit, P., Bhanarkar, A.D., Bertok, I., Borken-Kleefeld, J., Cofala, J., Heyes, C., Kiesewetter, G., Klimont, Z., Liu, J. and Majumdar, D. (2017). Managing future air quality in megacities: A case study for Delhi. *Atmospheric environment, 161*, 99–111.

Appannagari, R. R. (2017). Environmental pollution causes and consequences: a study. *North Asian International Research Journal of Social Science & Humanities, 3*(8), 151–161.

Bates, S., Candès, E., Lei, L., Romano, Y., & Sesia, M. (2023). Testing for outliers with conformal p-values. *The Annals of Statistics, 51*(1), 149–178.

Bhawal Mukherji, S., Sekiyama, M., Mino, T., & Chaturvedi, B. (2016). Resident knowledge and willingness to engage in waste management in Delhi, India. *Sustainability, 8*(10), 1065.

Bikkina, S., Andersson, A., Kirillova, E. N., Holmstrand, H., Tiwari, S., Srivastava, A. K., ... & Gustafsson, Ö. (2019). Air quality in megacity Delhi affected by countryside biomass burning. *Nature Sustainability, 2*(3), 200–205.

Bray, C. D., Battye, W. H., & Aneja, V. P. (2019). The role of biomass burning agricultural emissions in the Indo-Gangetic Plains on the air quality in New Delhi, India. *Atmospheric Environment, 218*, 116983.

Carslaw, D.C. and K. Ropkins (2012). openair — an R package for air quality data analysis. *Environmental Modelling & Software, 27–28*, 52–61.

Chatterji, A. (2021). Air pollution in Delhi: Filling the policy gaps. *Massachusetts Undergraduate Journal of Economics, 17*(1).

Chen, Y., Wild, O., Ryan, E., Sahu, S.K., Lowe, D., Archer-Nicholls, S., Wang, Y., McFiggans, G., Ansari, T., Singh, V. and Sokhi, R.S. (2020). Mitigation of PM 2.5 and ozone pollution in Delhi: a sensitivity study during the pre-monsoon period. *Atmospheric Chemistry and Physics, 20*(1), 499–514.

Chowdhury, S., Dey, S., Di Girolamo, L., Smith, K. R., Pillarisetti, A., & Lyapustin, A. (2019). Tracking ambient PM2. 5 build-up in Delhi national capital region during the dry season over 15 years using a high-resolution (1 km) satellite aerosol dataset. *Atmospheric Environment, 204*, 142–150.

Dumka, U. C., Kaskaoutis, D. G., Tiwari, S., Safai, P. D., Attri, S. D., Soni, V. K., ... & Mihalopoulos, N. (2018). Assessment of biomass burning and fossil fuel contribution to black carbon concentrations in Delhi during winter. *Atmospheric Environment, 194*, 93–109.

Dutta, A., & Jinsart, W. (2022). Air pollution in Delhi, India: It's status and association with respiratory diseases. *Plos one, 17*(9), e0274444.

Gajbhiye, M. D., Lakshmanan, S., Aggarwal, R., Kumar, N., & Bhattacharya, S. (2023). Evolution and mitigation of vehicular emissions due to India's Bharat Stage Emission Standards: A case study from Delhi. *Environmental Development, 45*, 100803.

Gargava, P., & Rajagopalan, V. (2016). Source apportionment studies in six Indian cities—drawing broad inferences for urban PM 10 reductions. *Air Quality, Atmosphere & Health*, *9*, 471–481.
Goel, V., Hazarika, N., Kumar, M., Singh, V., Thamban, N. M., & Tripathi, S. N. (2021). Variations in Black Carbon concentration and sources during COVID-19 lockdown in Delhi. *Chemosphere*, *270*, 129435.
Gulia, S., Kaur, S., Mendiratta, S., Tiwari, R., Goyal, S. K., Gargava, P., & Kumar, R. (2022). Performance evaluation of air pollution control device at traffic intersections in Delhi. *International Journal of Environmental Science and Technology*, *19*(2), 785–796.
Guo, H., Kota, S. H., Sahu, S. K., & Zhang, H. (2019). Contributions of local and regional sources to PM2. 5 and its health effects in north India. *Atmospheric Environment*, *214*, 116867.
Guttikunda, S. K., & Calori, G. (2013). A GIS based emissions inventory at 1 km × 1 km spatial resolution for air pollution analysis in Delhi, India. *Atmospheric Environment*, *67*, 101–111.
IBEF (2021). *Indian Real Estate Industry Report*. India Brand Equity Foundation.
Jethva, H., Torres, O., Field, R. D., Lyapustin, A., Gautam, R., & Kayetha, V. (2019). Connecting crop productivity, residue fires, and air quality over northern India. *Scientific Reports*, *9*(1), 16594.
Kazi, Z., Filip, S., & Kazi, L. (2023). Predicting PM2. 5, PM10, SO2, NO2, NO and CO Air Pollutant Values with Linear Regression in R Language. *Applied Sciences*, *13*(6), 3617.
Kumar, P. G., Lekhana, P., Tejaswi, M., & Chandrakala, S. (2021). Effects of vehicular emissions on the urban environment: a state of the art. *Materials Today: Proceedings*, *45*, 6314–6320.
Lin, C. C., Chiu, C. C., Lee, P. Y., Chen, K. J., He, C. X., Hsu, S. K., & Cheng, K. C. (2022). The adverse effects of air pollution on the eye: A review. *International Journal of Environmental Research and Public Health*, *19*(3), 1186.
Luthra, A., Chaturvedi, B., & Mukhopadhyay, S. (2023). Air pollution, waste management and livelihoods: Patterns of cooking fuel use among waste picker households in Delhi. *Geographical Review*, *113*(2), 229–245.
Marrapu, P., Cheng, Y., Beig, G., Sahu, S., Srinivas, R., and Carmichael, G. R. (2014). Air quality in Delhi during the Commonwealth Games, *Atmospheric Chemistry and Physics*, 14(19), 10619–10630, https://doi.org/10.5194/acp-14-10619-2014
Mogno, C., Palmer, P. I., Marvin, M. R., Sharma, S., Chen, Y., & Wild, O. (2023). Road transport impact on PM2.5 pollution over Delhi during the post-monsoon season. *Atmospheric Environment: X*, *17*, 100200.
Parveen, N., Siddiqui, L., Sarif, M. N., Islam, M. S., Khanam, N., & Mohibul, S. (2021). Industries in Delhi: Air pollution versus respiratory morbidities. *Process Safety and Environmental Protection*, *152*, 495–512.
TERI & ARAI, (2018), Strong policies needed to clean Delhi NCR air by 2025: ARAI & TERI report, https://www.teriin.org/press-release/strong-policies-needed-clean-delhi-ncr-air-2025-arai-teri-report

UNEP. n.d. *Open Waste Burning Prevention*. Paris: Climate and Clean Air Coalition. https://www.ccacoalition.org/en/activity/open-waste-burning-prevention

Yan, H., Ding, G., Li, H., Wang, Y., Zhang, L., Shen, Q., & Feng, K. (2019). Field evaluation of the dust impacts from construction sites on surrounding areas: a city case study in China. *Sustainability, 11*(7), 1906.

CHAPTER 18

A Systematic Review on Intelligent Mobile Beacon Systems

Applications and Features

Nurul Aida Muhamad Noor
and Mohammad Fadhli Asli

18.1 INTRODUCTION

The advances of networking technology have also revolutionized many digital technologies in this century with big data, Internet of Things, social media, and many more. Wireless sensors and devices now can be manufactured at affordable costs, thus enabling further embedding into connected systems for society. One of these wireless sensors is beacon, a Bluetooth-enabled device that sends signals to all smart devices within proximity range continuously. This beacon uses low-power transmit signal to connect with smart devices in range, allowing interaction like pushing notifications and registering presence. Mobile beacon technology has been observed to be applied to many instances such as indoor localization, proximity detection, and activity sensing.

Several prior works have examined the advances of mobile beacon technologies with regards to Internet of Things and marketing applications (Allurwar et al., 2016; Jeon et al., 2018). However, these works are focused on the applications of mobile beacon systems with limited discussions on their components and their notification interaction features. There is yet to

DOI: 10.1201/9781003400387-18

be specific work on identifying and organizing the components of mobile beacon applications and their notifications features.

In this chapter, a systematic review to examine the recent works on intelligent mobile beacon systems and applications is presented. This review focuses on gathering overview insights on recent advances and applications of mobile beacon systems and its notification features. Recent literatures on intelligent mobile beacon systems from quality scholarly databases were collected and examined. Reported mobile beacon's characteristics and features in recent works including the components, interaction features, domain applications, and limitations are systematically investigated. This chapter specifically offers the following contributions:

1. Classification of mobile beacon system components based on different functions.
2. Classification of proximity-based notifications in mobile beacon systems based on different interactions.
3. Summarization of emerging trends of intelligent mobile beacon systems across major sectors.
4. Summarization of challenges in existing intelligent mobile beacon systems.

The contents of this chapter are organized as follows: Section 18.2 introduces the fundamentals of intelligent mobile beacon systems and applications. Section 18.3 then describes the review method and process used in the systematic review including the review questions, search process, selection criteria, and data extraction. Section 18.4 then presents and discusses the results of systematic review based on the defined review questions. Finally, Section 18.5 concludes this chapter by highlighting significant findings and future work.

18.2 BACKGROUND OF INTELLIGENT MOBILE BEACON SYSTEMS

This section introduces the fundamental working and current overviews of intelligent mobile beacon systems and applications. Understanding the fundamentals and current state of intelligent mobile beacon allows the reader to learn its mechanisms, advantages, and practical applications.

18.2.1 Fundamentals of Intelligent Mobile Beacon Applications

Intelligent mobile beacon applications are currently at the forefront of technological advancements across many domains. Beacon technology, which refers to the use of small devices equipped with Bluetooth low energy (BLE) technology, has gained significant interest in recent years (Jeon et al., 2018). Mobile beacon applications then leverage these devices to deliver location-based services and proximity-based notifications to users' smartphones (Seo et al., 2019).

Intelligent mobile beacon systems are gaining increasing interest in response to the growing market demand for location-based services and personalized experiences. Beacon technology has been regarded as a promising solution to these demands and shows rapid adoption across various industries. Market studies have also highlighted the increasing market demand for beacon technology and its applications. For instance, a prior study has projected a substantial market growth for beacon technology, attributing it to the rising adoption of proximity marketing and indoor navigation systems (Adkar et al., 2018). Furthermore, another survey report underscores the significance of beacon technology in the context of personalized experiences and targeted marketing (Fortune Business Insights, 2020).

Beacon technology offers many advantages toward existing processes or tasks including enhancing user engagement, creating personalized experiences, and improving process efficiency. Leveraging the potential of beacon technology, intelligent mobile beacon systems offer valuable facilitation capabilities like indoor navigation, targeted marketing, asset tracking, and real-time monitoring (Karabtcev et al., 2019; Yamamoto et al., 2018). These systems evidently demonstrated their usefulness in many sectors including retail, healthcare, transportation, and tourism, leading to increased interest in their development and utilization (Van De Sanden et al., 2019; Spachos & Plataniotis, 2020; Huang et al., 2020).

18.2.2 Overview of Mobile Beacon Systems and Applications

Many notable trends and usage of mobile beacon systems and applications have been observed over recent years. For instance, the integration of beacon technology in smart retail systems has revolutionized the way customers interact with stores, enabling personalized recommendations, seamless checkouts, and location-based promotions (Alzoubi et al., 2021). These intelligent retail applications eventually facilitate enhanced customer engagement and optimized shopping experience.

In the healthcare sector, mobile beacon systems have been deployed to improve patient care and safety. For example, a prior study explored the use of mobile beacons in hospitals to track medical equipment, ensuring their availability when needed and reducing time wasted in searching for essential resources (Akbarzadeh et al., 2021). Similarly, in the context of elderly care, another study investigated the utilization of intelligent mobile beacon systems to enhance the safety and well-being of older adults (Li et al., 2020). By providing real-time location tracking and emergency alerts, these applications have the potential to improve the quality of care and mitigate risks.

Another domain where mobile beacon systems have shown promise is transportation. Intelligent beacon applications are increasingly being employed to facilitate efficient parking management systems (Park, 2021). By guiding drivers to available parking spaces and streamlining payment processes, these applications contribute to reduced congestion and enhanced user convenience. Furthermore, mobile beacon systems have been used in public transportation to provide passengers with real-time updates on routes, schedules, and delays, improving the overall commuter experience (Ferreira et al., 2020).

While existing survey and review articles have provided valuable insights into the applications and functionalities of mobile beacon systems, there are certain limitations and uncovered aspects that necessitate further investigation. For instance, some studies have primarily focused on specific application domains or provided a broad overview without delving into the specificities of intelligent mobile beacon applications across diverse domains (Jeon et al., 2018; Spachos & Plataniotis, 2020; Yang et al., 2020). In addition, the advances of mobile beacon technologies, emerging use cases, and integration with other intelligent systems may not have been discussed in existing literature.

18.3 REVIEW METHODOLOGY

This section describes the methods for reviewing the related literature on intelligent mobile beacon systems including the review questions, review phases, and analysis method.

18.3.1 Review Questions

Four review questions were defined to guide the examinations of recent applications using this technology.

RwQ1: What are the major component types of intelligent mobile beacon systems?

RwQ2: What are the types of proximity-based notification features in the existing intelligent mobile beacon systems?

RwQ3: What are the recent trends for intelligent mobile beacon systems and applications?

RwQ4: What are the remaining challenges or limitations of existing intelligent mobile beacon systems and its notification features?

The purpose of **RwQ1** is to identify notable characteristics of major components inside the reported intelligent mobile beacon systems. The classification of these components will provide deeper insights on the inner working of mobile beacon systems for improving future development. Next, **RwQ2** aims to classify the types of proximity-based notification features employed by the existing mobile beacon systems. Understanding the type of notification features can be a great development factor in designing interactivity of the mobile beacon systems. **RwQ3** then investigates the recent trends for intelligent mobile beacons systems and applications across various industries. The emerging trends can show the current and predicted development directions for this technological application. Finally, **RwQ4** aims to inform readers of the remaining challenges and limitations faced by the existing mobile beacon systems. Insights into these challenges and limitations can help future research to coordinate their efforts in advancing intelligent mobile beacon systems.

18.3.2 Protocol and Phases of the Review

This study uses the PRISMA review strategy (Moher et al., 2009) to ensure a rigorous and systematic approach to the review process. The detailed workflow of the review process is illustrated in Figure 18.1.

18.3.3 Inclusion and Exclusion Criteria

To ensure the inclusion of relevant studies, three key criteria were established in considering study selection for review. Table 18.1 shows the description and justification for the defined key criteria.

18.3.4 Information Sources and Search Process

A comprehensive search for relevant articles was carried out via online scholarly databases for manual analysis. The selected online databases for this review included ACM Digital Library, IEEE Xplore, Scopus, Google Scholar, SpringerLink, and ScienceDirect. The search strategy utilized the

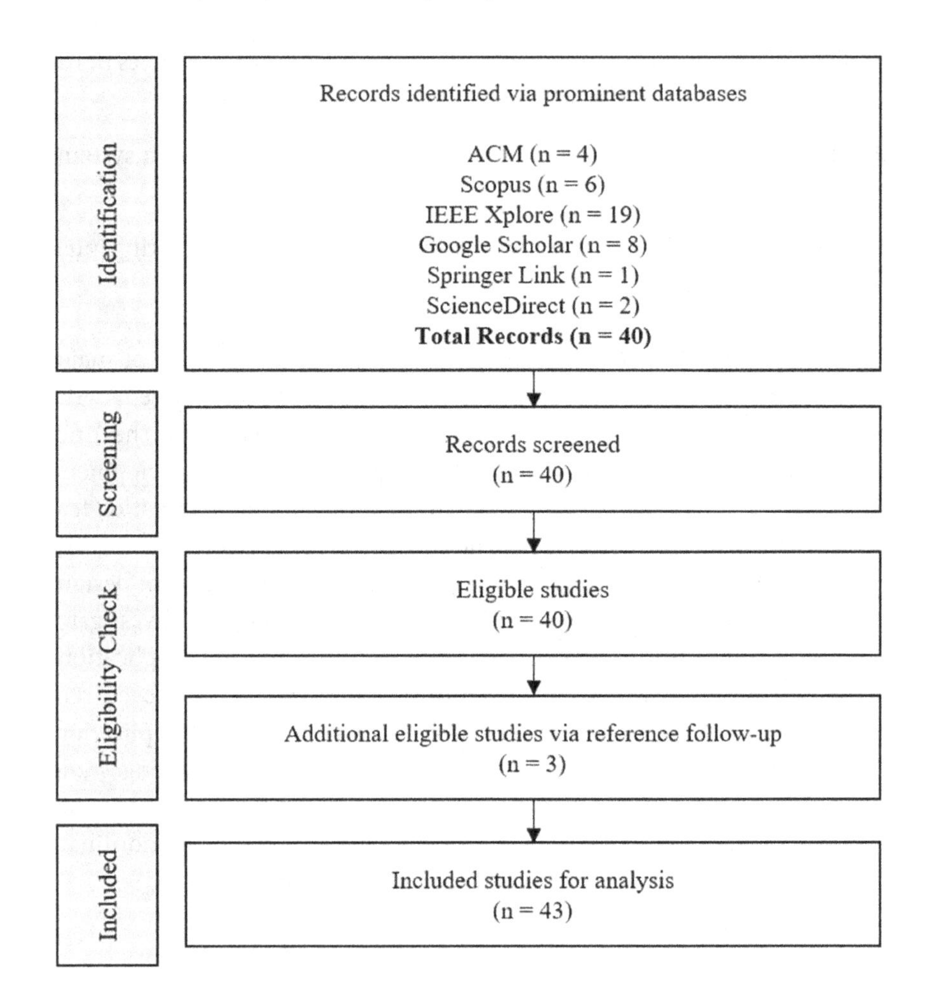

FIGURE 18.1 Workflow of systematic review in this study.

TABLE 18.1 Inclusion Criteria for Selecting Studies for Review.

Criteria	Justification
The study is written in English	English is the universal language and mostly used in academic writing
The study is a journal article or conference proceeding	This review investigates the application and features of intelligent mobile beacon systems. The selected forms should contain most of the necessary review information
Additional references from journal article or conference proceeding	Some additional studies offer information on the type of mobile beacon systems and notification features

following keywords "beacon technology" OR "mobile notification" AND "engagement" to gather articles with relevance to the topic. Each article obtained from the search then underwent selection process with regards to key inclusion criteria.

18.3.5 Study Selection and Review Process

The selected articles were searched based on key inclusion criteria and examined with regards to the defined review questions. The information extracted from the selected articles was carefully organized and compiled into a tabular format. Subsequently, a qualitative analysis was carried out by assigning keywords to each extracted information based on their relevance in addressing the review questions. This review extracted the following information from each study:

- Article details including authors, year of publication, and type of article.
- The type of components prevalent in intelligent mobile beacon systems.
- The type of proximity-based notification features used in the intelligent mobile beacon systems.
- Domain of application of the recent intelligent mobile beacon systems.
- Open issues and limitations of the existing intelligent mobile beacon systems.

18.4 RESULTS AND DISCUSSIONS

This section presents and discusses the results of the systematic review based on defined review questions and processes. Analysis results for each review question will be presented first, followed by discussions that eventually highlight the findings.

18.4.1 Result for RwQ1: Components

Referring to Table 18.2, eight studies were found that highlighted the use of beacon device (Zaim & Bellafkih, 2016; Menon et al., 2017; Ng et al., 2017; Adkar et al., 2018; Berka et al., 2018; Antoniou-Kritikou et al., 2019; Lin et al., 2019; Marín et al., 2021). Moreover, there are 13 studies that demonstrated the use of beacon applications (Akinsiku & Jadav, 2016;

TABLE 18.2 Instances of Highlighted Components in Recent Intelligent Mobile Beacon Systems.

Components	Number of Studies
Beacon device	8
Beacon app	13
Beacon sensor	6
Beacon tracker	9

Kaur & Maheshwari, 2016; Shinotsuka et al., 2016; AlBraheem et al., 2017; Concepción-Sánchez et al., 2017; Pugaliya et al., 2017; Oziom et al., 2019; Hasan & Hasan, 2020; Schwebel et al., 2021; Yamagami et al., 2021; Alabduljabbar, 2022; Đurđević et al., 2022; McGuirt et al., 2022). Next, there are six works that discuss the use of beacon sensors (Baig & Jilani, 2017; Wu et al., 2018; Dong et al., 2019; Srisura & Thakiguchi, 2020; Kong et al., 2021; Park, 2021). Lastly, nine studies were found that include the discussions on the beacon tracker (Cui et al., 2016; Boric et al., 2018; Hidayat & Simalango, 2018; Ferreira et al., 2020; X.-Y. 1Lin et al., 2015; Saraswat & Garg, 2016; Saranya et al., 2018; Surendran & Rohinia, 2019; Pai et al., 2020). The distinction between the highlighted components is explained in the following discussions.

18.4.2 Result for RwQ2: Notification Features

As shown in Table 18.3, this study discovered five works that have instances of receive details notification type among the mobile beacon systems (Kemble et al., 2016; Zaim & Bellafkih, 2016; Wanniarachchi et al., 2017; Hasan & Hasan, 2020; Pai et al., 2020). There are 11 studies that show instances of receive information notification (Kaur & Maheshwari, 2016; AlBraheem et al., 2017; Adkar et al., 2018; Hidayat & Simalango, 2018; Antoniou-Kritikou et al., 2019; Barthe et al., 2022; Shinotsuka et al., 2016; Saranya et al., 2018; Wu et al., 2018; Lin et al., 2019; Oziom et al., 2019). Moreover, eight studies were found that have instances of alert notification

TABLE 18.3 Instances of Notification Features in Recent Intelligent Mobile Beacon Systems.

Type of Notification	Number of Studies
Receive details	5
Receive information	11
Alert	8
User responsive	3

features (Akinsiku & Jadav, 2016; Baig & Jilani, 2017; Concepción-Sánchez et al., 2017; Menon et al., 2017; Ng et al., 2017; Pugaliya et al., 2017; Berka et al., 2018). Lastly, there are three studies with user responsive notification features (Yamagami et al., 2021; McGuirt et al., 2022; Đurđević et al., 2022). The characterizations of each notification type are further elaborated in the following discussions.

18.4.3 Result for RwQ3: Trends

Based on Figure 18.2, among reported work from 2016 until 2022, this study discovered 19 studies under the retail and marketing sector category (Lin et al., 2015; Kemble et al., 2016; Shinotsuka et al., 2016; Zaim & Bellafkih, 2016; Baig & Jilani, 2017; Ng et al., 2017; Pugaliya et al., 2017; Wanniarachchi et al., 2017; Adkar et al., 2018; Saranya et al., 2018; Oziom et al., 2019; Surendran & Rohinia, 2019; Chu et al., 2020; Kong et al., 2021; Alabduljabbar, 2022; Barthe et al., 2022; Đurđević et al., 2022; McGuirt et al., 2022). Next, there are 19 numbers of works that leverage mobile beacon systems under the active environment sector category (Akinsiku & Jadav, 2016; Saraswat & Garg, 2016; Kaur & Maheshwari, 2016; Concepción-Sánchez et al., 2017; Menon et al., 2017; AlBraheem et al., 2017; Boric et al., 2018; Hidayat & Simalango, 2018; Wu et al., 2018; Berka et al., 2018; Dong et al., 2019; Ferreira et al., 2020; Hasan & Hasan, 2020; Pai et al., 2020; Srisura & Thakiguchi, 2020; Marín et al., 2021; Park,

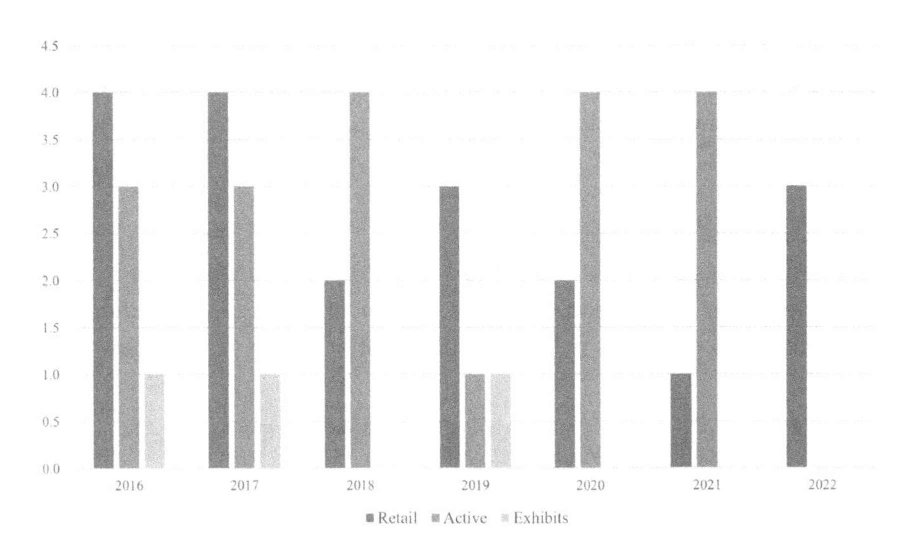

FIGURE 18.2 Yearly trends of recent intelligent mobile beacon systems across major sectors.

2021; Schwebel et al., 2021; Yamagami et al., 2021). Lastly, three works were found under the exhibitions sector category (Cui et al., 2016; Ng et al., 2017; Antoniou-Kritikou et al., 2019). The emerging trends from this result are discussed further in the following discussions.

18.4.4 Result for RwQ4: Challenges and Limitations

Based on the findings indicated in Table 18.4, this study found nine works that highlighted inaccuracy issues due to dampening signals by external obstacles (Lin et al., 2015; Akinsiku & Jadav, 2016; Berka et al., 2018; Hasan & Hasan, 2020; Kong et al., 2021; Alabduljabbar, 2022; Đurđević et al., 2022). Furthermore, there are seven works that demonstrate the current beacon's close-proximity inconvenience due to the limited beacon broadcast range (Saraswat & Garg, 2016; Pugaliya et al., 2017; Boric et al., 2018; Ferreira et al., 2020; Schwebel et al., 2021; Yamagami et al., 2021; McGuirt et al., 2022). Subsequently, this study found seven works that deliberated on the challenges of the real-world use cases of mobile beacon technology (Kemble et al., 2016; AlBraheem et al., 2017; Baig & Jilani, 2017; Chu et al., 2020 Menon et al., 2017; Wu et al., 2018; Lin et al., 2019). The characterizations of each limitation are further deliberated in the following discussions.

18.4.5 Components of Intelligent Mobile Beacon Systems and Notifications

Based on the results in Table 18.2, this study further examines and classifies the working components in a mobile beacon application's environment. The characterization of each major components is described as follows.

18.4.5.1 Beacon Device

The term "beacon" is commonly used to denote the utilization of devices facilitating user interaction by deploying beacon devices at specific locations, employing various principles such as attachment to walls, entrance

TABLE 18.4 Instances of Reported Limitations in Recent Intelligent Mobile Beacon Systems

Limitations	Number of Studies
Signal inaccuracy	9
Close-proximity inconvenience	7
Practical real-world use	7

or exit doors, objects, and so forth. This deployment enables successful connection between the user's Bluetooth-enabled smart device and the beacon. To illustrate, a beacon device positioned in proximity to an artwork would transmit a push notification to the visitor's smart device when they approach the artwork's location where the beacon is deployed (Ng et al., 2017). Subsequently, another study (Pugaliya et al., 2017) involved placing a beacon device in a shopping mall that upon the user's approach or departure from the store, would send offers or information about specific products to their smartphone. In addition, a prior work (Antoniou-Kritikou et al., 2019) emphasized the significance of selecting suitable areas or locations for beacon placement. For example, areas where users commonly gather or traverse within and around a building with their Bluetooth-enabled smartphones facilitate active communication between the device and the beacon within range, enabling the transmission of informational content to their smartphones.

18.4.5.2 Beacon Application

A beacon application can be defined as the utilization of two distinct devices, namely a beacon device and a mobile application, to facilitate user interaction through the exchange of notifications with the user's smartphone device via embedded BLE technology. A smartphone equipped with BLE functionality actively scans for nearby beacons within its range, and upon detection of a beacon signal, the smartphone receives a notification visible to the user, subsequently redirecting them to the associated mobile application. For instance, one study presents an example of a beacon application that employs iBeacon broadcasting in the BeaSmart mobile application, enabling users to navigate indoor environments to locate specific items (Akinsiku & Jadav, 2016). Furthermore, another study demonstrates the deployment of a beacon system on construction sites, serving as an alarm mechanism by promptly notifying users to steer clear of hazardous areas (Kong et al., 2021). When near the beacon, the application signals an alert notification, updating the user about the current conditions within that location or zone.

18.4.5.3 Beacon Sensor

A beacon sensor, serving as a proximity sensor, can be strategically deployed either indoors or outdoors to detect the presence of a user through their mobile device's capability to identify nearby beacon deployments. A notable instance is when a driver enters a parking space with their smartphone mobile app, the beacon sensor detects the driver's presence and offers recommendations for nearby available parking spots (Srisura & Thakiguchi, 2020). Furthermore, beacon sensors find diverse applications across various domains. For instance, one study (Baig & Jilani, 2017) demonstrates the utilization of iBeacon in conjunction with wearable sensors worn by elderly patients to monitor their movement, enabling family members to track their whereabouts within indoor and outdoor spaces. This capability enhances patient safety by facilitating swift response from nearby medical emergency centers in the event of a potential incident. In addition, beacon sensors also play a role in the marketing industry, as highlighted by prior work (Kemble et al., 2016). Employing iBeacon technology, a designated region around a showroom is established, and when users enter this region, the beacon sensor identifies their proximity and subsequently delivers targeted information about showcased vehicles through push notifications.

18.4.5.4 Beacon Tracker

Beacon trackers employ location-based services to accurately determine the user's current position and deliver relevant information via notifications. These trackers find application in both indoor and outdoor environments, encompassing buildings and open spaces. To track individuals in such settings, the beacon tracker relies on receiving signals within a specific proximity to calculate the distance. A notable instance of this component usage is by prior work (Ferreira et al., 2020), wherein BLE beacons are deployed inside vehicles and at station stops to track passengers throughout their entire journey, ensuring their safe arrival at the destination. Another notable use case involves employing a beacon tracker for monitoring student attendance and delivering pertinent details, such as assignment links, through notifications when students come into close range of the beacon (Saraswat & Garg, 2016). Moreover, the beacon tracker plays a pivotal role in safeguarding local cultural artifacts. By leveraging iBeacons, the tracker can detect the presence of visitors and enforce a specific distance to be maintained from the protected items, thus preserving their integrity (Cui et al., 2016).

18.4.6 Types of Notifications in Intelligent Mobile Beacon Systems

Based on the results in Table 18.3, this study further examines and classifies the notification features in mobile beacon applications based on its different interaction functions. The characterization of each type of notification features are described as follows.

18.4.6.1 Receive Details

This study defines the term "receive details" as a proximity-based notification mechanism or protocol designed to receive and deliver a bundle of information to recipients, including users or administrators. An exemplary instance of this notification type is mentioned by prior work (Wanniarachchi et al., 2017), which employs a specific Advanced Message Queuing Protocol to receive selected information details pertaining to approaching customers during insurance engagement. Furthermore, this notification type finds extensive application in the retail and marketing industries. For instance, a Geomarketing system implemented within a supermarket assists shoppers as they navigate the shopping arena by collecting relevant information and providing details about discounted products, offers, or promotions (Zaim & Bellafkih, 2016). The concept of proximity-based notifications is also embraced by smart airports, as elucidated in previous study (Pai et al., 2020). Upon entering the airport premises, beacons deployed throughout the facility send a range of offers to users, while a Wi-Fi beacon-based system triggers location-based actions on passengers' smartphones, enabling them to receive detailed flight information such as flight details and gate numbers. These notifications are effectively pushed to users' smartphones by the Wi-Fi beacon, enhancing the overall airport experience.

18.4.6.2 Receive Information

The concept of "receive information" aligns closely with that of "receive details," wherein a beacon transmits a notification to the user's smartphone when they come within the range of the beacon. In the case of "receive information," the notification serves to inform the user about specific details they may encounter upon entering the beacon range. An example highlighting this type is shown in one study (AlBraheem et al., 2017) where an advertising application sends notifications to users' smartphones when Bluetooth is enabled and they are present on public transportation or in other public spaces equipped with beacon infrastructure. Moreover, to enhance customer engagement, another study (Saranya et al., 2018) explores the placement of beacons on counters, enabling the system to send notifications to users during instances of long queues, providing updates on the current situation at the respective counters. Another instance of "receive information" notifications is presented in prior work (Kaur & Maheshwari, 2016), which focuses on the implementation of a smart tourist application. When users approach specific locations within various tourist spots around the city, notifications are sent to their smartphones, providing relevant information such as historical context, images, and other related data about the places they visit.

18.4.6.3 Alert

An "alert" represents a specific type of notification employed to prompt users with important tasks, reminders, or messages requiring immediate attention. Upon receiving an alert notification, users promptly open the notification and take appropriate actions accordingly. A notable study (Ng et al., 2017) demonstrates this notification type in an art gallery setting, where notifications are sent as alerts to users standing near or in front of specific artworks. The content of these notifications enables users to engage with the exhibited artwork by providing opportunities to write comments, view related videos, and more. Another example pertaining to the use of this notification type is in the context of Alzheimer's disease care (Surendran & Rohinia, 2019). Caretakers and healthcare professionals can monitor the movements of patients through alert notifications. When the patients move within indoor spaces or depart from their designated locations, an alert notifies the responsible individuals. Furthermore, this notification type also described in a smart application that incorporates an alert system to facilitate efficient work management for employees (Menon et al., 2017). By notifying employees of their tasks and deadlines when they enter the proximity region, the alert notifications assist in ensuring timely task completion.

18.4.6.4 User Responsive

User responsiveness refers to a notification type that requires the user's active engagement before accessing the linked content. This notification prompts the user with options to either accept or reject the notification. An instance of this notification type described within the context of smart retail (Đurđević et al., 2022), when users of the SimplyTastly mobile application come within proximity of a beacon, they receive a notification on their smartphone along with a prompt to accept or decline the beacon promotion. In addition, this type of notification is also implemented in the V-Wall app for evacuation guidance (Yamagami et al., 2021). To display the virtual wall map on their phones for guidance to a safe location, users are required to press the response button on the V-Wall app's settings screen, which is triggered when the app receives signals from beacons.

18.4.7 Recent Trends of Mobile Beacon Applications

Based on the results in Figure 18.2, this study further examines and summarizes the recent trends of intelligent mobile beacon systems across major sectors described as follows.

18.4.7.1 Retail and Marketing

This study clusters several sectors under the retail and marketing category, i.e., retails, insurances, healthcare, wildlife preservations, and constructions. Based on the Figure 18.2, from 2016 until 2022, beacon technology was constantly used in this category, specifically in the retails, marketings, and healthcare. In 2016 and 2017, increasing studies in this category are observed with four studies related to it. In contrast, in the year 2021, only two studies highlighted the use of beacon technology under this category. In 2019 and 2022, the application trend is mainly observed under this category. As for 2018 and 2020, the number of studies pertaining the retail and marketing category has reduced to two studies, shifting it into second-highest trend for those years.

18.4.7.2 Active Environment

This trend category encompasses the beacon systems uses in active environments like education, traffic safety, parking systems, workplaces, and forestry. The trend for active environment is prevalent from 2016 until 2021. However, by 2022 there are no work that discusses intelligent mobile beacon systems for this category. In 2018 and 2021, the active environment sector records the highest trend for both years, with four studies highlighting the use of intelligent beacon technology. In 2016 and 2017, the trend for this category started to dwindle and reduced to second spot for both years. Year 2019 shows the lowest instance of intelligent mobile beacon systems work in this category with only one study.

18.4.7.3 Exhibitions

The exhibitions category mainly covers the intelligent mobile beacon systems for art galleries and museums. Albeit the limited applications in this category, its prevalent use is evident within the recent years. This study discovers at least one new study under this category in 2016, 2017, and 2019. More mobile beacon systems for exhibitions are expected in the future in line with the increasing interests toward interactive and immersive exhibition experience.

18.4.8 Challenges and Limitations of Existing Intelligent Mobile Beacon Systems

Based on our results in Table 18.4, this study further examines and characterizes the highlighted limitations and challenges of the existing intelligent mobile beacon systems described as follows.

18.4.8.1 Signal Inaccuracy

Inaccurate signals can arise in both indoor and outdoor settings due to various factors, thereby impacting the performance of beacon–receiver interactions. The issue of signal inaccuracies is addressed in nine articles. For instance, a study (Saranya et al., 2018) reported disruptions in information transfer due to physical barriers obstructing Bluetooth signals within the supermarket environment. Similarly, in the context of hazard alarm systems, previous work (Kong et al., 2021) highlights the need for repeating experiments due to the occurrence of signal inaccuracies resulting from environmental factors such as temperature, humidity, and physical obstacles like walls that obstruct sensor signals. Furthermore, another study (Berka et al., 2018) discusses a navigation application designed for blind pedestrians. The article emphasizes that configuring beacons accurately becomes challenging in certain locations characterized by high signal distortions, particularly when multiple users are concurrently utilizing the application, leading to signal interference.

18.4.8.2 Close-Proximity Inconvenience

Close-proximity inconveniences may arise when users encounter various challenges during their interaction with beacons. A good example of this issue is highlighted in a study (Boric et al., 2018). For users to have their attendance recorded automatically, they are required to physically approach a beacon deployed at a specific location, confirming the proper transmission of signals without errors. Furthermore, a prior work (Ferreira et al., 2020) emphasizes that difficulties can arise when multiple users relying on the same mobile application attempt to receive signals from a beacon in proximity, resulting in potential interference and reducing the effectiveness of the beacon's proximity functionality. Another scenario involves a mobile application designed for evacuation purposes, as discussed by a recent study (Yamagami et al., 2021). The application provides a navigation map that guides users through virtual walls during evacuation, which is delivered through the installed mobile app. However, an inconvenience arises in this context as users only receive beacon notifications based on their walking speed. The speed of their movement and the positioning of their smartphones can impact the beacon's ability to effectively reach them during the evacuation process.

18.4.8.3 Practical Real-World Use

A notable limitation of mobile beacon technology lies in the practical application of this technology in real-world scenarios. While some applications of beacon technology with proximity-based notifications have been tested, many of the results are still in their early stages or have not undergone thorough evaluation in real-world settings. For instance, a study (Chu et al., 2020) introduces a combination of popular social media platforms with beacon technology to create a social media-based tracking system aimed at enhancing work efficiency in the healthcare industry. However, further research and evaluation are required to fully assess the effectiveness of this system. Moreover, the use of beacons in wildfire monitoring systems has been explored (Wu et al., 2018), with the beacon's role being to push notifications to visitors regarding relevant information, such as wildfire detection. Future system implementations are expected to incorporate additional functionalities, including the integration of solar energy charging systems. These examples highlight the ongoing development and potential of beacon technology, while underscoring the need for further research and evaluation to fully understand and optimize its practical applications in real-world contexts.

18.5 CONCLUSIONS

This chapter presented a comprehensive review of recent intelligent mobile beacons systems to offer a better understanding of the usage of mobile beacon technology and proximity-based notifications in various industries within the context of intelligent systems and informatics. The review encompasses an overview of mobile beacon systems and proximity-based notifications, serving as a foundation for comprehending the fundamental concepts of this technology within intelligent systems. The classification and discussion of mobile beacons and proximity-based notifications shed light on recent types of applications, considering their relevance to intelligent systems.

Furthermore, emerging trends in intelligent mobile beacon systems and its notification features are analyzed within the framework of intelligent systems and informatics. Notable trends under the major categories like retails and marketings, active environments, and exhibitions are discussed. The retails and marketings category encompasses intelligent marketing strategies, insurance applications, and intelligent healthcare systems. The active environment category covers intelligent education systems, parking solutions, and zone managements. Lastly, the exhibitions category explores the integration of intelligent systems in art galleries, museums, and tourism applications.

The present chapter also summarizes the challenges and limitations of the existing intelligent mobile beacons systems. These challenges include addressing signal inaccuracy, overcoming inconveniences in close-proximity situations, and considering the real-world use cases of intelligent mobile beacons. By highlighting these challenges, future researchers and practitioners can focus their efforts on addressing the remaining limitations to realize the full potential of mobile beacon technology.

In conclusion, this review offers insights into the current advances of intelligent mobile beacon systems and its notification features with regards to intelligent systems and informatics. The identified trends and challenges pave the way for further research and development in intelligent applications, where mobile beacons play a crucial role. Future endeavors should focus on refining the beacon system deployment processes, addressing signal inaccuracies, and optimizing the functionality and performance of mobile beacons.

REFERENCES

Adkar, N., Talele, A., Mundhe, C., & Gunjal, A. (2018). Bluetooth Beacon Applications in Retail Market. *2018 International Conference on Advances in Communication and Computing Technology (ICACCT)*, 225–229.

Akbarzadeh, O., Baradaran, M., & Khosravi, M. R. (2021). IoT-Based Smart Management of Healthcare Services in Hospital Buildings During COVID-19 And Future Pandemics. *Wireless Communications and Mobile Computing, 2021*, 1–14.

Akinsiku, A., & Jadav, D. (2016). BeaSmart: A Beacon Enabled Smarter Workplace. *NOMS 2016-2016 IEEE/IFIP Network Operations and Management Symposium*, 1269–1272.

Alabduljabbar, R. (2022). An IoT-Aware System for Managing Patients' Waiting Time Using Bluetooth Low-Energy Technology. *Computer Systems Science and Engineering, 40*(1), 1–16.

AlBraheem, L., Al-Abdulkarim, A., Al-Dosari, A., Al-Abdulkarim, L., Al-Khudair, R., Al-Jasser, W., & Al-Angari, W. (2017). Smart City Project Using Proximity Marketing Technology. *2017 Intelligent Systems Conference (IntelliSys)*, 177–183.

Allurwar, N., Nawale, B., & Patel, S. (2016). Beacon For Proximity Target Marketing. *Int. J. Eng. Comput. Sci, 15*(5), 16359–16364.

Alzoubi, H. M., Alshurideh, M., & Ghazal, T. M. (2021). Integrating BLE Beacon Technology with Intelligent Information Systems IIS for Operations' Performance: A Managerial Perspective. *The International Conference on Artificial Intelligence and Computer Vision*, 527–538.

Antoniou-Kritikou, I., Economou, C., Flouda, C., & Karioris, P. (2019). Innovative Mobile Tourism Services: The Case of 'Greek at the Hotel'. In *Strategic Innovative Marketing and Tourism* (pp. 801–808). Springer.

Baig, M. M. B., & Jilani, M. T. (2017). An iBeacon Based Real-time Context-Aware E-Healthcare System. *2017 First International Conference on Latest Trends in Electrical Engineering and Computing Technologies (INTELLECT)*, 1–5.

Barthe, G., Viti, R. D., Druschel, P., Garg, D., Gomez-Rodriguez, M., Ingo, P., Kremer, H., Lentz, M., Lorch, L., & Mehta, A., & others. (2022). Listening to Bluetooth Beacons for Epidemic Risk Mitigation. *Scientific Reports*, *12*(1), 1–14.

Berka, J., Balata, J., & Mikovec, Z. (2018). Optimizing The Number of Bluetooth Beacons with Proximity Approach at Decision Points for Intermodal Navigation of Blind Pedestrians. *2018 Federated Conference on Computer Science and Information Systems (FedCSIS)*, 879–886.

Boric, M., Vilas, A. F., & Redondo, R. P. D. (2018). Automatic Attendance Control System Based on BLE Technology. *ICETE (1)*, 455–461.

Chu, E. T.-H., Lin, K.-H., Chen, S.-Y., Hsu, J., & Wu, H.-M. (2020). SBOT: A Social Media Based Object Tracking System. *2020 Indo–Taiwan 2nd International Conference on Computing, Analytics and Networks (Indo-Taiwan ICAN)*, 132–137.

Concepción-Sánchez, J., Caballero-Gil, P., Suárez-Armas, J., & Álvarez-Diaz, N. (2017). Mobile Application for Elderly Assistance in Public Transport. *Proceedings of the 1st International Conference on Internet of Things and Machine Learning*, 1–7.

Cui, B., Zhou, W., Fan, G., & He, Z. (2016). System Design for Local Culture Protection Based on Smart Item. *2016 11th International Conference on Computer Science & Education (ICCSE)*, 144–147.

Dong, H. J., Abdulla, R., Selvaperumal, S. K., Duraikannan, S., Lakshmanan, R., & Abbas, M. K. (2019). Interactive On Smart Classroom System Using Beacon Technology. *International Journal of Electrical & Computer Engineering (2088-8708)*, *9*(5).

Đurđević, N., Labus, A., Barać, D., Radenković, M., & Despotović-Zrakić, M. (2022). An Approach to Assessing Shopper Acceptance of Beacon Triggered Promotions in Smart Retail. *Sustainability*, *14*(6), 3256.

Ferreira, M. C., Dias, T. G., & Cunha, e (2020). Is Bluetooth Low Energy Feasible for Mobile Ticketing in Urban Passenger Transport? *Transportation Research Interdisciplinary Perspectives*, *5*, 100120.

Fortune Business Insights. (2020). *Beacon Market Size, Share & Industry Analysis*. https://www.fortunebusinessinsights.com/industry-reports/beacon-market-100142

Hasan, R., & Hasan, R. (2020). Towards Designing a Sustainable Green Smart City Using Bluetooth Beacons. *2020 IEEE 6th World Forum on Internet of Things (WF-IoT)*, 1–6.

Hidayat, M. A., & Simalango, H. M. (2018). Students Attendance System and Notification of College Subject Schedule Based on Classroom Using iBeacon. *2018 3rd International Conference on Information Technology, Information System and Electrical Engineering (ICITISEE)*, 253–258.

Huang, Y., Fansheng, K., & Yifei, H. (2020). Applying Beacon Sensor Alarm System for Construction Worker Safety in Workplace. *IOP Conference Series: Earth and Environmental Science*, *608*(1), 12036.

Jeon, K. E., She, J., Soonsawad, P., & Ng, P. C. (2018). BLE Beacons For Internet of Things Applications: Survey, Challenges, And Opportunities. *IEEE Internet of Things Journal*, *5*(2), 811–828.

Karabtcev, S. N., Khorosheva, T. A., & Kapkov, N. R. (2019). BLE Beacon Interaction Module and Mobile Application in The Indoor-Navigation System. *2019 International Science and Technology Conference "East Conf"*, 1–6.

Kaur, M. J., & Maheshwari, P. (2016). Smart Tourist for Dubai City. *2016 2nd International Conference on next Generation Computing Technologies (NGCT)*, 30–34.

Kemble, P., Mane, R., Salokhe, D., Mane, P., Davare, S., & Kadam, A. V. (2016). *An iPhone Application for Car Showroom with Deployment of Beacon*.

Kong, F., Ahn, S., Seo, J., Kim, T. W., & Huang, Y. (2021). Beacon-Based Individualized Hazard Alarm System for Construction Sites: An Experimental Study on Sensor Deployment. *Applied Sciences*, *11*(24), 11654.

Li, J.-W., Chang, Y.-C., Xu, M.-X., & Huang, D.-Y. (2020). A Health Management Service With Beacon-Based Identification for Preventive Elderly Care. *Journal of Information Processing Systems*, *16*(3), 648–662.

Lin, X.-Y., Ho, T.-W., Fang, C.-C., Yen, Z.-S., Yang, B.-J., & Lai, F. (2015). A Mobile Indoor Positioning System Based on iBeacon Technology. *2015 37th Annual International Conference of the IEEE Engineering in Medicine and Biology Society (EMBC)*, 4970–4973.

Lin, C.-Y., Jhu, G.-Y., & Siao, S. (2019). An Innovative Queue Management System Based on Bluetooth Beacon Technology. *2019 IEEE International Conference on Computational Science and Engineering (CSE) and IEEE International Conference on Embedded and Ubiquitous Computing (EUC)*, 427–430.

Marín, J. S. P., Cabrera, J. J. F., Garzôn, N. V. O., & Mora, H. R. C. (2021). Virtual Learning Environments Based on Beacons Devices. *2021 IEEE Fifth Ecuador Technical Chapters Meeting (ETCM)*, 1–6.

McGuirt, J. T., Gustafson, A., Ammerman, A. S., Tucker-McLaughlin, M., Enahora, B., Moore, C., Dunnagan, D., Prentice-Dunn, H., & Bedno, S. (2022). EatWellNow: Formative Development of a Place-Based Behavioral "Nudge" Technology Intervention to Promote Healthier Food Purchases Among Army Soldiers. *Nutrients*, *14*(7), 1458.

Menon, S., George, A., Mathew, N., Vivek, V., & John, J. (2017). Smart workplace—Using iBeacon. *2017 International Conference on Networks & Advances in Computational Technologies (NetACT)*, 396–400.

Moher, D., Liberati, A., Tetzlaff, J., Altman, D. G., Altman, D., Antes, G., Atkins, D., Barbour, V., Barrowman, N., Berlin, J. A., Tovey, D., & Tugwell, P. (2009).

Preferred Reporting Items for Systematic Reviews and Meta-Analyses: The PRISMA Statement. *PLoS Medicine, 6*(7). https://doi.org/10.1371/journal.pmed.1000097

Ng, P. C., She, J., & Park, S. (2017). Notify-and-interact: A Beacon-Smartphone Interaction for User Engagement in Galleries. *2017 IEEE International Conference on Multimedia and Expo (ICME)*, 1069–1074.

Oziom, M., Bachl, S., Wimmer, C., & Grechenig, T. (2019). GroFin: Enhancing In-Store Grocery Shopping with A Context-Aware Smartphone App. *Proceedings of the 18th International Conference on Mobile and Ubiquitous Multimedia*, 1–11.

Pai, V., Castelino, K., Castelino, A., & Gonsalves, P. (2020). Smart Airport System Using Beacon Technology. *International Research Journal of Engineering and Technology, 7*(6), 7552–7560.

Park, S. (2021). D-Park: User-Centric Smart Parking System Over BLE-Beacon Based Internet of Things. *Electronics, 10*(5), 541.

Pugaliya, R., Chabhadiya, J., Mistry, N., & Prajapati, A. (2017). Smart Shoppe Using Beacon. *2017 IEEE International Conference on Smart Technologies and Management for Computing, Communication, Controls, Energy and Materials (ICSTM)*, 32–35.

Saranya, K., Fathima, S. S. A., & Ismail, M. N. A. (2018). Enhancing Customer Engagement Using Beacons. *Int. J. Recent Technol. Eng, 7*(4S), 390–393.

Saraswat, G., & Garg, V. (2016). Beacon Controlled Campus Surveillance. *2016 International Conference on Advances in Computing, Communications and Informatics (ICACCI)*, 2582–2586.

Schwebel, D. C., Hasan, R., Griffin, R., Hasan, R., Hoque, M. A., Karim, M. Y., Luo, K., & Johnston, A. (2021). Reducing Distracted Pedestrian Behavior Using Bluetooth Beacon Technology: A Crossover Trial. *Accident Analysis & Prevention, 159*, 106253.

Seo, S., Jeon, K. E., & Park, S. (2019). Reliable Mobile-Proximity Interaction Mechanism Based on BLE Beacon-Initiated Notification. *2019 IEEE 90th Vehicular Technology Conference (VTC2019-Fall)*, 1–5.

Shinotsuka, Y., Ishida, T., Takahagi, K., Iyobe, M., Akatsu, N., Sugita, K., Uchida, N., & Shibata, Y. (2016). Proposal Of a Zoo Walk Navigation System for Regional Revitalization. *2016 19th International Conference on Network-Based Information Systems (NBiS)*, 307–310.

Spachos, P., & Plataniotis, K. (2020). BLE Beacons In the Smart City: Applications, Challenges, And Research Opportunities. *IEEE Internet of Things Magazine, 3*(1), 14–18.

Srisura, B., & Thakiguchi, T. (2020). Beacon Proximity Based Service to Find Nearby Parking Spaces. *International Journal of Electrical and Electronic Engineering & Telecommunications, 9*(5), 373–379.

Surendran, D., & Rohinia, M. (2019). BLE Bluetooth Beacon Based Solution to Monitor Egress of Alzheimer's Disease Sufferers from Indoors. *Procedia Computer Science, 165*, 591–597.

Van De Sanden, S., Willems, K., & Brengman, M. (2019). In-Store Location-Based Marketing With Beacons: From Inflated Expectations to Smart Use in Retailing. *Journal of Marketing Management, 35*(15–16), 1514–1541.

Wanniarachchi, V. U., Fernando, O. N. N., & Fernandopulle, Y. (2017). Application of Beacon Technology for Enhanced Customer Engagement in Insurance Companies. In *SIGGRAPH Asia 2017 Mobile Graphics & Interactive Applications* (pp. 1–3).

Wu, H.-T., Chen, J.-K., & Tsai, C.-W. (2018). Wildfire Monitoring and Guidance System. *2018 27th Wireless and Optical Communication Conference (WOCC)*, 1–3.

Yamagami, M., Kuroki, H., Ikeoka, H., & Nakamichi, N. (2021). Evacuation Guidance Experiment using V-Wall App. *2021 IEEE 10th Global Conference on Consumer Electronics (GCCE)*, 699–702.

Yamamoto, D., Tanaka, R., Kajioka, S., Matsuo, H., & Takahashi, N. (2018). Global Map Matching Using BLE Beacons for Indoor Route and Stay Estimation. *Proceedings of the 26th ACM SIGSPATIAL International Conference on Advances in Geographic Information Systems*, 309–318. https://doi.org/10.1145/3274895.3274918

Yang, J., Poellabauer, C., Mitra, P., & Neubecker, C. (2020). Beyond Beaconing: Emerging Applications and Challenges of BLE. *Ad Hoc Networks*, *97*, 102015.

Zaim, D., & Bellafkih, M. (2016). Bluetooth Low Energy (BLE) Based Geomarketing System. *2016 11th International Conference on Intelligent Systems: Theories and Applications (SITA)*, 1–6.

CHAPTER 19

Enhancing the Use of Simulation Software with Artificial Intelligence Methods for Clean Energy System Performance

Saiful Azmi Husain, Adam Ahmad,
and Samsul Ariffin Abdul Karim

19.1 INTRODUCTION

Energy plays an important role in the development of a country, and demands for energy are always increasing due to growth in population and industrialization around the world. Increase in energy demand and growth in the population continuously deplete the resources of non-renewable energy such as coal, petroleum, and nuclear power. Moreover, they pose two major challenges: the first one being the need to satisfy global energy demand whilst reducing or keeping the greenhouse gas emission and environmental pollution to the minimum, and the second one is to provide equal energy access to people and countries all over the globe, since current energy access is neither universal nor guaranteed (Singh & Singal, 2017; Quaranta, 2018). According to the World Energy Data (2021), fossil fuels remain the major contributor to the world electricity generation

DOI: 10.1201/9781003400387-19

by 2020 accounting to 61.3%, whereas the remainders are from renewable resources. It is a well-known fact that the usage of fossil fuels emits greenhouse gases such as carbon dioxide, consequently causing environmental pollutions and greenhouse effect. Human activities over the past decades have raised the concentration of greenhouse gases in the atmosphere and these gases trap the sun's energy and prevent its heat from escaping into space, thus warming the planet (greenhouse effect) and causing climate change (UNESCO 2018). The impacts of climate change can affect labor and operations, cause detrimental damages to physical assets, and leave adverse effects on supply chain, distribution chain, consumers, and the communities on which companies depend (Gardiner David & Associates, 2012).

The ever increasing energy demands continuously deplete fossil fuel resources worldwide and the effects of carbon emissions to the atmosphere on climate change has raise global awareness on reducing non-renewable energy generations and environmental protections (Vine, 2008). These goals can be achieved by adopting different environmental sustainability strategies including reducing dependency on fossil fuels and use of alternative energy sources with minimal negative environmental impacts for large-scale energy production and even stand-alone systems to meet and satisfy the continuously increasing global energy demand (Zhou et al., 2010). Examples of alternative energy sources, or also known as, renewable energy resources include solar, wind, hydropower, tidal, wave, biomass, geothermal, etc. These renewable energies are inexhaustible and offer many environmental benefits over traditional energy source (Ozturk et al., 2009); however, their technologies are relatively less competitive than traditional electric energy conversion systems, mainly due to their intermittent productivity and the relatively high maintenance cost (Banos et al., 2011). For example, renewable energy resources such as solar and wind on a large scale have enabled energy sectors to produce meaningful electricity but a common drawback of solar and wind power systems, is their power generation reliability depends on weather conditions and climatic changes, which are unpredictable and the variations of wind and solar energy may not able to meet the demands for energy in time (Yang et al., 2008).

Hydropower is potentially the cleanest, yet an efficient source of clean energy and arguably the most versatile and hassle-free technology for power generation (Singh & Singal, 2017). As stated in Sustainable Development Goal 7 (SDG 7 or Global Goal 7), established by the United Nations General Assembly in 2015, it aims to "Ensure access to affordable, reliable, sustainable and modern energy for all." One of the target indicators is to promote access to research, technology, and investments in clean energy. Hence hydropower generation lays down an interesting option for handling the challenges posed by the growths of energy demand and population. Hydropower has been estimated to contribute about 70% of the planet's electricity through renewable energy sources in 2016 (IRENA, 2018) as cited in Ciric (2019). Large hydropower plant can potentially meet the global energy demand; however, due to the need of large dams, the usage of large hydro plants may generate some adverse effects on the ecosystems such as flooding of large areas upstream, interruption of the river longitudinal connectivity, changing in the hydrological regime and sediment transport processes of rivers, and, sometimes, social impacts (Singh & Singal, 2017). Nonetheless, hydropower plants remain as the planet's most contributory renewable energy source, providing about 19% of electrical power production worldwide, and it continues to be installed globally, especially in emerging countries (Paish, 2002; Quaranta, 2018). A work of literature state that the wave energy in the entire ocean can potentially generate 8.0×10^3 to 80×10^3 TWh/year and provide about 15 to 20 times more available energy per square meter than either wind or solar (Muetze and Vining (2006)), Malik (2011). However, despite its huge potential, wave energy is far from ready for commercial viability (Polinder et al., 2004). Polinder et al. (2004) describes the commonly used concept of wave energy conversion systems, although many of the ocean wave technologies are still in the development stage with a lot of researchers still investigating to improve and optimize the devices, and unfortunately no clear engineering solutions have yet been provided (Malik (2011)). Meanwhile, advancement in turbine technologies may eventually offer opportunity to greater power generation from ocean, especially from its tidal current using the tidal stream designs. As of today, there exist two main variations for tidal energy conversion systems, which are the tidal stream systems that converts kinetic energy of current, and barrages which make uses of potential energy (Blanco Ilzarbe et al. (2013)). Malik (2011) state that the marine and tidal energies potentially possess about 5 TW globally. The kinetic energy of water (from ocean, tidal and etc.) collectively can be classified as hydrokinetic energy.

The main objectives of the studies are:

a. To discuss on the use of ANSYS simulation software that apply AI methods to automatically find simulation parameters in order to improve speed and accuracy simultaneously in the product design for clean energy system. One of the product design being simulated is for the hydrokinetic turbine system, typically used to drive power production from natural flow of a river.

b. To discuss the importance of this simulation software with AI enhancement in predicting whether or not the product design really works in the real world, by taking hydrofoil selection and turbine creation, geometrical configurations, mesh generation and boundary conditions for numerical simulation settings into considerations.

c. To discuss the importance of ANSYS simulation software in teaching and learning, especially in advanced engineering courses for higher education level.

d. Limitations and challenges of this study and how to move forward.

This chapter is organized as follows. In Section 19.1, we give an overview of non-renewable energies and their environmental impacts to climate change as well as renewable energies and clean energies, giving their advantages, limitations and challenges. Section 19.2 discusses the ANSYS simulation software with artificial intelligence (AI) methods. Section 19.3 is dedicated to results and discussion of using ANSYS simulation software with AI methods enhancement on hydrokinetic turbine system. Limitation and challenges will also be presented in this section. Section 19.4 discusses the advantages of using ANSYS simulation software for teaching and learning in advanced engineering courses for higher education level. Finally, Section 19.5 presents the conclusion and future works.

19.2 ANSYS SIMULATION SOFTWARE

ANSYS Inc. is an American multinational company with its headquarters based in Canonsburg, Pennsylvania. ANalysis of SYStems or in short form ANSYS develops and markets Computer Aided Engineering (CAE) and multi-physics engineering simulation software for product design, testing, and operation and offers its products and services to customers worldwide, including NASA (NASA press release (2022)). ANSYS Mechanical finite element analysis software is used to simulate computer models of structures, electronics, or machine components for analyzing the strength, toughness, elasticity, temperature distribution, electromagnetism, fluid flow, and other attributes (ANSYS official website). ANSYS is used to determine how a product will function with different specifications, without building test products or conducting crash tests (ANSYS Official Website). Most ANSYS simulations are performed using the ANSYS Workbench system, which is one of the company's main products (ANSYS Official Website). Typically, ANSYS users break down larger structures into small components that are each modeled and tested individually (Houser, Mark (1995)). A user may start by defining the dimensions of an object, and then adding weight, pressure, temperature and other physical properties. The ANSYS software simulates and analyzes movement, fatigue, fractures, fluid flow, temperature distribution, electromagnetic efficiency and other effects over time (Nakasone (2006)).

ANSYS uses AI/machine learning (ML) methods to automatically find simulation parameters to improve speed and accuracy simultaneously, and guide early product optimization efforts to help engineers quickly find the best design space for their product design, based on thousands of parameters. ANSYS can use augmented simulation to speed up the simulation by factors of 100× by training neural networks via data-driven or physics-informed methods. Advanced simulation technology, enhanced with AI/ML, is underpinning the engineering design process (ANSYS Official Website) (Figure 19.1).

19.3 ANSYS SIMULATION ON HYDROKINETIC TURBINE SYSTEM

19.3.1 Hydrokinetic System

The emergence of hydrokinetics as a renewable and clean energy solution opens new possibilities for regions where installations of hydropower is impossible. Hydrokinetic system was originally developed to overcome the numerous problems associated with water reservoir (Kusakana & Vermaak, 2013) and their technologies are designed to be deployable directly without significantly altering its environment (Khan et al., 2009). The typical design of hydrokinetic technology brings about several advantages over conventional hydropower including:

- Dam-free power generation
- Preservation of the natural flow of the river
- Better preservation of the ecosystem and natural habitat.

Being a relatively new energy solution, there are still ambiguity in the definition of its technology classes, field of applications and their

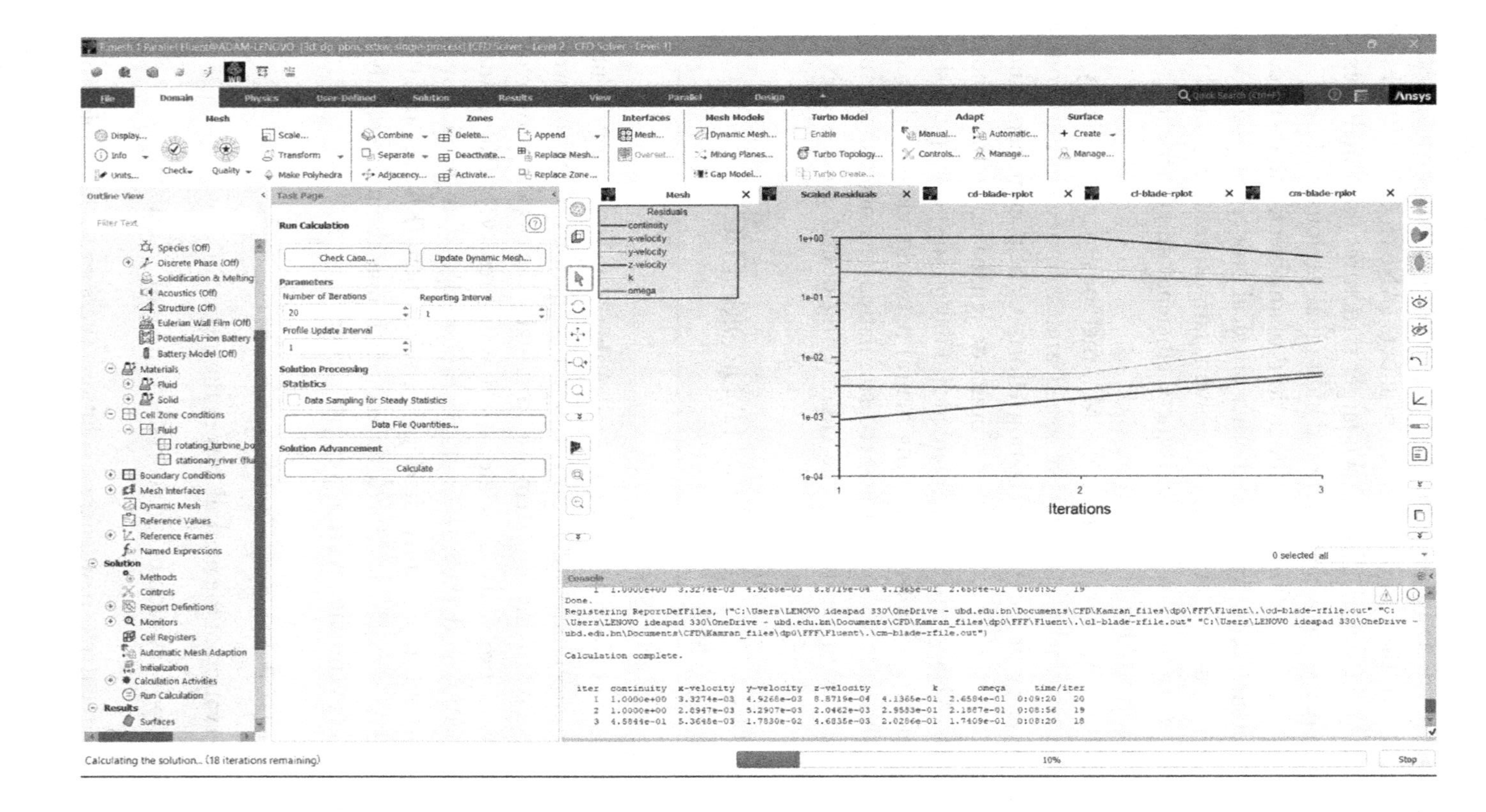

FIGURE 19.1 ANSYS simulation platform.

conversion concepts. Generally, the term hydrokinetic is used to describe both river and tidal applications and depending on their working principles, the current existing hydrokinetic systems can be organized into two main classes: Wave Energy Conversion (WEC) and Current Energy Conversion (CEC) systems (Yuce & Muratoglu, 2014).

19.3.2 Current Energy Conversion (CEC) System

The process of CEC system utilizes the kinetic energy available in river streams, tidal currents, canals, and other man-made waterways to rotate a turbine for electricity generation without impounding or diverting the flow of water resources (Inc, 2006). One of the earliest technologies that exploit water current as sources of energy is the watermill. A typical watermill system consists of a simple water wheel used to drive a mechanical process such as grinding, rolling, and hammering. Historically, watermills have been installed at fast-flowing rivers for food, textile and paper production, amongst other applications (Tanier-Gesner et al., 2014, as cited in (Ibrahim et al., 2021)). The terminology used to describe current energy converter varies in the works of literature throughout the years, it is sometimes referred to as "Water Current Turbine (WCT)", "Free Flow/Stream Turbine", "Zero Head Hydro Turbine", "In-stream Hydro Turbine" or just simply with the term "Hydrokinetic Turbine". Depending on its areas of application, similar technologies generally are identified as "River Current Energy Conversion System (RCECS)" for river and artificial waterway applications, "Tidal In-Stream Energy Converter (TISEC)" for tidal applications, and "Marine Current Turbine" for ocean application. Other common albeit misleading terms used to identify the system include "Watermill", "Water Wheel", or just as "Water Turbine" (Khan et al., 2009).

A CEC system follows a familiar structure design. The kinetic energy of water stream is converted into mechanical energy as it rotates the blades of the hydrokinetic turbine. The hydrokinetic turbine is connected to a generator by a shaft. The generator, usually of a permanent magnet synchronous type, is coupled with a gearbox unit which constitutes for the conversion of mechanical energy into electrical energy. Generated electricity is then distributed to power electrical loads or in a battery (Salleh et al. 2019).

19.3.3 Wave Energy Conversion (WEC) System

A WEC system extract wave energy by creating a system of reacting forces, in which two or more bodies move relative to each other. The primary types of the system include oscillating wave columns (OWC), overstopping devices, attenuators, terminators and point-absorbers (RESOLVE Inc., 2006).

19.3.4 ANSYS Using AI Enhancement Simulation on Hydrokinetic Turbine System

19.3.4.1 Methodology: Hydrofoil Selection and Turbine Creation for the Product Design

The product design of the hydrokinetic turbine begins by determining the sizing of the rotor which factors into the cost of the rotor. Normally, the cost of a rotor increases with an increase in its size, therefore it is much preferable to obtain an efficient rotor instead of a very large rotor (Chica et al., 2015). Followed by, the modelling of the rotor blades which starts with deciding the shape of the blade cross-section. The cross-section geometry of a hydrokinetic turbine blade is normally referred to as a hydrofoil, which typically a similar design as the airfoil for a wind turbine blade. But in contrast, most hydrofoils differ by having a higher curvature at the upper surface than at the lower surface. The efficiency and performance of a hydrokinetic turbine depends on the geometry of the hydrofoil used on the blade which can be composed of one or more hydrofoil types. In general, thinner hydrofoils have better efficiency than thick hydrofoils but relatively thicker hydrofoils should be employed near the blade root than at the blade tip to achieve a balance of efficient and structurally strong blades. For hydrokinetic turbines, the blades sections should be relatively thicker than that of a wind turbine to withstand the hydrodynamic forces acting on the hydrokinetic blades that are greater than the aerodynamic forces on the wind turbine (Ahmed, 2012).

The turbine is modeled using a computer aided design (CAD) software called SOLIDWORKS 2021. To generate the hydrofoil model into Solidworks, the software requires the calculated coordinates of the hydrofoil which may be easily taken from the airfoil database available online. To automatically produce the hydrofoil model into Solidworks, the selected hydrofoil coordinates must be rearranged and reformatted in three separate columns for x, y, and z coordinates before transferring the said coordinates into Solidworks. The operational and geometrical description of the 3-straight-bladed horizontal axis turbine can be found in Table 19.1.

After generating the turbine geometry, the next step involves determining the suitable geometry that is compatible with running simulations on a computational fluid dynamics solver in ANSYS. According to the work of

TABLE 19.1 Geometrical Parameters of Horizontal Axis Hydrokinetic Turbine

Parameter	
Blade profile	Asymmetrical NACA2421
Mass [g]	3237.05
Number of blades	3
Pitch angle, β [°]	0, 3, 6, 9, 12, 15, 18
Radius [mm]	500
Swept area [mm^2]	785,398
Shaft radius [mm]	100

Schleicher et al. (2013), the conditions for ANSYS (that uses AI methods) to be successful is to have two sub-domains in the computational system, which makes the simulation result with better accuracy and improved computational time. For the hydrokinetic turbine system, by using the said settings, enables calculations of velocity and pressure fields in the channel sub-domain to be updated from an absolute frame of reference whereas the calculation for the turbine sub-domain to be taken from a relative frame of reference to the rotation rate of the rotor. The computational domain for the Computational Fluid Dynamics (CFD) simulations is created using ANSYS SpaceClaim.

After all the geometric models are built, the next step involves the discretization of the computational domain in a finite number of elements in which governing equations (such as turbulent kinetic energy and specific dissipation equations) are solved, in which these equations are already embedded in ANSYS. The accuracy of the computational simulations results is dependent on the mesh generation that users provided for the analysis (Schleicher et al., 2013). For this hydrokinetic turbine system, an unstructured tetrahedral mesh was applied on the channel sub-domain together with a structured prism with tetrahedral meshing on the turbine sub-domain. It is important to implement both large and small elements for the mesh construction. Typically, smaller elements should be employed in the regions of importance such as in the near-wall regions to improve the accuracy of the solution, however the challenge is that, its usage increasingly put more burdens on the computational time and resources for the completion of the solution and so larger elements should be placed in regions of less importance to help ease the computational process. In this context, to avoid having a too large number of small elements, prism layers are used to generate enough fine mesh on the near-wall regions of interest and provide enough small numerical values, i.e. non-dimensional parameters, which defines the distance of the first mesh node of the adjacent cell to the non-slip wall of a solid surface which are essential for resolving the boundary layer flow (La′ın et al., 2019).

The quality of the mesh plays an important role in ensuring that a simulation performs as accurately as possible, which can be measured according to the skewness and the orthogonal quality of the generated mesh. A good quality mesh is expected to have a skewness value in between 0.25 and 0.50 and an orthogonal quality value close to 1. Whereas a bad mesh has a skewness value close to 1 and inversely, an orthogonal value close to 0. Although, to ensure that a simulation may performs as it is expected, a minimum orthogonal quality for all types of cells should at least has a value of 0.01. (Al-Dabbagh & Yuce, 2018).

The mesh for the computational domains is generated through ANSYS Meshing and the overall mesh of the computational domain can be seen in Figure 19.2.

Tetrahedron elements are chosen for the discretization of the channel and turbine sub-domains. This meshing employs a body of influence meshing technique to influence the sizing of the elements in the overall mesh. The mesh elements are larger near the surfaces of the channel subdomain and it gradually becomes smaller as the mesh approaches the

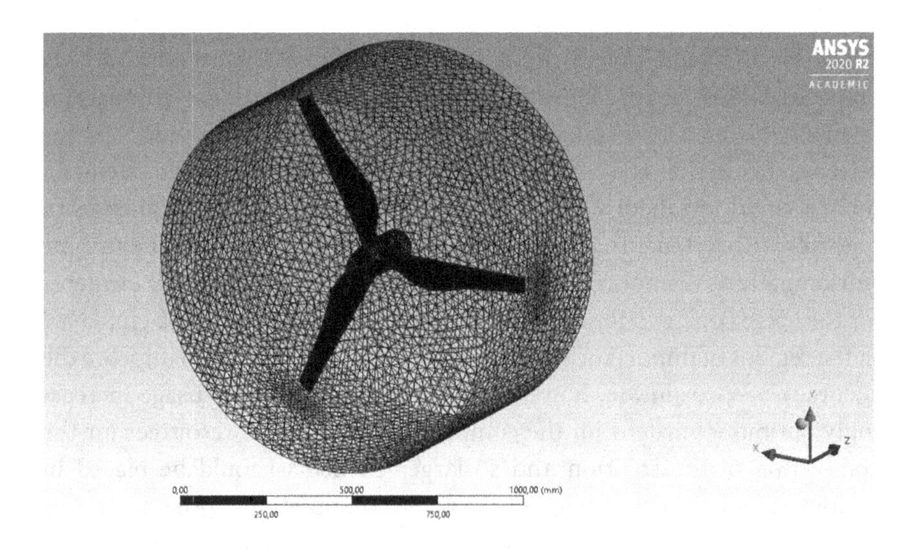

FIGURE 19.2 Overall mesh of the computational domain.

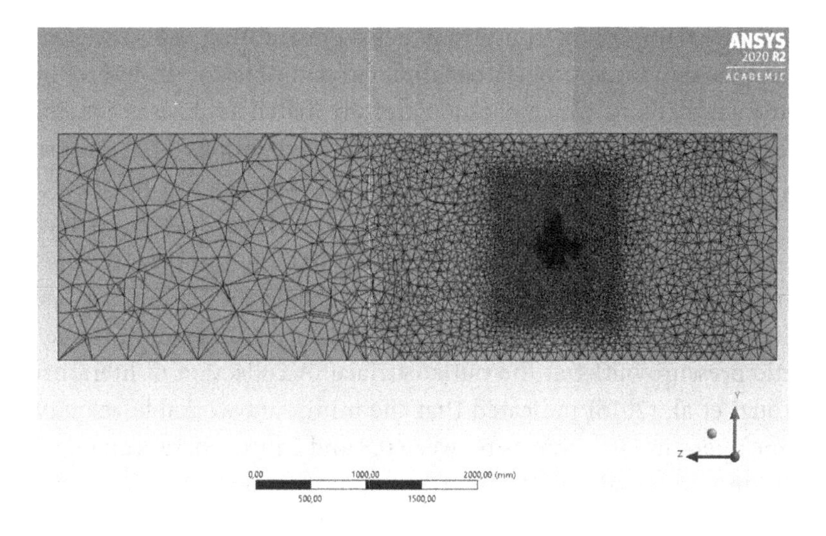

FIGURE 19.3 The cross-section of the computational domain mesh.

center of the rotational domain as can be seen in Figure 19.3. The inflation function of the ANSYS Meshing is used to employ the prism layers surrounding the turbine. Figure 19.4 depicts the prism layers generated along a cross section of the blade profile. Multiple meshes were prepared, with their prismatic layers before the simulations are performed.

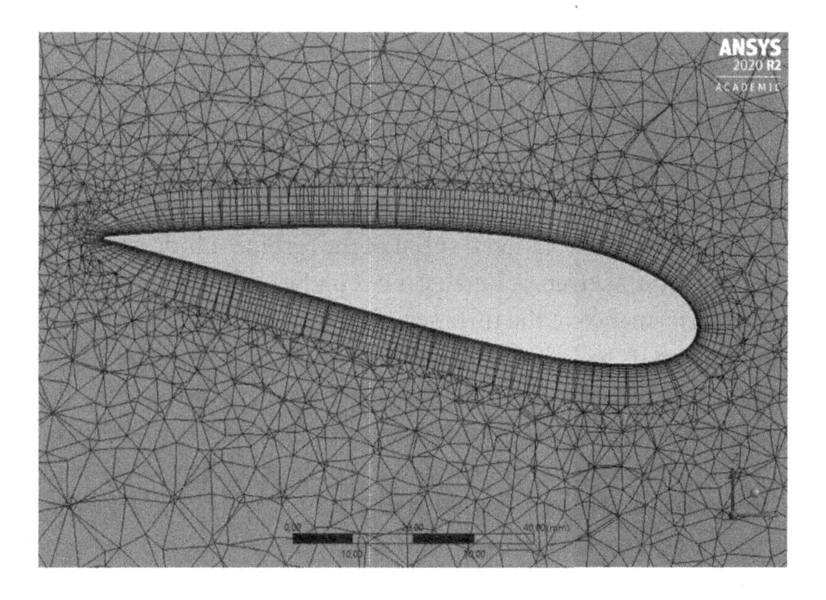

FIGURE 19.4 Details of the prism layers surrounding the blade profile.

One other important conditions before proceeding with the simulation, is to assign the appropriate boundary conditions for the hydrokinetic turbine simulations, which should carefully match as close as possible to the experimental conditions. The boundary conditions hold important roles which factor for the flow distribution, hence affecting the forces and torques acting on the turbine and the generated power coefficient, among other results (La′ın et al., 2019). Some of the common settings for boundary conditions in the hydrokinetic turbine simulation is to assign a velocity inlet with the desired current velocity at the inlet surface and a static pressure outlet at the outlet surface. A collection of literature by Behrouzi et al. (2016) indicated that the minimum workable velocity for turbine operation in a river is between 0.8 and 1 m/s, whereas an expected optimum current speed for a marine application is at least 2 m/s. Other beneficial boundary conditions that may be employed for the hydrokinetic turbine simulation are described below:

- Symmetry: This boundary condition overlooks the effects of free surface.
- Free slip wall: Enables the tangential motion allowing for the simulation of the free surface without deformation.
- No slip wall: This condition sets the relative velocity at the solid wall as stationary.

19.3.4.2 Results and Discussion from ANSYS Simulation Using AI Enhancement on Hydrokinetic Turbine System

In the present study, ANSYS simulation using AI enhancement via ANSYS Fluent was carried out given the initial flow velocity of 1 m/s, to drive the turbine and obtain the rotational speed of the turbine sub-domain, namely the tip speed ratio, λ, given by Equation 19.1 of the turbine, in order to investigate the performance of the turbines of various fixed blade pitch angles β beginning from 0° to 18° at a regular increment of 3° per case. Using AI methods, the results will generate the power generation from the effect of tip speed ratio as well as the optimum tip speed ratio. The AI methods will automatically generate the pressure distribution and flow analysis of the blade hydrofoil at specified blade length and blade tip. These factors are important to investigate the factors affecting the lift and the drag production of the blade, and therefore the efficiency of the hydrokinetic turbine.

The torque of the turbine was numerically evaluated using AI methods in ANSYS Fluent, where torque calculation was defined from each

FIGURE 19.5 Geometry of the simulated Turbine.

surface of the turbine geometry seen in Figure 19.5. The torque output of the various turbines is then illustrated in Figure 19.6 and more detailed results can be observed in Table 19.2. The estimated power generation of the turbine is depicted in Figure 19.7 and their values are recorded in Table 19.3. Since the power output and torque output of turbines are directly proportional, both the torque output and power output graphs show a similar curved distribution for all the turbines under the same testing conditions. The tip speed ratios of a turbine rotor may affect the drag and lift forces production which are the two important parameters in hydrokinetic turbine performances. The torque generation and power output of the turbine can be seen to increase as the rotational speed of the turbine increases until it obtains the maximum power output at the optimum tip speed ratio. When a turbine reaches its optimum tip speed ratio, it experiences the greatest ratio of lift to drag forces which lead to the peak performance of the turbine. However, beyond the optimum tip speed ratio, the torque generation of turbine declines in quasilinear fashion as the rotational speed of the turbine increases for a specific water flow speed. Lower performances are observed in turbines with a

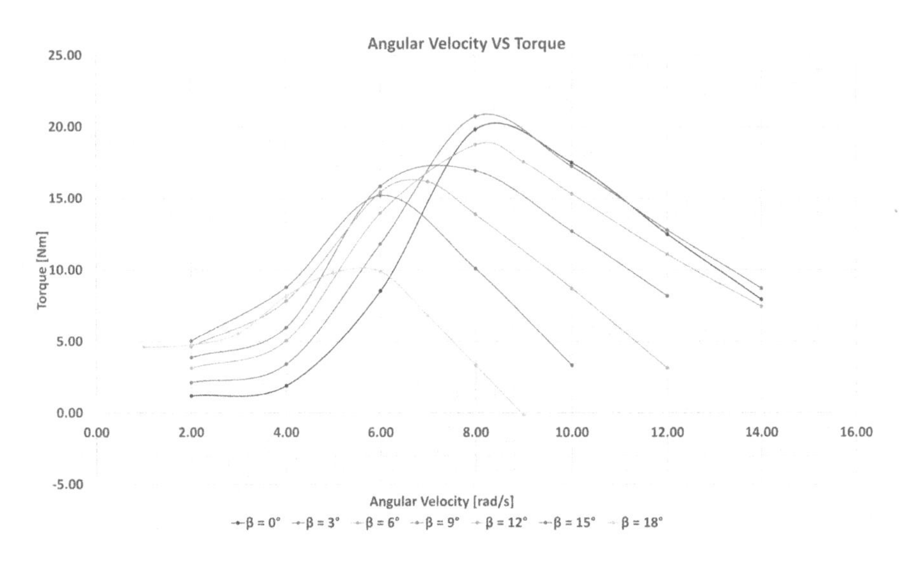

FIGURE 19.6 Angular velocity ω vs torque output of turbines at various blade pitch angle β.

rotational speed below its optimum tip speed ratio which is normally due to small lift generation and greater drag production is observed in a turbine operating at a higher tip speed ratio which consequently lead to smaller torque production in a turbine (Abuan & Howell, 2019; Ali & Jang, 2021).

The efficiency of the turbine can be evaluated by comparing the power coefficient against its operating tip speed ratio by calculating using the Cp equation defined in Equation 19.2. The results are shown in Figure 19.8 and the estimated Cp values are listed in Table 19.4. According to Figure 19.8, the maximum Cp value is measured below 0.592 (Betz's limit,

TABLE 19.2 Torque Distribution Table of Turbines with Different β Values, Rotating at Different ω Values

T [Nm]		ω [rad/s]						
		2	4	6	8	10	12	14
β [°]	0	1.19	1.89	8.54	19.8	17.5	12.5	7.94
	3	2.12	3.40	11.8	20.7	17.3	12.8	8.72
	6	3.12	5.06	14.0	18.7	15.3	11.1	7.50
	9	3.87	6.00	15.9	16.9	12.7	8.19	
	12	4.64	7.85	15.4	13.9	8.72	3.13	
	15	5.04	8.80	15.2	10.1	3.31		
	18	5.42	9.41	12.4	5.62			

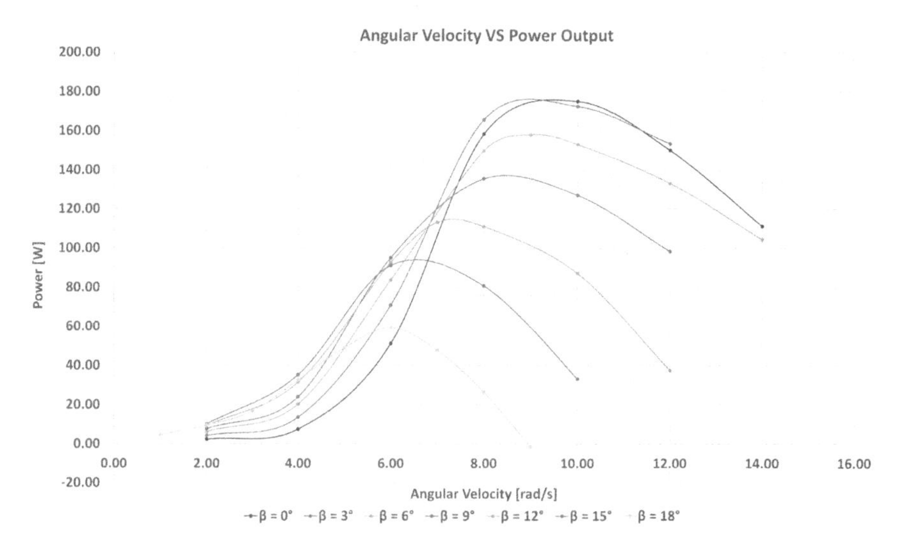

FIGURE 19.7 Angular velocity ω vs power output of turbines at various blade pitch angle β.

TABLE 19.3 Power Distribution Table of Turbines with Different β Values, Rotating at Different ω Values

P [W]		ω [rad/s]						
		2	**4**	**6**	**8**	**10**	**12**	**14**
β [°]	**0**	2.37	7.55	51.3	158	174	150	111
	3	4.24	13.6	70.9	166	172	153	122
	6	6.25	20.2	83.8	150	153	133	104
	9	7.75	24.0	95.1	136	127	98.3	
	12	9.28	31.4	92.7	111	87.2	37.5	
	15	10.1	35.2	91.2	80.8	33.1		
	18	10.9	37.6	74.3	45.0			

TABLE 19.4 Power Coefficient, *Cp* Distribution Table of Turbines with Different β Values, Rotating at Different ω Values

C_p		λ						
		1	**2**	**3**	**4**	**5**	**6**	**7**
β [°]	**0**	0.01	0.02	0.13	0.40	0.45	0.38	0.28
	3	0.01	0.03	0.18	0.42	0.44	0.39	0.31
	6	0.02	0.05	0.21	0.38	0.39	0.34	0.27
	9	0.02	0.06	0.24	0.35	0.32	0.25	
	12	0.02	0.08	0.24	0.28	0.22	0.1	
	15	0.03	0.09	0.23	0.21	0.08		
	18	0.03	0.10	0.19	0.11			

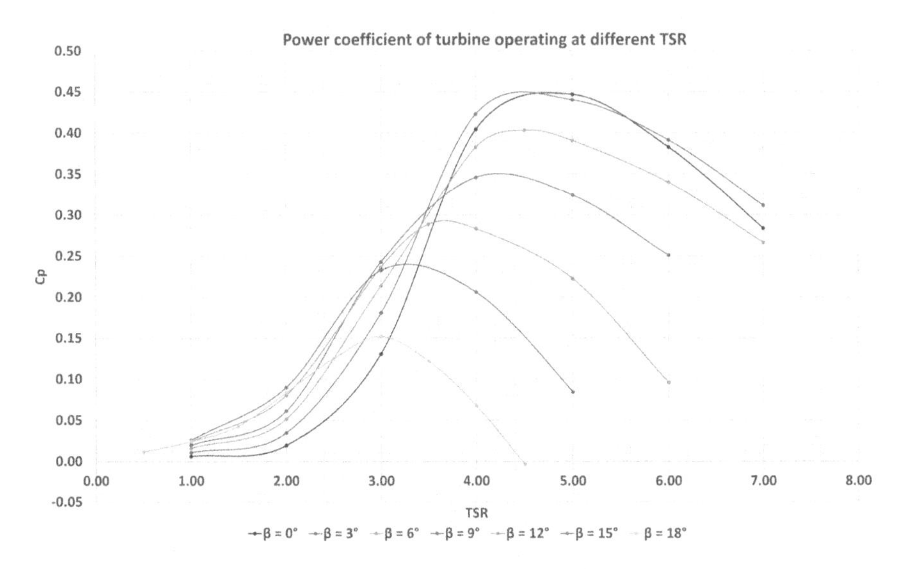

FIGURE 19.8 Tip speed ratio λ vs power coefficient Cp of turbines at various blade pitch angle.

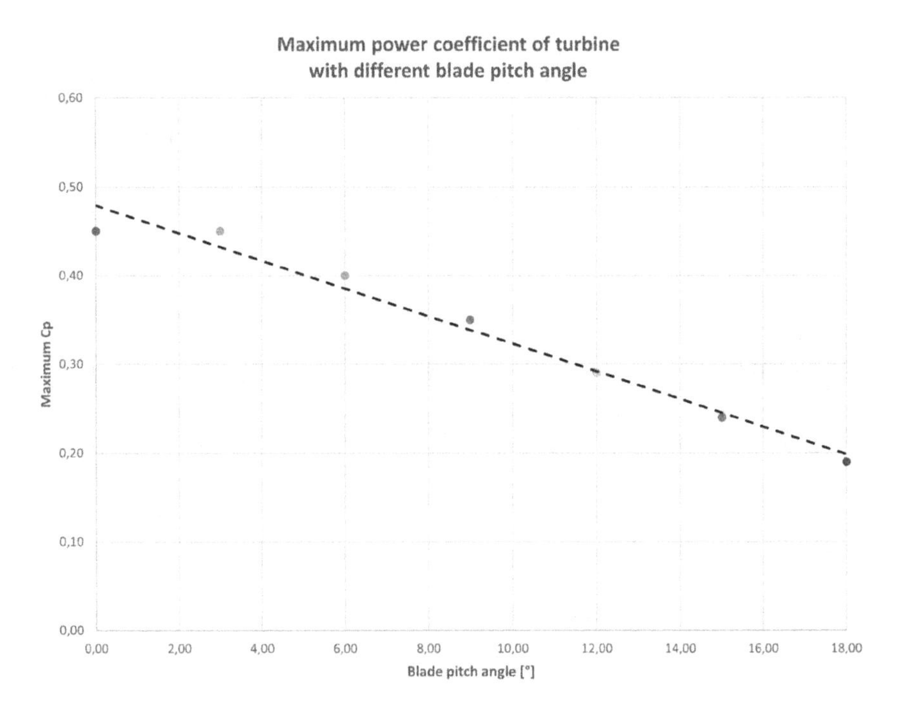

FIGURE 19.9 Trend of power coefficient of turbines at different blade pitch angle β.

i.e. there exists a limit to the proportion of extractable kinetic energy from the flowing fluid relative to its maximum available kinetic energy in the fluid) for all the turbines under the same testing conditions.

However, it can be observed that with increasing fixed blade pitch angle, the maximum power coefficient of the turbine decreases as can be seen in Figure 19.9. Therefore, a turbine with a low blade pitch angle may provide a better option than a turbine with a high blade pitch angle when compared by its power efficiency alone.

The limitation of this research study is the overall time needed to run all the simulations, which are also increasing exponentially for a given small interval increment (blade pitch angle β).

19.4 APPLICATIONS OF ANSYS IN HIGHER EDUCATION CURRICULUM

Applications of ANSYS has currently been implemented in the form of ANSYS GRANTA EduPack (ANSYS Official Website) which helps educators in higher education to engage and inspire students who are taking engineering courses and supports students' learning with these packages throughout the engineering curriculum. Teachers and lecturers can also support existing teaching materials and create new courses as well as applying innovative teaching approach, using this ANSYS GRANTA EduPack. These packages can also promote interdisciplinary learning and project work among group of students.

19.5 CONCLUSION & FUTURE WORK

This study has presented the performance of horizontal axis hydro (water)-kinetic turbine system across a range of blade pitch angles in a steady flow with a water velocity of 1 m/s. From these results, using AI methods enhancement, the best scenario for achieving the greatest efficiency in the power generation for this type of turbine is to employ the blades with a low pitch angle. A turbine with a fixed blade pitch angle of 3° has shown the greatest power production among its different variations of a turbine and hence provides the best efficiency with a maximum Cp = 0.44 at optimum tip speed ratio, $\lambda = 4.5$. At low blade pitch angle, turbines can be operated at a higher optimum tip speed ratio which increases the maximum lift to drag ratio of the blade and therefore better efficiency. Hence, using AI enhancement, the ANSYS simulation can help the engineers quickly explore and predict whether or not the product design works feasibly in the real world. Thus, this simulation software is not only important for research but also beneficial for teaching and learning in higher education, in which students, especially engineering students could explore their products design in more depths using ANSYS GRANTA EduPack. Future work is recommended to validate the results from ANSYS simulation software by designing the same product design hardware in experimental settings.

REFERENCES

Abuan, B. E., & Howell, R. J. (2019). The performance and hydrodynamics in unsteady flow of a horizontal axis tidal turbine. Renewable Energy, 133, 1338–1351.

Ahmed, M. R. (2012). Blade sections for wind turbine and tidal current turbine applications-current status and future challenges. International Journal of Energy Research, 36, 829–844.

Al-Dabbagh, M. A., & Yuce, M. I. (2018). Simulation and comparison of helical and straight-bladed hydrokinetic turbines. International Journal of Renewable Energy Research, 8, 504–513.

Ali, S., & Jang, C. M. (2021). Effects of tip speed ratios on the blade forces of a small h-darrieus wind turbine. Energies, 14, 1–19.

ANSYS Official Website: https://www.ansys.com/products/structures/ansys-mechanical#:~:text=Ansys%20Mechanical%20is%20a%20finite,hydrodynamic%2C%20explicit%2C%20and%20more.

Banos, R., Manzano-Agugliaro, F., Montoya, F. G., Gil, C., Alcayde, A., & Gomez, J. (2011). Optimization methods applied to renewable and sustainable energy: A review.

Behrouzi, F., Nakisa, M., Maimun, A., & Ahmed, Y. (2016). Global renewable energy and its potential in Malaysia: A review of hydrokinetic turbine technology. Renewable and Sustainable Energy Reviews, 62, 1270–1281.

Blanco Ilzarbe, J. M., Malpartida, J. G., & Chandro, E. R. (2013). Recent patents on geothermal power extraction devices. Recent Patents on Engineering, 7, 2–24.

Chica, E., Perez, F., Rubio-Clemente, A., & Agudelo, S. (2015). Design of a hydrokinetic turbine. WIT Transactions on Ecology and the Environment, 195, 137–148.

Ciric, R. M. (2019). Review of techno-economic and environmental aspects of building small hydroelectric plants: A case study in Serbia. Renewable Energy, 140, 715–721.

Gardiner David & Associates, L. (2012). Physical Risks from Climate Change: A guide for companies and investors on disclosure and management of climate impacts, (p. 32).

Ibrahim, W. I., Mohamed, M. R., Ismail, R. M., Leung, P. K., Xing, W. W., & Shah, A. A. (2021). Hydrokinetic energy harnessing technologies: A review. Energy Reports, 7, 2021–2042.

IRENA (2018). Renewable Power Generation Costs in 2017, A Report by the International Renewable Energy Agency, Abu Dhabi, UAE. 2018. Technical Report.

Khan, M. J., Bhuyan, G., Iqbal, M. T., & Quaicoe, J. E. (2009). Hydrokinetic energy conversion systems and assessment of horizontal and vertical axis turbines for river and tidal applications: A technology status review.

Kusakana, K., & Vermaak, H. J. (2013). Hydrokinetic power generation for rural electricity supply: Case of South Africa. Renewable Energy, 55, 467–473.

La'ın, S., Contreras, L. T., & Lopez, O. (2019). A review on computational fluid dynamics modeling and simulation of horizontal axis hydrokinetic turbines. Journal of the Brazilian Society of Mechanical Sciences and Engineering, 41. https://doi.org/10.1007/s40430-019-1877-6.

Malik, A. Q. (2011). Assessment of the potential of renewables for Brunei Darussalam. 15, 427–437.

Muetze, A., & Vining, J. G. (2006). Ocean wave energy conversion: A survey. In Conference Record - IAS Annual Meeting (IEEE Industry Applications Society) (pp. 1410–1417). volume 3.

Nakasone, Y. (2006). Engineering Analysis With ANSYS Software. Butterworth-Heinemann. Oxford Burlington, MA. *ISBN978-0-7506-6875-0.*

NASA Press Release (2022) https://www.nasa.gov/press-release/nasa-awards-contract-for-modeling-simulation-capabilities-to-ansys

Ozturk, M., Bezir, N. C., & Ozek, N. (2009). Hydropower-water and renewable energy in Turkey: Sources and policy. Renewable and Sustainable Energy Reviews, 13, 605–615.

Paish, O. (2002). Small hydro power: Technology and current status. Renewable and Sustainable Energy Reviews, 6, 537–556.

Polinder, H., Damen, M. E., & Gardner, F. (2004). Linear PM generator system for wave energy conversion in the AWS. IEEE Transactions on Energy Conversion, 19, 583–589. doi:10.1109/TEC.2004.827717.

Quaranta, E. (2018). Stream water wheels as renewable energy supply in flowing water: Theoretical considerations, performance assessment and design recommendations. doi:10.1016/j.esd.2018.05.002.

Inc, R. E. S. O. L. V. E. (2006). Proceedings of the Hydrokinetic and Wave Energy Technologies Technical and Environmental Issues Workshop. Washington D.C. October 26-28, 2005.

Salleh, M. B., M. Kamaruddin, N., & Mohamed-Kassim, Z. (2019). Savonius hydrokinetic turbines for a sustainable river-based energy extraction: A review of the technology and potential applications in Malaysia. Sustainable Energy Technologies and Assessments, 36.

Schleicher, W. C., Riglin, J. D., Kraybill, Z., & Gardner, G. (2013). Design and Simulation of a Micro Hydrokinetic Turbine. Proceedings of the 1st Marine Energy Technology Symposium METS13, (pp. 1–8).

Singh, V. K., & Singal, S. K. (2017). Operation of hydro power plants-a review. doi:10.1016/j.rser.2016.11.169.

Tanier-Gesner, F., Stillinger, C., Bond, A., Egan, P., & Perry, J. (2014). Design, build and testing of a hydrokinetic H-Darrieus turbine for developing countries. In IEEE Power and Energy Society General Meeting. volume 2014-October.

UNESCO (2018). Getting the Message Across: Reporting on Climate Change and Sustainable Development in Asia and the Pacific. A Handbook for

Journalists. UNESCO: https://diraj.org/resources/getting-the-message-across-reporting-on-climate-change-and-sustainable-development-in-asia-and-the-pacific/

Vine, E. (2008). Breaking down the silos: The integration of energy efficiency, renewable energy, demand response and climate change. Energy Efficiency, 1, 49–63. doi:10.1007/s12053-008-9004-z.

World Energy Data (2021). World Final Energy. https://www.worldenergydata.org/world-final-energy/.

Yang, H., Zhou, W., Lu, L., & Fang, Z. (2008). Optimal sizing method for stand-alone hybrid solar-wind system with LPSP technology by using genetic algorithm. Solar Energy, 82, 354–367.

Yuce, M. I., & Muratoglu, A. (2014). Hydrokinetic energy conversion systems: A technology status review. Renewable and Sustainable Energy Reviews, 43, 72–82.

Zhou, W., Lou, C., Li, Z., Lu, L., & Yang, H. (2010). Current status of research on optimum sizing of stand-alone hybrid solar-wind power generation systems. Applied Energy, 87, 380–389.

CHAPTER 20

Intelligent Systems of Computing and Informatics

An Extension

Samsul Ariffin Abdul Karim

20.1 CONCLUSION

Intelligent Systems of Computing and Informatics (ISCI) is a new paradigm via novel framework that can be used to cater the need of IR4.0, SDGs by 2020, and Carbon Zero Net by 2050 in many aspects in our daily lives. For instance, machine learning and big data analytics can be used in Teaching and Learning (T&L) especially on course delivery and automated marking and grading system. This will be enhancing the learning process through online platform. The students can receive instant feedback and the grading is obtained in quick manner. This will ensure the students can improve their understanding as well as score and perform better in their respective assessments. Machine learning also plays an important role to the special students especially in improving their abilities to read, hear, and speak. Besides, the automated process is a crucial task in the manufacturing processes [1–5]. PETRONAS is also in the process to automate their processes especially in oil and gas exploration and reducing carbon emissions [4]. Figure 20.1 shows the proposed ISCI framework that has been developed in the present studies.

DOI: 10.1201/9781003400387-20

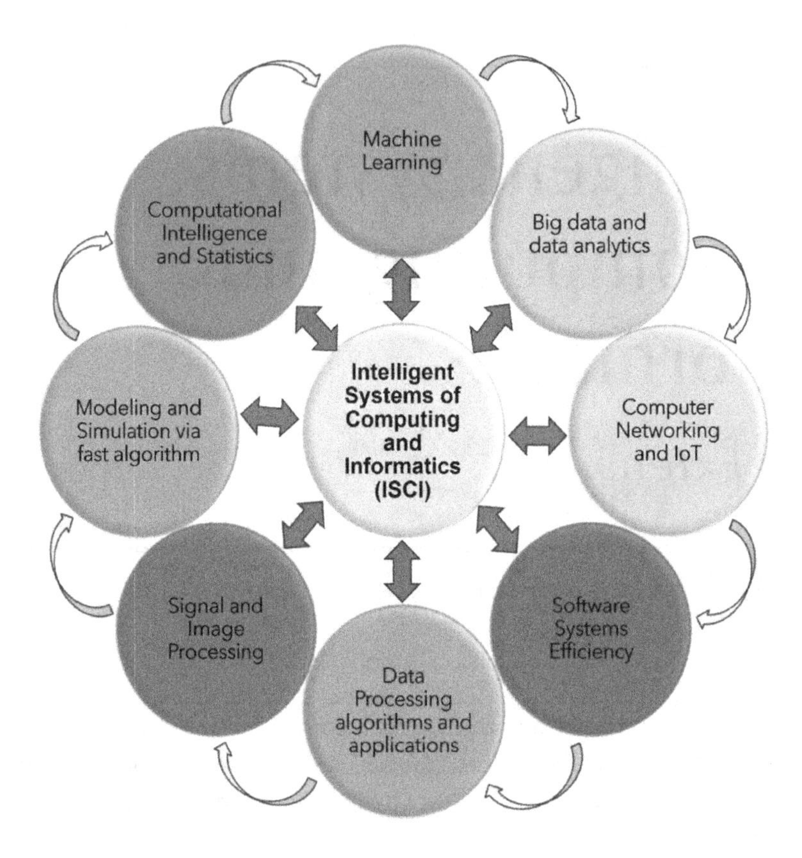

FIGURE 20.1 Intelligent systems of computing and informatics (ISCI) framework.

20.2 FUTURE EXTENSION

There are many potential applications that can be derived and developed based on the ISCI framework. Some possible future extensions are summarized below:

- Building upon the foundation laid by the noninvasive microwave system (MiS), the study aims to enhance the accuracy and applicability of fat depth assessment. Further investigation could involve delving into advanced machine learning techniques, such as ensemble methods or deep learning, to optimize the predictive capabilities of the models. In addition, exploring alternative dimensionality reduction techniques beyond PCA and K-means, like t-SNE or Lasso regression, could provide novel insights into feature selection and model enhancement. Moreover, the research could extend its scope to incorporate other relevant variables that might impact fat depth estimation, leading to a more comprehensive predictive framework. By continuously refining methodologies and embracing emerging technologies, this chapter's future endeavors promise to contribute to the advancement of fat depth measurement in the realm of animal agriculture.

- Investigating the other features that can improve detection accuracy while simultaneously decreasing the false positive rate (FPR) and increasing the true positive rate (TPR). Moreover, exploring deep learning methods like convolutional nural networks (CNNs), recurrent neural networks (RNNs), and long short-term memory (LSTM) models shows potential for advancing the field of malware detection further.
- More multidisciplinary cooperations are interested in gaining insight into the analysis of electroencephalogram (EEG) data to further understand the human mind. Hence EEG data analysis manifests as a dynamic field that has been developing quickly. EEG signals are often complex and contain noisy data, which require advanced processing techniques. Some possible directions might involve the development of more sophisticated methods for denoising, artifact removal, feature extraction, and feature selection. Machine learning approaches, including swarm intelligence techniques, could play a significant role in improving the accuracy and efficiency of these processes. Regardless promising results found in this study, in the future more swarm intelligence techniques need to be tested on different datasets to determine the quality of selected features in mining human emotions.
- In future endeavors, it is crucial to give primary importance to improving document feature vectors, term-document matrices, and the semantic connections between words. This can be achieved by conducting comprehensive analysis involving both human insights and large text datasets. This strategy will significantly enhance the efficiency and precision of the analysis process.
- Another compelling direction involves delving into more intricate SVM kernels, such as nonlinear kernels like polynomial or Gaussian radial basis function (RBF), which could unlock even greater predictive potential. Expanding the dataset to include additional relevant

features, such as nutritional aspects or breed characteristics, may further enhance the model's precision. As the field of machine learning evolves, experimenting with ensemble methods or hybrid models could lead to unprecedented predictive accuracy. The incorporation of interpretability techniques, such as SHAP values or feature importance analysis, could provide valuable insights into the model's decision-making process. By harnessing these innovative approaches and staying attuned to emerging methodologies, the chapter's future trajectory is poised to redefine the landscape of lamb carcass fat depth prediction.

- Given the challenges faced in face detection at specific distances, future research could focus on optimizing the CNN model to handle a wider range of distances effectively. This could involve refining the network architecture, incorporating additional training data with varying distances, and exploring techniques like depth estimation to improve accuracy across different distances. Integrate the attendance tracking system with broader educational analytics platforms. This could provide insights into attendance patterns, student engagement, and correlations between attendance and academic performance. The T&L delivery and assessments can be improved further.
- Higher learning institutions are exposed to phishing attacks, endangering sensitive data and reputation. Therefore, a comprehensive awareness program, including posters, infographics, videos, and seminars, will educate and empower staff to identify and thwart phishing attempts. Higher learning institutions should conduct awareness programme periodically using suitable means to protect their organisation from attacks on human or users. This is very crucial for all institutions facing the current challenges as well as to fulfil the Education 4.0 objectives.
- Intelligence Random Forest Application in Developing Regression Model has made significant strides in revolutionizing fat depth prediction through the utilization of machine learning techniques. However, there are several potential future directions that can be explored to further enhance and expand upon the findings of this study. For instance, exploring various algorithms such as gradient boosting, support vector regression, or neural networks could potentially lead to even more precise predictions. In this study, the focus

was primarily on the application of machine learning algorithms to the given dataset. However, further investigations into feature engineering techniques may uncover additional variables or transformations that can enhance the predictive power of the models. Feature selection methods, dimensionality reduction techniques, or incorporating domain-specific knowledge could be explored to refine the input variables.

- A database must be created to map the user hardware's MAC address to the authentication database in the server. Network speed should be considered as the number of users increases. The standard operating procedure for user (facial image) registration must be designed to comply with the Personal Data Protection Act. After the user logs in, the average eye blinking computation time could also be considered. This approach will enhance the security and the safety of the users and the data itself.
- Object detection and recognition in low-light conditions is a very crucial tasks in many aspects of 4IR. The evaluation of the latest deep neural network algorithms, the enhancement of the dataset, and the development of mobile applications as testing modules are all promising areas of research that could lead to significant advances towards intelligence systems.
- The proposed EDGSOR algorithm to solve nonlinear problems significantly have improved the performance of existing iterative methods. Some possible extension is by leveraging optimization and data-driven approaches, researchers can unlock new levels of accuracy, efficiency, and understanding in predicting and controlling complex diffusion phenomena. This research direction has the potential to revolutionize applications ranging from environmental remediation to advanced materials design.
- Further studies that may be carried out are implementing more complex functions to find the roots to solve arbitrary nonlinear equations. Furthermore, it may also be possible to upgrade a method that has a relatively low error from the current one. The method that has been developed also needs to be reviewed whether it can be implemented in the problem of finding the optimum value of a function.
- The implementation of different interpolation method applies to the data sets such as spline or other new interpolation methods.

Beside that utilizing different or new numerical method would also improve the finding. This will ensure data processing will be as fast and as accurate as possible so that the efficiency of predicting is improve and without any lagging.

- The future direction of succinate and lactate production in metabolic engineering holds exciting prospects with the potential for numerous applications in biotechnology and industry. First, future efforts will likely focus on engineering metabolic pathways to increase the efficiency of succinate and lactate biosynthesis. This could include the design of novel enzymes or the optimization of existing ones. Second, advances in fermentation processes, including bioreactor design and process control, will play a crucial role in increasing the production of succinate and lactate. Continuous fermentation, metabolic flux analysis, and online monitoring will likely be employed for real-time optimization. Lastly, the integration of metabolic modeling and synthetic biology will become more prominent. Furthermore, the predictive modeling can be used to guide the design of engineered strains, and synthetic biology toward an automated predictive maintenance.
- Another possible further study includes Internet of Things (IoT) architecture on multiple variable speed air conditioning to gain and analyse variety of data. Independent pressure sensor and collect the refrigerant pressure data to study the relationships and gain important insights. This will enable the consumers to reduce electricity consumption as well as reducing CO_2 emissions toward Carbon Net Zero by 2020.
- In terms of processing the large data set recorded at the various meteorological station and later it can be analyzed using the Openair package which can be used in the Comprehensive R Archive Network (CRAN). The air quality data recorded in the various meteorological stations across the world can also be used by this Openair package for further analysis.
- Future directions of the beacon and its technology in teaching and learning toward 4IR and SDG No 4: Quality Education by 2030. Intelligent interactive environment and personalized immersive educational experience are one of the many potential applications offered by this exciting technology.

- The future direction of using ANSYS simulation software, which uses artificial intelligence methods to help engineers in designing optimal new or existing product design for the clean energy system, specifically, the emergence of hydrokinetics as a renewable & clean energy solution. This ANSYS simulation software opens new possibilities for regions where installations for hydropower are impossible. The typical design of hydrokinetic technology such as hydrokinetic turbine system brings about several advantages over conventional hydropower including dam-free power generation, preservation of the natural flow of the river and better preservation of the ecosystem and natural habitat.
- Another possible extension is the implementation of the proposed Intelligent Systems of Computing and Informatics (ISCI) to cater scattered data interpolation problems arising in many science and engineering applications. For instance, in medical imaging, where the data are very huge, deep learning can be integrated into existing scattered data interpolation scheme to produce efficient method to reconstruct highly accurate interpolating surfaces. These surfaces are smooth and can be used to represent the geometric shape of the objects.

Overall, the future of the ISCI looks bright especially to cater the objectives of SDGs, 4IR, and Carbon Net Zero by the year 2030 until 2050.

ACKNOWLEDGMENT

This research is fully supported by the Ministry of Higher Education (MOHE), Malaysia for the financial support received in the form of a research grant: **[FRGS/1/2023/ICT06/UMS/02/1]** (New Scattered Data Interpolation Scheme Using Quasi Cubic Triangular Patches for RGB Image Interpolation and fruits quality inspection) and Universiti Malaysia Sabah. Special thanks to Faculty of Computing and Informatics, Universiti Malaysia Sabah for the computing facilities support that made the completion of the book possible.

REFERENCES

1. https://sdgs.un.org/goals (Retrieved on 22 August 2023).
2. https://mia.org.my/knowledge-centre-resources/digital-economy/ (Retrieved on 22 August 2023).

3. https://www.ekonomi.gov.my/sites/default/files/2021-07/National-4IR-Policy.pdf (Retrieved on 22 August 2023).
4. https://www.petronas.com/sustainability/getting-to-net-zero (Retrieved on 22 August 2023).
5. https://www.sustainabilitymatters.net.au/content/sustainability/article/net-zero-carbon-neutral-carbon-negative-what-do-they-mean-exactly–1597925690 (Retrieved on 22 August 2023).

Index

Q

R

S

T

Printed in the United States
by Baker & Taylor Publisher Services